African Natural Plant Products Volume II: Discoveries and Challenges in Chemistry, Health, and Nutrition

ACS SYMPOSIUM SERIES **1127**

African Natural Plant Products Volume II: Discoveries and Challenges in Chemistry, Health, and Nutrition

H. Rodolfo Juliani, Editor
Rutgers University
New Brunswick, New Jersey

James E. Simon, Editor
Rutgers University
New Brunswick, New Jersey

Chi-Tang Ho, Editor
Rutgers University
New Brunswick, New Jersey

Sponsored by the
ACS Division of Agriculture and Food Chemistry, Inc.

American Chemical Society, Washington, DC

Distributed in print by Oxford University Press

Library of Congress Cataloging-in-Publication Data

African natural plant products volume II: Discoveries and challenges in chemistry, health, and nutrition / H. Rodolfo Juliani, editor, James E. Simon, editor, Chi-Tang Ho, editor.

p. cm. -- (ACS symposium series ; 1127)

Includes bibliographical references and index.

ISBN 978-0-8412-2804-7 (alk. paper)

1. Materia medica, Vegetable--Africa. 2. Essences and essential oils--Africa. 3. Medicinal plants--Africa. 4. Aromatic plants--Africa. I. Juliani, H. Rodolfo. II. Simon, James E. III. Ho, Chi-Tang, 1944-

RS181.A37 2009

615'.321--dc22

2009031670

The paper used in this publication meets the minimum requirements of American National Standard for Information Sciences—Permanence of Paper for Printed Library Materials, ANSI Z39.48n1984.

Distributed in print by Oxford University Press

PRINTED IN THE UNITED STATES OF AMERICA

Foreword

The ACS Symposium Series was first published in 1974 to provide a mechanism for publishing symposia quickly in book form. The purpose of the series is to publish timely, comprehensive books developed from the ACS sponsored symposia based on current scientific research. Occasionally, books are developed from symposia sponsored by other organizations when the topic is of keen interest to the chemistry audience.

Before agreeing to publish a book, the proposed table of contents is reviewed for appropriate and comprehensive coverage and for interest to the audience. Some papers may be excluded to better focus the book; others may be added to provide comprehensiveness. When appropriate, overview or introductory chapters are added. Drafts of chapters are peer-reviewed prior to final acceptance or rejection, and manuscripts are prepared in camera-ready format.

As a rule, only original research papers and original review papers are included in the volumes. Verbatim reproductions of previous published papers are not accepted.

ACS Books Department

Contents

Preface xi

Traditional Medicines from Africa

1. ***Euphorbia tirucalli* L. (Euphorbiaceae) – The Miracle Tree: Current Status of Knowledge** 3
   Julius Mwine, Patrick Van Damme, Bernadetta Rina Hastilestari, and Jutta Papenbrock

2. **Marula [*Sclerocarya birrea* (A.Rich) Hochst]: A Review of Traditional Uses, Phytochemistry, and Pharmacology** 19
   Gerhard Prinsloo and Renée A. Street

3. ***Piper guineense* (Piperaceae): Chemistry, Traditional Uses, and Functional Properties of West African Black Pepper** 33
   H. Rodolfo Juliani, A. R. Koroch, L. Giordano, L. Amekuse, S. Koffa, J. Asante-Dartey, and J. E. Simon

Chemistry , Pharmacognosy, and Validation of Traditional Medicines

4. ***Hypoxis hemerocallidea* (African potato): A Botanical Whose Time Has Come?** 51
   David R. Katerere

5. ***Pycnanthus angolensis*: Bioactive Compounds and Medicinal Applications** 63
   Kelsey Gustafson, Qing-Li Wu, Julie Asante-Dartey, and James E. Simon

6. **Prospects of *Croton membranaceus* for Prostate Health** 79
   A. A. Appiah, W. A. Asomaning, I. V. Oppong, G. Duker-Eshun, J. Clement, L. K. N. Okine, G. Achel, O. Gyampo, S. Adjei, A. K. Nyarko, F. A. Aboagye, and D. A. Edoh

7. **Utilization of Medicinal Plants and Their Products in the Treatment and Control of Disease in Fish** 93
   Albert P. Ekanem and Austin I. Obiekezie

8. **Rooibos: Effect on Iron Status in South African Adults at Risk for Coronary Heart Disease** 103
   Jeanine L Marnewick, Irma Venter, Fanie Rautenbach, Henry Neethling, and Maritha Kotze

9. **Secoiridoid Glucosides from Fraxinus Excelsior with Effects on LPS-Induced Nitrite Production in RAW 264.7 Macrophages and Human Cancer Cell Lines** 115
Naisheng Bai, Kan He, Marc Roller, Ching-Shu Lai, Xi Shao, Min-Hsiung Pan, and Chi-Tang Ho

10. **Nutritional Value of Fonio (*Digitaria exilis*) from Senegal** 127
Adolfina R. Koroch, H. Rodolfo Juliani, and James E. Simon

Quality Control of African Natural Plant Products

11. **African Nightshades and African Eggplants: Taxonomy, Crop Management, Utilization, and Phytonutrients** 137
Ray-Yu Yang and Chris Ojiewo

12. **Quality Characteristics of Shea Butter, *Vitellaria paradoxa*** 167
Hee Seung Nahm, H. Rodolfo Juliani, and James E. Simon

13. **Guidelines for Quality Control during Preclinical Testing of African Traditional Medicines** 185
Renee A. Street, Gerhard Prinsloo, and Nceba Gqaleni

Applications and Commercialization of African Natural Plants Products as Foods and for Health and Nutrition

14. ***Hibiscus sabdariffa*: Phytochemistry, Quality Control, and Health Properties** 209
Tom Villani, H. Rodolfo Juliani, James E. Simon, and Qing-Li Wu

15. **Increasing Micronutrient Availability from Food in Sub-Saharan Africa with Indigenous Vegetables** 231
Ray-Yu Yang, Sahrah Fischer, Peter M. Hanson, and J. D. H. Keatinge

16. **Morphologic and Biochemical Diversity of Peasant's Baobab Tree (*Adansonia digitata* L.) Morphotypes in Senegal** 255
Diaminatou Sanogo, Babacar Diedhiou, Abdoulaye Diop, Haby Sanou, Mactar Thiome, Yaye Kène Gassama, and Emmanuel Bassène

17. **The Role of African Indigenous Plants in Promoting Food Security and Health** 273
Lorraine S. Cordeiro

18. **An Assessment of the Essential Oil and Aromatic Plant Industry with a Focus on Africa** 289
Ramu Govindasamy, Surendran Arumugam, and James E. Simon

Editors' Biographies 323

Indexes

Author Index 327

Subject Index 329

Preface

The wealth of African biodiversity and indigenous knowledge of plants and their products continue to create valuable opportunities for development while enriching the lives and cultures of those using these plants. Many African plant products are already enjoyed on a regular basis in foods, beverages, and as spices, fragrances and medicines as well as used in religious and cultural traditions. Such plants not only provide sustenance, but many can improve the health and nutrition while others can create real income generating opportunities for both the collectors, growers as well as the processors and distributors. Given the serious issues of malnutrition and poverty, and the increasingly acceptable link between the health and nutritional status of a person and the effectiveness of prescribed medicines, such as retrovirals, Africa has the opportunity and the obligation of examining its indigenous resources for their contribution to improving the health and nutrition of its citizens. Many African plants exhibit both promising medicinal activity and are nutrient rich yet remain underutilized. Africa, a continent so rich in natural resources is poised to provide *new plants and products* to the world while strengthening and increasing its own consumption and use of indigenous fruits, vegetables, greens and herbs, spices, medicinal and aromatic plants. This can be accomplished while improving income generation and economic activities at the local and village levels and in an environmentally sustainable manner as well, given the increasing importance in the protection of the environment.

The American Chemical Society Symposium book African Natural Plant Products was originally conceived as a vehicle to present scientific discoveries, challenges and to create a dialogue focused on African natural products, an area still very underexplored as a vehicle to benefit the African people. Yet, as we began on our journey, we and in concert with our scientific colleagues, who have contributed to this series, realized early on the remarkable diversity and range of African plants and products being used for a wide variety of applications (foods, flavorings, medicine, nutrition and health). This series is to provide a scientific stimulus for greater research, enhanced collaboration, and confirmation and/or validation on the uses and importance of African natural plant products, particularly those steeped in a rich traditional history.

The Volume II "Discoveries and Challenges in Chemistry, Health and Nutrition" is a new installment of an international effort to provide a communication platform for scientists to share their interest in African plants and products. The book seeks to promote the identification of new uses and applications that can contribute to the development of the African continent, as the value of plant uses emerges from the interaction of the rich biodiversity of the African ecosystems with societies and cultures. We will continue to expand the focus to include health and nutritional considerations in addition to the core natural product chemistry and continue to present new findings that we hope others will build upon and develop.

For Volume II, 75 authors from four continents have written and contributed the eighteen chapters, the majority of them were African (52%), followed by American (35%), Asian (8%) and European (5%). The book has been organized in four sections that represent the main areas of current research (traditional knowledge, chemistry, validation of traditional medicines, quality, commercial uses and applications). The section "Traditional medicines from Africa" provides research findings and reviews on traditional and new uses of some emerging Sub-Saharan plants with a focus on phytochemistry and pharmacology. Section 2 "Chemistry, Pharmacognosy and Validation of Traditional Medicines" provide case studies of African medicinal plants with latest findings on scientific validation of traditional uses and recent discoveries on the chemistry and biological activities. The section "Quality Control of African Natural Plant Products" seeks to assess effectiveness and safety of African plant products in support of commercialization efforts. The last section "Applications and Commercialization of African Natural Plants Products as Foods and for Health and Nutrition" highlights applications and commercialization aspects of plants.

As the editors, we acknowledge and want to highlight the contributions from each of the authors. We hope the publication of this second volume will continue to promote scientific research in African natural plant products and collaborations within higher education and research institutions, government, and private sector with each other and with the international community.

We are particular thankful to the United States Agency for International Development (USAID) and to the Millennium Corporate Account-Namibia, whose funding of several of our ongoing projects in Africa has contributed indirectly in the preparation of this new volume. We likewise are thankful to the U.S. National Institute of Health (NIH) and the United States Department of Agriculture (USDA) whose funding of several of our domestic research projects helped development a number of analytical protocols and approaches that facilitated several research studies on African plant products. We also express our gratitude to the American Chemical Society for their strong support and encouragement that led to the publication of this second volume. We thank Tim Marney and Mary Calvert of the ACS for their persistence and support.

We are indebted to our close colleagues and friends of the programs Agri-Business in Sustainable Natural African Plant Products (ASNAPP), Excellence in Higher Education for Liberian Development (EHELD), People, Rules and Organizations Supporting the Protection of Ecological Resources (PROSPER) for their dedication and promotion of the different facets of natural plant products sector for sustainable economic development and the preservation of the forest environment in the African continent.

H. Rodolfo Juliani
Associate Director, New Use Agriculture & Natural Plant Products Program
Department of Plant Biology and Plant Pathology
School of Environmental and Biological Sciences
Rutgers, The State University of New Jersey
59 Dudley Road, New Brunswick, New Jersey 08901

James E. Simon
Director, New Use Agriculture and Natural Plant Products Program
Department of Plant Biology and Plant Pathology
School of Environmental and Biological Sciences
Rutgers, The State University of New Jersey
59 Dudley Road, New Brunswick, New Jersey 08901

Chi-Tang Ho
Department of Food Science
School of Environmental and Biological Sciences
Rutgers, The State University of New Jersey
65 Dudley Road, New Brunswick, New Jersey 08901

Traditional Medicines from Africa

Chapter 1

Euphorbia tirucalli L. (Euphorbiaceae) – The Miracle Tree: Current Status of Knowledge

Julius Mwine,[1] Patrick Van Damme,[2,3] Bernadetta Rina Hastilestari,[4] and Jutta Papenbrock[*,4]

[1]Faculty of Agriculture, Uganda Martyrs University, Nkozi, P.O. Box 5498, Kampala, Uganda
[2]Faculty of Bioscience Engineering, Tropical and Subtropical Laboratory of Plant Agronomy and Ethnobotany, University of Ghent, Coupure Links 653, 9000 Ghent, Belgium
[3]ICRAF, POBox 30677, Nairobi, Kenya
[4]Institute of Botany, Herrenhäuserstr. 2, D-30419 Hannover, Germany
[*]E-mail: Jutta.Papenbrock@botanik.uni-hannover.de.

Euphorbia tirucalli L. is an important tree *Euphorbia* species, known for its many uses in the tropics. Endemic to tropical eastern Africa where it often grows wild, it is usually planted for boundary demarcation but also as a live fence around compounds, shrines and kraals. Due to its rather unique combination of CAM stems and C3 leaves it has the ability to withstand extreme aridity and still produce a lot of biomass, whereas it withstands high herbivore pressure basically because the presence of caustic latex. *E. tirucalli* indeed contains a white latex which is vesicant and rubifacient, but also known to be a remedy against many ailments. However, most of its medicinal features are mostly informal reports from folk medicine and there appears to be little medical/laboratory analysis to validate them. In this review, we attempt to explore the current knowledge status about *E. tirucalli* in relation to its classification, chemical content and functions, and the extent to which modern research has been able to validate them. It was found that although a great deal has been done to analyze its chemical composition (bark, roots and latex), and potential for biofuel production, little is available on validation of its potential application for medicinal purposes, yet it continues to be used in traditional and alternative medicine on a daily basis. As a result, more empirical research is called for to clarify the current situation and potentials.

Introduction

Euphorbia tirucalli L. belongs to the genus *Euphorbia*, one of the 8,000 species within family Euphorbiaceae. It is a shrub or a small tree that can grow up to 7 – 12 m high. It is present in most (sub)tropical areas with a propensity for (semi-arid areas. Its pencil-like branches have inspired its vernacular name, the pencil-tree. *E. tirucalli* is generally evergreen since its stems and branches remain green for a number of years; eventually the main trunk, and older branches get a brown bark appearance, however. The tree is rarely fed on by herbivores, nor has it many natural enemies. It bears a white poisonous latex, which accounts for its low herbivore pressure and medicinal features. Most of these features are shared by a number of other, so-called coralliform euphorbias occurring in its centre of origin, e.g. *E. intisy*, or *E. fiha* (or as it is known by its synonymous *E. plagiantha*). However, *E. tirucalli* is rather unique because it can grow very fast, and produce a lot of biomass even under very marginal soil and extreme climatic conditions. This can at least partly be explained by its rather unique combination of CAM stems and C3 leaves which allow it to assimilate CO_2 during 24 h per day. And apparently it does this better than some of its closest relatives in the subtribe Euphorbiinae.

The species is easily vegetatively propagated through stem cuttings. It quickly spread from Africa into the other continents which have subtropical to tropical climates. It became naturalized in most of them as evidenced by the wide variety of local names in different countries (Table I). The species epithet is derived from tiru-calli, a local name given to the plant on the Malabar Coast of India. There is a hypothesis that the plant came to that part of the world with Portuguese travelers who had traveled from Portugal to East Africa (*1*) and then further to India.

Table I. Vernacular Names for *E. tirucalli* in Different Languages. Source: Agroforestry online data base (www.worldagroforestrycentre.org)

Language	*Name*
Amharic	Kinchib
Arabic	Knjil
English	Finger euphorbia, Indian spurge tree, Milk bush, Naked lady, Pencil-tree, Rubber euphorbia
Philipine	Bali bali

Continued on next page.

Table I. (Continued). Vernacular Names for *E. tirucalli* in Different Languages

Language	*Name*
French	Arbre de saint Sebastien, Euphorbe effilé, garde maison, tirucalli
Malay	Kayu patah, tentulang, Tulang, Tulang-tulang
Somali	Dana
Spanish	Alfabeto chino, Antena, Esqueleto, Palito, Aveloz
Swahili	Mtupa mwitu, Mwasi, Utupa
Thai	Khia cheen, Khia thian
Ugandan	Kakoni (luganda), Oruyenje (runyankole)
Vietnamese	San h(oo)xanh, X(uw) (ow)ng c(as).

Classification

In the binomial system (USDA plants data base seen at www.plants.usda.gov), *Euphorbia tirucalli* L. belongs to:

Kingdom: Plantae
Division: Magnoliophyta
Class: Magnoliopsida
Order: Malpighiales
Family: Euphorbiaceae
Genus: *Euphorbia*
Species: *E. tirucalli*
Binomial name: *Euphorbia tirucalli*

According to the APG II system (*2*), *Euphorbia tirucalli* L. belongs to:

Cladus: Eukaryota
Regnum: Plantae
Cladus: Angiospermae
Cladus: Eudicots
Cladus: Core eudicots
Cladus: Rosids
Cladus: Eurosids I
Ordo: Malpighiales
Familia: Euphorbiaceae
Subfamilia: Euphorbioideae
Tribus: Euphorbieae
Subtribus: Euphorbiinae
Genus: *Euphorbia*
Species: *Euphorbia tirucalli*

Figure 1. Twig of E. tirucalli showing position of leaves. Courtesy: Van Damme (3).

While classification from regnum to species is clear, no literature is available on lower subdivisions such as subspecies, varieties and others. However, referring to his 18 accessions/samples of *E. tirucalli* collected from different countries, Van Damme (*1*) reports that there are some differences in vegetative structure in terms of relative dimensions of stems, leaves, general plant habit and growth rate, although no conclusive bio-chemical/molecular tests have been made to evaluate genetic differences that would be a basis of infra-species classification. According to Van Damme (pers. comm.), differences between specimens are minor, except maybe for specimens initially obtained from US, New York State, whose young, growing pencils/branches are yellow-red in colour while the older ones are green. Other accessions, such as those from Morocco and Senegal form thick, short and stocky whorls at branching which remain relatively closed, giving them a closed broom appearance even after opening up. On the other hand, specimens from Burundi, Kenya and Rwanda tend to form thin, elongate and more-open-at-base whorls. The former specimens show a higher canopy formation tendency than the latter. Ugandan specimens have similar appearances as those from Burundi, Kenya and Rwanda. Also, as communicated by Van Damme above, young pencils of samples obtained from US (New York State), are bright yellow-red in colour and stand out from others which are green (Figure 1). Up till now there have been no further comparisons made of the specimens for other features such as height, rate of growth, flowering, fruiting and other factors. What is certain, however, is that these variations are not just environmental differences since the specimens described hereabove were raised as cuttings under the same greenhouse conditions for over twenty years, whereby they remain true-to-type during that whole period (Geirnaert, pers. comm.). Whether these differences are big enough to allow one to distinguish between subdivisions within the species still remains to be confirmed. More research is required to analyze more specimens and using genetic methodologies to establish whether there is a need to subdivide the species into its lower taxonomic forms.

Plant Description

Morphology of *E. tirucalli* has been extensively studied by Van Damme (*1*, *4*, *5*). *E. tirucalli* is an unarmed shrub or small tree that can grow 4 to 12 m high with an adult trunk diameter of 15 to 20 cm. Its younger branches are evergreen, longitudinal, succulent, about 7 mm thick and are usually produced in whorls, rarely single, giving it a broom-like structure. Branches usually end in smaller pencil-like twigs, dull green to red green in colour, with fine white striations and produced in whorls of 2 to 6. Its young stem is green and photosynthetically active, with grooves which in fact are small furrow-like structures containing stomata through which CO_2/O_2 exchanges are operated and that are thus protected from extreme environmental conditions. Stem stomatal frequency is estimated at 12 per mm^2 in grooves on older stem parts, while it may reach 40 per mm^2 on a younger stem. Older stems become rough, brown and lose their photosynthetic ability with age.

Leaves are few, simple, scattered, entire, alternate, oblanceolate, about 1.3 to 2.5 cm long and 2 cm wide but broadest beyond the middle. They are present only at the tips of young growing branchlets. They have glandular, minute, dark brown stipules and are quickly deciduous ((*1*, *4*) Figure 1).

According to the same author, inflorescences are pedicel-less and appear as yellowish heads in clusters of 2 to 6 cymes. *E. tirucalli* is dioecious. Male cymes produce a dense cluster of flowers, whereas female cynthia only contain a few female flowers. Cynthia have cup-shaped single sex involucres. Male involucres have bracteoles which are linear with plumose apices. Stamens are usually single per stalk and are about 4.5 mm long. Occasionally, an aborted female flower is present. In female involucres, the perianth is distinct and 3-lobed existing below a tomentose ovary which is lobular and about 0.5 mm long. The ovary is joined at the base with thickened bifid apices. Occasionally, a female flower exists within the involucres. Each involucre bears five independent nectaries that produce nectar and therefore flowers are insect-pollinated.

Fruits are glabrescent capsules on a tomentose pedicel, yellowish red when ripe and fall off easily. Seeds are ovoid, 3.5 x 2.8 mm in size, smooth, buff-speckled and with a dark brown ventral line. A caruncle exists which is about 1 mm across (*4*).

Ecology and Distribution

E. tirucalli is probably the best-known and most widespread of all tree *Euphorbia* species (*6*). According to the same author, the plant's origin is not known but Van Damme (*3*), and Schmelzer and Gurib-Fakim (*7*) believe it originated from tropical East Africa and is endemic in countries such as Angola, Eritrea, Ethiopia, Kenya, Malawi, Mauritius, Rwanda, Senegal, Sudan, Tanzania, Uganda and Zanzibar. The same authors intimate that the tree is currently widely distributed in southern Europe, Asia and the Americas, having been repeatedly introduced there for its ornamental and medicinal features.

The largest distribution of succulent *Euphorbia* is in the drier regions of southern and eastern Africa (*8*). *E. tirucalli* combines high drought and salinity tolerance with low-input requirements (*9*). As presented hereabove, it is clear that *E. tirucalli*'s drought tolerance has connection with its photosynthetic system. Most succulent Euphorbiaceae have been listed as species with Crassulacean Acid Metabolism (CAM) (*10, 11*). Interestingly, Van Damme (*1*) reports that *E. tirucalli* performed C3 in leaves and CAM pathways in stems.

E. tirucalli can survive in a wide range of habitats. Van Damme (*1*) states that the plant can grow under conditions in which most crops and other trees cannot grow. They include: tropical arid areas with low rainfall, on poor eroded soils, saline soils and high altitudes up to 2000 m, but that it cannot survive frost. Its distribution is therefore limited by low temperatures. He goes on to say that this combination of the C3 and CAM photosynthetic pathways could probably be the reason for its survival in hardier conditions. To clarify this, the same author explains that CAM pathway entails a high carboxylic acid accumulation (than if the plant would be a C3 plant) at night, raising the osmotic potential of the plant which increases its salt tolerance. Also like other succulent plants, *E. tirucalli* stores extra water in the parenchyma and vacuoles which can be used to dilute salt ions entering the plant and as a reserve for survival in dry conditions.

As a crop, *E. tirucalli* can be grown in a variety of areas since it is tolerant to a variety of conditions. Propagation is by cuttings (from any part of the aboveground material) which form roots easily and quickly form a bush. Experiments show that *E. tirucalli* can be grown at a density of 10,000 to 20,000 plants per hectare at a spacing of 1 m x 1 m whereas it coppices excellently at 20 to 30 cm height ((*12*), cited at www.hort.purdue.edu). In many tropical areas, *E. tirucalli* grows wild often in abandoned sites of homesteads and kraals where they sometimes form thick woody vegetation tending towards a forest.

Chemical Composition

E. tirucalli contains white milky latex in all its parts, including the roots. According to Kapaczewski (*13*), the latex contains about 28% solid matter whose composition is: 21 to 27% water-soluble substances, 59 to 63% resin-soluble substances and 12 to 14% rubber-like substances. The chemical composition of the different parts of the plant has been extensively studied and a variety of chemical compounds have been isolated from them (Table II). This great variety of listed chemical substances reveals the complexity of *E. tirucalli* latex and may explain most of its functions. For example, low herbivore pressure, poisonous nature, pesticidal features and medicinal characteristics may all be attributed to this chemical constitution.

Table II. Main Groups of Chemical Substances in the Tree Sap Latex of *E. tirucalli*

Chemical substances	*Source*	*References*
Campesterol, stigmasterol, betasitosterol, isofucosterol, cycloartenol (Sterol)	Stem callus	(*14*)
Cyclotirucanenol (triterpene)	Latex	(*15*)
Diterpene ester	Latex	(*16*)
Euphol and beta-amyrin (triterpenoids)	Stem callus	(*14*)
Steroid	Latex	(*17*)
Taraxerane triterpene	Stem bark	(*18*)
Tirucalicine (diterpene)	Latex	(*19*)

Pests and Disease

There is a tendency to believe that *E. tirucalli* has no pests and diseases because of its poisonous latex. However, a few pests including *Meloidogyne incognita* (*20*), *Botrytis* spp. (*21*) and the weed *Cuscuta* spp. (*1*) have been reported. The latter author notes that an infestation by *Botrytis* spp. causes the plant stem and roots to rot especially in warm and humid conditions. He reports that a combination of *Meloidogyne* spp. and *Botrytis* spp. infestations can wipe out a whole field in a short time.

Uses of *E. tirucalli*

Traditional Medicines

Possibly due to a great variety of chemical substances found in *E. tirucalli* tissues (Table II), medical folklore literature of different parts of the world, especially tropical and subtropical areas where it is endemic, is tainted with its curative abilities (Table III). According to Schmelzer and Gurib-Fakim (*7*) and Van Damme (*3*), in East Africa, latex is used in cases of sexual impotence, warts, epilepsy, toothache, hemorrhoids, snake bites, for the extraction of ecto-parasites and cough among others. In Peninsular Malaysia, a poultice of the roots or stems is applied to nose ulceration, hemorrhoids and swellings. Root scrapings mixed with coconut oil are taken for stomach ache.

In India, Kumar (*22*) notes that it is an unavoidable plant in many traditional homesteads whereas it is used as a remedy for ailments such as: spleen enlargement, asthma, dropsy, leprosy, biliousness, leucorrhoea, dyspepsia, jaundice, colic, tumors, and bladder stones. He further says that vesicant and rubifacient though it is, its latex is emetic in large doses but a purgative in small doses and applied against toothaches, earaches, rheumatism, warts, cough, neuralgia and scorpion bites. The same author points out that its branch and root decoctions are administered for colic and gastralgia while ashes are applied as a caustic to open abscesses.

Table III. The Use of *E. tirucalli*

Function	*Detail*	*Part of plant*	*Country*	*Reference*
Traditional medicine	Wart, epilepsi, sexual impotence, toothache, hemorrhoids, snake bites	Latex	East Africa	(*7*)
	Colic and gastralgia	Branch and root	India	(*22*)
	Cancer, cancroids, ephitheliomas, sarcomas, tumors, wart	Latex	Brazil	(*12*)
	Aching bones, nose ulcers, hemorrhoids	Root	Indonesia	(*12*)
Ornamental				www.iuc-nredlist.org
Glue and adhesive		Latex (resin)		(*12*)
Afforestration and reforestration	To achieve soil conservation	Whole plant	Tanzania, Kenya,	(*24, 25*)
Mosquitoes pesticide	*Aedes aegypti* and *Culex quinquefasciatus*			(*26*)
Anti-bacterial	*Staphylococcus aureus*			(*27*)
Anti-moluscs	*Lymnaea natalensis*			(*28*)
Anti-parasitic nematodes				(*29*)

Duke (*12*) and Van Damme (*3*) mention that in Brazil, *E. tirucalli* is used against cancer, cancroids, epitheliomas, sarcomas, tumors, and warts although they argue that this may have no scientific basis since the same tree is known to be co-carcinogenic. In Malabar (India) and the Moluccas, latex is used as an emetic and anti-syphilitic while in Indonesia, the root infusion is used for aching bones while a poultice of roots or leaves is used to treat nose ulcers, hemorrhoids and extraction of thorns. Wood decoctions are applied against leprosy, and hands and feet paralysis following childbirth (*12*). The same author states that in Java, the plant latex is used to cure skin ailments and bone fractures. In a US patent, *E. tirucalli* extracts have been claimed to be active against prostate cancer (US 2003/0171334 A1, http://patft.uspto.gov).

Ornamental

E. tirucalli has increasingly become popular as an ornamental plant. Potted plants are placed in offices and homes but can also be grown in lawns. It is preferred for its ease of maintenance and beautiful evergreen pencil-like branches which factors have increased its international trade resulting into a wide distribution in areas where it was not endemic.

Source of Energy

E. tirucalli is looked at as a potential source of biodiesel as it can produce lots of biomass and grow in marginal areas unfit for production of other crops. There has been increasing attention on biofuel production in order to reduce over-dependence on fossil fuel (*30*). While agreeing with such a venture, Eshel *et al.* (*31*) warn that emphasis should be put on non-food sources (such as *E. tirucalli*) to avoid hunger that can result from the use of food crops as a source of biodiesel.

E. tirucalli stores abundant amounts of latex in so-called laticifers (*32, 33*). This latex can be processed into biofuel. According to earlier findings and projections from small field experiments, the hydrocarbon of the latex would be able to produce the equivalent of 15 (range: 10 to 50) barrels of oil per acre (or 37 barrels per ha; (*28*)).

Calvin (*34*, *35*) and Duke (*12*) report that latex of *E. tirucalli* contains petroleum-like hydrocarbons, largely C_{30} triterpenoids (Table II), which on cracking yield high-octane gasoline. These postulations were validated by Calvin (*36*) and although they were found to be true, extraction projects to this time have never materialized. *E. tirucalli*'s latex contains large amounts of steroles and triterpenes (*37*), which can also be used for biofuel (*17*).

Several chromatographic and spectroscopic methods can be used to analyze and characterize the latex of *E. tirucalli*. Uchida *et al.* (*38*) conducted characterization of the latex of *E. tirucalli* using GC-MS. They got the steroid, triterpenoid and diterpenoid profile of *E. tirucalli*. Expression analysis of genes encoding for enzymes involved in sterole synthesis was done through EST screening (*39*). The intermediate product of sterole and terpenoid synthesis in the mevalonate pathway is squalene (*40*). The squalene synthase gene of *E. tirucalli* has been cloned and characterized (*14*).

Apart from biofuel production, methane (biogas) can also be generated from *E. tirucalli* material. Rajasekaran *et al.* (*41*) and Van Damme (*5*), considering its reported high biomass production rates and ease with which the harvested material ferments, note that it is a potential source of methane/biogas. Sow *et al.* (*42*) experimentally demonstrated that *E. tirucalli* produces biomass that is well-suited for biogas generation under thermophilic conditions especially using chopped material. Based on productivity estimations from field research carried out near Lake Baringo (Kenya) with 80,000 plants per hectare yielding 20 dry metric tons per year, they estimated *E. tirucalli*'s potential annual methane production (with a continuous digester), at around 3,000 m^3, equivalent to approximately 3,000 l of biofuel. The same authors estimated production of about 100 metric tons of compost per year as a byproduct.

Other forms of energy associated with high *E. tirucalli* biomass, are fuelwood and charcoal. Van Damme (*1*, *3*) names the provision of charcoal and fuelwood among its traditional uses. He further explains that the plant's ability to grow in semi-arid areas, which are devoid of forests, makes it one of the few alternatives for fuelwood production in such situations. For the same reason, it has been recommended for commercial fuelwood production projects for purposes of woodlot restocking in semi-arid parts of Kenya (*43*). Mahiri (*44*) points out that *E. tirucalli* is preferred for this purpose due to its fast growth rate, high productivity, quick acclimatization to an area and ease with which it dries.

Table IV. Annual Energy Yields After Conversion to Electricity of *E. tirucalli* (Et), *Jatropha curcas* (Jc), and Oil Palm (Considering Small-Scale Communal Electrification Technology)

	Et oil	*Et gasifier*	*Et biogas*	*Jc (oil + biogas from press cake*	*Oil palm (only oil taken)*
yield/ha/yr	2 MT dry	30 MT dry	8,250 m^3	1,500 kg dry seeds /ha	1,950 liter/ha
electricity	6,600 kWh	3,000 kWh	11,880 kWh	2,330 kWh	5,850 kWh
level technology	very high[1]	high	low	medium	medium
production cost	very high	medium	medium	high	medium

[1] Oil from latex is included to give an indication of yields but the technology is too complicated to apply in communal bio-energy systems .Source: Franken (*45*).

As conversion of terpenoids of *E. tirucalli* into biofuel is only beneficial when fossil petroleum prices would reach US $150 to $200, it would seem to be preferable to converse the whole biomass into methane (*7*). Conversion of *E. Tirucalli*'s latex, sugar and cellulose into biogas through anaerobic digestion can yield high output figures. Franken (*45*) compared the energy yield of *E. tirucalli* with that from other energy plants such as *Jatropha curcas* and oil palm (*Elaeis guineensis*) (Table IV).

In summary, the latex hydrocarbon is largely a C_{30} triterpenoid which, after cracking, yields high octane gasoline (*46*). The costs for extraction, however, are enormous and oil quality will be low; upgrading to fuel quality leads to considerable losses. Taking this into consideration, about 2,200 liters of fuel oil equivalent could be harvested per hectare per year requiring advanced chemical technology against very high costs. When used to generate electricity through a gasoline generator, this amount could generate 6,600 kWh (at 3 kWh/liter) per ha per year.

Source of Rubber

E. tirucalli is reported for possessing hydrocarbon polymers that are used for manufacturing rubber substitutes. Several researchers point out that its latex is an emulsion of terpenes and resins in water, which can easily be transformed into rubber at low cost (*3*, *34*, *47*). The same authors further report that during the Second World War, its latex was used in South Africa to develop a rubber substitute which proved unprofitable as due to its high resin content, it could not yield high quality rubber. Also, due to the strong fixative power of the resin, it has for long been used on the East African coast in local gum manufacture, for fastening knife-blades to wood handles and spear-heads to shafts (*3*). In the same view, Murali and Mwangi (*23*) notes that the resin produces comparably good wood-based glue and adhesives whereas with a few modifications, it would compete favourably with other commercial resins.

Conservation and Agroforestry

Due to its favourable agronomic features such as drought resistance, *E. tirucalli* is used in semi-arid areas to carry out afforestation and re-forestation for purposes of achieving soil conservation. Van Damme points out that such plants can be used as a soil cover in places where other plants (even grasses) cannot grow (*1*). Involvement of *E. tirucalli* has been mentioned in successful reforestation and conservation programs in: Tanzania (*23*), Kenya (*24*, *43*), and Sri Lanka (*48*) among others. It has also featured in agroforestry programs (*49–51*) as a hedge plant or as an intercrop.

Other related uses of *E. tirucalli* include: boundary demarcation (*3*, *52*), live fencing around compounds and kraals (*3*, *53*, *54*), cultural connotations e.g. as a sign of starting a new home in Luo culture of East Africa (*44*) and as a windbreak in semi-arid areas (*49*). Simons (*54*) points out that the plant plays these roles due to its latex toxicity and hence low herbivore pressure.

Pesticides

E. tirucalli latex has been reported to have pesticidal features against such pests as aphids (*Brevicoryne brassicae*) (*55*), mosquitoes (*Aedes aegypti* and *Culex quinquefasciatus*) (*56*), micro-organisms such as bacteria (*Staphylocococcus aureus*) (*26*) and molluscas (*Lymneae natalensis* (*28*) and *Biomphalaria gabrata* (*57*) among others). Siddiqui *et al.* (*29*) report a dose-dependent latex toxicity to parasitic nematodes such as *Haplolaimus indicus*, *Helicotylenchus indicus* and *Tylenchus filiformis in vitro*, with increasing exposure period although some nematodes like *Meloidogyne* spp. are known to attack the plant.

The latex is also reported to be a hunters' tool applied in local fishing and arrow poisoning in tropical Africa (*58*). The pesticidal feature has been validated by Kumar (*22*) and Tiwari and Singh (*57*). Although the plant is generally mentioned as a pesticidal plant, scanty experimental work has been performed to confirm this.

Disuses

A number of disuses have also been mentioned. Associated with its vesicant and rubifacient features, *E. tirucalli* latex is reported to cause conjunctivitis (*59–62*) when it accidentally gets in contact with eyes. Eke *et al.* (*63*) report that symptoms range from mild epithelial keratoconjunctivitis to severe keratitis with stromal oedema, epithelial sloughing, and anterior uveitis which usually heal in 2 to 7 days but can also result into permanent blindness. They advise that it should be handled with caution.

Research also shows that *E. tirucalli* is co-carcinogenic. Roe (*64*) observed that papillomas and malignant tumors were elicited in mice treated with acetone extracts of *Euphorbia* lattices. Mizuno *et al.* (*65*) reports a high incidence of Burkitt's lymphoma - a latent Epstein-Barr virus (EBV) malignancy in East Africa where *E. tirucalli* is endemic. EBV causative factors were detected in soil and drinking water (where *E. tirucalli* grows) implying that people living in such areas run a high cancer risk. The findings have been further clinically validated in rats (*66–69*) some of which developed full blast lymphomas. However, folklore reports anti-cancer treatment by the latex (*70*), and there are scientific indications that it may modulate myelopoiesis and enhance resistance against tumor bearing (*71*), both of which are suggestive of a cancer cure.

E. tirucalli is known to be an irritant to herbivores and due to its nasty and acrid features, most herbivores learn to avoid it. Howes (*72*) and Simons *et al.* (*54*) point out that this is one of the reasons why it is a good live fencing material.

Conclusively, *E. tirucalli* is a miracle tree. This is undoubtedly expressed by the vast number of uses cited. Evidently, quite a lot has been done on exploration of its chemistry (*14*, *15*, *18*) and evaluation of its potential as an energy plant (*35*, *41*). However, most of the medicinal uses mentioned have been left to folklore and need validation. For example, in spite of the vast number of ailments it is reported to cure e.g. (*3*, *26*, *73*), to our knowledge, no substance of pharmaceutical importance has so far been obtained from it. Also scanty literature has been cited on validation of other functions like the reported insecticidal, nematicidal, piscicidal and molluscicidal features. This calls for more research/laboratory investigation, in order to establish scientific authenticity of these important functions and to ascertain with confidence, that *E. tirucalli* is a wonder plant for modern science.

Further to application of *E. tirucalli* for biofuel, more research should be needed to know effective plantation system and management, genetic improvement to yield higher production, minimum nurient requirement to grow and protection of potential diseases that will decrease the yield, and energy transformation coefficient.

References

1. Van Damme, P. L. J. In *Combating Desertification With Plants*; Schlissel, A, Pasternak, D., Eds.; Kluwer Academic Pub.: Norwell, MA, 1999; pp 169−187.
2. Stevens, P. F. *Angiosperm Phylogeny Website*, Version 9; June 2008; http://www.mobot.org/MOBOT/research/APweb/.
3. Van Damme, P. *Afr. Focus* **1989**, *5*, 176–193.
4. Van Damme, P. *Meded. Fac. Landbouwwet., Univ. Gent* **1985**, *37*, 1369–1382.
5. Van Damme, P. *Afr. Focus* **1990**, *6*, 19–44.
6. Gildenhuys, S. *Euphorbia W.* **2006**, *2*, 9–14.
7. *Plant Resources of Tropical Africa*; Schmelzer, G. H., Gurib-Fakim, A., Eds.; PROTA Foundation, Wagenigen, Netherlands/Backhuys Publishers, Leiden, Netherlands/CTA, Wageningen, Netherlands, 2008; pp. 296−301.
8. Bruyns, P. V.; Mapaya, R. J.; Hedderson, T. *Taxon* **2006**, *55* (2), 397–420.
9. Janssen, M. J.; Keutgen, N.; Pohlan, J. *J. Agric. Rural Dev. Trop. Subtrop.* **2009**, *110*, 39–47.
10. Webster, G. L.; Brown, W. V.; Smith, B. N. *Taxon* **1975**, *24*, 27–33.
11. Batanouny, K. H.; Stichler, W.; Ziegler, H. *Oecologia* **1991**, *87*, 565–569.
12. Duke, J. A. *Handbook of Energy Crops*; Purdue University Centre for New Crops and Plant Products, 1983; www.hort.purdue.educ (accessed on 1 March 2009).
13. Kapaczewski, W. *J. Sci. Chem. Biol.* **1947**, *20*, 924–928.
14. Uchida, H.; Ohyama, K.; Suzuki, M.; Yamashita, H.; Muranaka, T.; Ohyama, K. *Plant Biotechnol.* **2010**, *27* (1), 105–109.
15. Khan, A. Q.; Ahmed, Z. *Chem. Sci.* **1988**, *43*, 1059–1062.
16. Khan, A. Q.; Malik, A. *J. Nat. Prod.* **1990**, *53*, 728–731.
17. Nielsen, P. E.; Nishimura, H.; Liang, Y.; Calvin, M. *Phytochemistry* **1979**, *18*, 103–104.
18. Rasool, N.; Khan, A. Q. *Phytochemistry* **1989**, *28*, 1193–1196.
19. Khan, A. Q.; Rasheed, T.; Malik, A. *Heterocycles* **1988**, *27*, 2851–2856.
20. Trivedi, P.; Sharma, C.; Datta, S. *Indian J. Nematol.* **1986**, *16*, 279–284.
21. Yanxia, D.; Chonggang, X.; Long, Y. *Plant Prot. (Beijing)* **2007**, *33*, 99–102.
22. Kumar, A. *Some Potential Plants for Medicine from India, Ayurvedic Medicines*; University of Rajasthan: Rajasthan, 1999; pp 1−12.
23. Murali, R.; Mwangi, J. G. In *International Conference on Domestication and Commercialization of Non-Timber Forest Products in Agro-systems*, Proceedings of an International FAO Conference held in Nairobi, Kenya, 19-23 February 1996; 1998.

24. Smith, W.; Meredith, T.; Johns, T. *Econ. Bot.* **1996**, *50*, 290–299.
25. Macharia, P. *Environ. Monit. Assess.* **2004**, *99*, 141–147.
26. Rahuman, A.; Gopalakrishnan, G.; Venkatesan, P.; Geetha, K. *Parasitol. Res.* **2008**, *102*, 867–873.
27. Lirio, L.; Hermano, M.; Fontanilla, M. *Pharm. Biol.* **1998**, *36*, 357–359.
28. Vassiliades, G. *Rev. Elev. Med. Vet. Pays. Trop.* **1984**, *37*, 32–34.
29. Siddiqui, M.; Alam, M.; Trivedi, P. In *Nematode Management in Plants*; Scientific Publishers, India, 2003; pp 173-174.
30. Prusty, B.; Chandra, R.; Azeez, P. *Nat. Prec.* **2009** doi:10.1038/npre.2008.2658.1.
31. Eshel, A.; Zilberstein, A.; Alekparov, C.; Eilam, T.; Oren, I.; Sasson, Y.; Valentini, R.; Waisel, Y. In *Proceedings of the 5th IASME WSEAS International Conference on Recent Advances in Energy & Environment* (EE 2010), 2010; pp 362−367.
32. Fahn, A. *Plant Anatomy*, 4th ed.;, Pergamon Press: New York, 1990; ISBN: 0-08-037491-3.
33. Uchida, H.; Yamashita, H.; Kajikawa, M.; Ohyama, K.; Nakayachi, O.; Sugiyama, R.; Yamato, K. T.; Muranaka, T.; Fukuzawa, H.; Takemura, M. *Planta* **2009**, *229*, 1243–1252.
34. Calvin, M. *Pure Appl. Chem.* **1978**, *50*, 407–425.
35. Calvin, M. Petroleum plantations for fuel and materials. *Bioscience* **1979**, *29*, 533–538.
36. Calvin, M. *Naturwissenschaften* **1980**, *67*, 525–533.
37. Nielsen, P. E.; Nishimura, H.; Liang, Y.; Calvin, M. *Phytochemistry* **1979**, *18*, 103–104.
38. Uchida, H.; Ohyama, K.; Suzuki, M.; Yamashita, H.; Muranaka, T.; Ohyama, K. *Plant Biotechnol.* **2010**, *27*, 105–109.
39. Kajikawa, M.; Yamato, K. T.; Kohzu, Y.; Sakata, R.; Fukuzawa, H.; Uchida, H.; Ohyama, K. *Plant Biotechnol.* **2004**, *21*, 349–353.
40. Shechter, I.; Guan, G.; Boettcher, B. R. In *Comprehensive Natural Products Chemistry: Isoprenoids Including Carotenoids and Steroids*; Cane, D. E., Ed.; Elsevier Science: Oxford, 1999; Vol. 2, pp 245–266.
41. Rajasekaran, P.; Swaminathan, K.; Jayapragasam, M. *Biol. Waste* **1989**, *30*, 75–77.
42. Sow, D.; Ollivier, B.; Viaud, P.; Garcia, J. L. *J. Appl. Microbiol. Biotechnol.* **1989**, *5*, 547–550.
43. Mahiri, I. *Land Degrad. Dev.* **2002**, *14*, 163–171.
44. Mahiri, I. *PLA Notes* **1998**, *31*, 4–8.
45. Franken, Y. J. http://www.fact-foundation.com, 2011.
46. Orwa, C.; Mutua, A.; Kindt, R.; Jamnadass, R.; Simons, A. *Agroforestree Database: A Tree Reference and Selection Guide*, Version 4.0; http://www.worldagroforestry.org/af/treedb, 2009.
47. Uzabakiliho, B.; Largeau, C.; Casadevall, E. *Phytochemistry* **1987**, *6*, 3041–3045.
48. Melvani, K. In *XIII World Forestry Congress*, Buenos Aires, Argentina, 18−23 October 2009.

49. Jama, B.; Njui, A.; Njenga, K. *Management and Utilization of Dryland Forests in Sub-Saharan Africa: The Role of Agroforestry*, VITRI/ETFRN/IUFRO-SPDC Workshop: Trees, Agroforestry And Climate Change In Dryland Africa, Hyytiälä, Finland, 30 June - 4 July 2003.
50. Long, A. J.; Nair, P. K. R. *New Forest.* **1999**, *17*, 145–174.
51. Mbwambo, L. *Drylands Agroforestry Workshop Held at ICRAF*, Nairobi-Kenya, 1 - 3 November 2004.
52. Kindt, R.; Van Damme, P.; Simons, A.; Beeckman, H. *Agroforestry Syst.* **2006**, *67*, 215–228.
53. Nascimento, V.; Sousa, L.; Alves, A.; Araújo, E.; Albuquerque, U. *Environ Dev. Sustainability* **2009**, *11*, 1005–1029.
54. Simons, T.; Schmelzer, G.; Omino, E. *Auxiliary Plants.*; Prota 9, PROTA Foundation, Nairobi, Kenya, 2004.
55. Mwine, J.; Van Damme, P. In *World Food System – A Contribution from Europe*, ETH Zurich, Switzerland, 14−16 September 2010.
56. Rahuman, A.; Gopalakrishnan, G.; Venkatesan, P.; Geetha, K. *Parasitol. Res.* **2008**, *102*, 867–873.
57. Tiwari, S.; Singh, A. *Chemosphere* **2006**, *64*, 36–42.
58. Neuwinger, H. D. *Toxicon* **2004**, *44*, 417–430.
59. Hsueh, K.; Lin, P.; Lee, S.; Hsieh, C. *J. Chin. Med. Assoc.* **2004**, *67*, 93–98.
60. Joshi, D.; Shingal, G. *J. Arm. Force India* **2008**, *64*, 293–294.
61. Scott, I.; Karp, C. *Brit. Med. J.* **1996**, *80*, 823.
62. Shlamovitz, G.; Gupta, M.; Diaz, J. *J. Emerg. Med.* **2009**, *36*, 239–241.
63. Eke, T.; Al-Husainy, S.; Raynor, M. K. *Arch. Ophthalmol.* **2000**, *118*, 13–16.
64. Roe, F. J. C.; Peirce, W. E. H. *Cancer Res.* **1961**, *21*, 338–345.
65. Mizuno, F.; Osato, T.; Imai, S.; Koizumi, S.; Aya, T.; Kinoshita, T.; Hirai, N.; Hirota, M.; Ohigashi, H.; Koshimizu, K. *AIDS Res.* **1986**, *2*, S151–S155.
66. Fürstenberger, G.; Hecker, E. *Biosciences* **1985**, *40*, 631–646.
67. Imai, S.; Sugiura, M.; Mizuno, F.; Ohigashi, H.; Koshimizu, K.; Chiba, S.; Osato, T. *Anticancer Res.* **1994**, *14*, 933–933.
68. MacNeil, A.; Sumba, O. P.; Lutzke, M. L.; Moormann, A.; Rochford, R. *Brit. J. Cancer* **2003**, *88*, 1566–1569.
69. Silva, A. C. P.; de Faria, D. E. P.; Borges, N. B. D.; de Souza, I. A.; Peters, V. M.; Guerra, M. D. *J. Ethnopharmacol.* **2007**, *110*, 154–159.
70. Cataluna, P.; Rates, S. M. K. *Second World Congress on Medicinal and Aromatic Plants for Human Welfare Wocmap-2: Pharmacognosy, Pharmacology, Phytomedicines, Toxicology*, WOCMAP, 1999; pp 289−295.
71. Valadares, M. C.; Carrucha, S. G.; Accorsi, W.; Queiroz, M. L. S. *Int. Immunopharmacol.* **2006**, *6*, 294–299.
72. Howes, F. *Kew Bull.* **1946**, *1*, 51–87.
73. Cataluña, P.; Rates, S. M. K. *Medicinal and Aromatic Plants*, ISHS Acta Horticulturae, 1997; pp 1−14.

Chapter 2

Marula [*Sclerocarya birrea* (A.Rich) Hochst]: A Review of Traditional Uses, Phytochemistry, and Pharmacology

Gerhard Prinsloo[1] and Renée A. Street[*,2]

[1]University of South Africa (UNISA), Department of Agriculture and Animal Health, Florida Campus, Florida, 1710, South Africa
[2]University of KwaZulu-Natal, Traditional Medicine Laboratory, Durban, 4041, South Africa
[*]E-mail: street@ukzn.ac.za.

Marula (*Sclerocarya birrea*) is a widespread species throughout the semi-arid deciduous savannas of much of sub-Saharan Africa. The tree has been a part of civilization since ancient times with use of all parts of the tree. The fruits are utilized for food, juice, jelly, jam and beer, the bark for medicinal purpose, the kernels for food and oil and the wood for fuel wood. There has been a wide interest in the medicinal uses of the tree since it has been utilized for centuries to treat diseases. Recent studies revealed hypoglaecemic activity as well as other positive attributes such as anti-inflammatory and antibacterial properties. There have been numerous studies conducted to elucidate its valued components which contribute to its medicinal properties. Several initiatives are exploiting the commercial value of the tree such as the production of internationally acclaimed alcoholic drinks such as Amarula liquer.

Sclerocarya birrea (marula) belongs to the family Anacardiaceae also known as the Mango family. The genus Sclerocarya has about five species found in southern Africa. The tree is widely distributed in Africa, but only the subspecies *caffra* is found in southern Africa (*1*). There is a real initiative to exploit the subspecies *caffra* in eastern and southern Africa, however, the subspecies *birrea* remains very much underutilized and less studied in western Africa (*2*). The Marula tree is a medium sized, single stemmed tree of up to 15 meters in height (Figure 1). The rough stem bark is flaky, with a mottled appearance due to the contrasting grey and pale brown patches. Male and female flowers appear separately, usually but not always on separate trees. The flowers are small, with red sepals and yellow petals (*1*). The fruit which is the size of a plum is pale yellow when ripe. The rounded slightly flattened fruit is about 30 mm in diameter and borne in late summer to mid-winter (Figure 1). The fruits are aromatic and edible, and are much sought after for their delicious pulp and edible nuts (*3*). Trees start to bear fruit at an average size of 42.8 cm circumference which relates to an approximate age of 19 years. About 92% of the fruit can be removed without impacting the current population profile (*4*). Fruit of 6 year old Marula seedlings have been recorded (*5*). The trees flower from September to November and bear fruit rom January to March (*6*).

Figure 1. Sclerocarya birrea (marula) tree.

Figure 1. Marula leaves and immature fruits.

Of all fruit trees indigenous to South Africa, the Marula has received most attention in terms of domestication and commercialization. It is known for its medicinal uses (*4*) but several products such as beer, juice, jam and jelly have also been developed from the mesocarp and successfully marketed, the most recent being a liqueur (*4*, *7*, *8*). The kernels are important as a food source and as flavouring and as a preservative for other foods, and the use of the wood for firewood and carvings (*4*). The hard stone of the fruit contains a small nut which is a sought after food item. These nuts are highly nutritious and contain valuable oil. Considerable skill is acquired to crack the hard shells to extract the nuts. The Tsonga people of southern Africa and Mozambique use the oil for cooking and as a moisturizer for women and as baby oil (*9*).

The plant is widely used in traditional medicine in Africa against many diseases and affections such as hypertension, dysentery, stomachache and gastroenteritis. The bark decoction has been used as an anti-cough remedy, the leaves, pulp and fruit and mistletoe used for hypertension and other affections. The bark and leaves are also used as antihyperglaecemic (*10*, *11*). It is also used for a prophylactic remedy against gangrenous rectitis, and the fruits for the destruction of ticks. The Zulus regard the fruit as a potent insecticide (*12*). Powdered bark is allegedly administered to pregnant women to regulate the sex of babies (*12*). In South Africa, diarrhea, dysentery and unspecified stomach problems are treated with the bark, which is believed to be of value in combating fever and in the treatment of malaria. It is also used as a general tonic. Decoctions of the bark or roots are taken orally or as enemas (*6*, *9*). Leaf infusions or decoctions are drunk for diabetes while chewing the fresh leaves and swallowing the astringent juice helps with indigestion (*13*).

Studies on Validation of the Medicinal Uses of *S. birrea*

A number of studies have been performed on *S. birrea* to validate the use and mechanisms of the different parts of the plant that are used. Among these are hypergleacemia, hypertension, anti-inflammatory, anti-bacterial and many more.

Hyperglaecemia

The stem-bark aqueous extract produced dose-dependent significant reductions in the blood glucose concentrations of both fasted normal and fasted diabetic 'test' rats (*12*). Optimum reduction in blood glucose levels were observed at a dose of 800 mg/kg p.o. The aqueous extract like chlorpromide – a sulphonylurea antidiabetic agent, produced significant reductions in the blood glucose levels of fasted and normal fasted STZ-treated diabetic rats. The dichloromethane (DCM):methanol (1:1) extract of *S. birrea* bark decreased the blood glucose level and increased plasma insulin levels in STZ-treated rats. The treatment significantly decreased the plasma glucose levels by 70%. The effect was ascribed to stimulation of insulin secretion or protection of the ß-cells from further deterioration (*11*). A significant improvement was seen in glucose tolerance during an oral glucose tolerance test in diabetic rats treated with the extract, but the organic root extract of *S. birrea* was unable to stimulate glucose utilization (*14*). The extract caused a marked increase in glucose utilization in Chang liver cells and C2C12 muscle cells (*11*).

Sclerocarya birrea stem-bark ethanolic (SBE) extract has hypoglaecemic properties. The effect is seen as the extract ameliorates kidney function and lowers blood glucose of rats. It conceptualizes the hypoglaecemic, renal and hypotensive effects of the extract. The hypoglaecemic, hypotensive and renal effects are however independent of each other. The hypoglaecemic effects might be associated with renoprotective and blood-lowering effects (*15*) .

The SBE extract mediates the antihyperglycemic effect similar to metformin. The study shows an extra pancreatic effect of SBE. STZ-induced diabetic rats have permanently destroyed pancreatic ß-cells, thus SBE may have some chemical constituents that are capable of reducing blood glucose levels via metformin mechanisms. The SBE extract shows a possible renal and cardiac protection (*15*) .

Antibacterial Activity

Antimicrobial activity ranged from 0.15-3 mg/ml (*16*). Most of the antibacterial activity was in highly non-polar fractions which were chloroform and carbon tetrachloride. Some activity was also found in the n–butanol fractions. According to the bioautograms there are only two major bacterial inhibiting compounds present in the leaf extracts. Based upon TLC analysis, the components in the bark and leaves differed, but in both extracts the Rf values of the inhibiting compounds were the same. That might be an indication of similar compounds in the extracts that are responsible for the activity of the extracts. The leaf extract was effective against *Staphyllococcus aureus, Pseudomonas aeruginosa, Escherichia coli* and *Enterococcus faecalis*.

Ethanolic extracts exhibited high antibacterial activity of < 0.1 mg/ml. The twig extract was the most active when compared to the leave and opercula extracts, with a total activity of 1609.1 ml/g against *Bacillus subtilis* and a MIC of 0.098 mg/ml. The leaf ethanolic extracts showed an activity of 0.390 mg/ml against *B. subtilis* (*17*). The twigs were twice as active as the leaves (*17*) and the leaves are 11 times less active than the stem bark extracts (*16*). Although the bark is probably preferred for use in medicinal preparations as it contains larger quantities of the antibacterial compounds (*16*), its use is unsustainable.

The methanol extract of Marula showed antibacterial activity against *E. faecalis* at 0.4 mg/ml and *S. aureus* at 0.1 mg/ml (*18*). The methanol extract also showed some anthelmintic activity at 0.5, 1 and 1.5 mg/ml (*18*) Recent studies have shown that the kernel oil is effective in inhibiting growth of one Gram-positive bacterial strain (*S. aueus)* and two Gram-negative strains *(P. aeruginosa* and *Salmonella choleraesuis*) (*19*).

Polydipsia

The increase in body weight of rats fed with *S. birrea* extract (300 mg/kg) was similar to those in the control groups in a hyperglaecemic study. The *S. birrea* extract possibly compensated for the energy supplied from food intake, or contributed to the metabolism. The untreated diabetic rats showed significant reduction in body weight. Treatment with *S. birrea* extracts also showed a reduced food intake compared to the untreated diabetic rats. Both metformin and *S. birrea* reduced polydipsia in this study. The extract also had a tendency to decrease the plasma cholesterol, triglyceride and urea (*11*).

Amoebic Dysentery

In one study where African medicinal plants were screened, *S. birrea* was one of four medicinal plants that showed the greatest activity against the causal organism of amoebic dysentery, *Entamoeba histolytica.* The other three plants being *Acoruscalamus, Albizia adianthifolia* and *Deinbollia oblongifolia* (*20*) .

Anti-Inflammatory Effect

Results indicate that both the aqueous and methanolic extracts of *S. birrea* stem-bark posses anti-inflammatory activity. Activity was relatively moderate in comparison to acetylacetic acid (ASA). The methanolic extract was also more active than the aqueous extract. No mechanism has been proposed for the mode of action (*12*). Petroleum ether and DCM extracts exhibited high COX-1 activity of 90.7-99.8 % and COX-2 activity of 69-92.6 % at concentrations of 250 μg/ml (*17*). Both the bark and the leaves showed comparable results (*17*).

Muscle Relaxation

The leaves are reported to have a relaxing effect on skeletal smooth muscles, but the leaf extract had no effect on the resting calcium (Ca) levels. The mode of action is believed to affect Ca signaling in the smooth muscles. The crude, aqueous, ethanolic and chloroformic extracts showed significant antagonistic effects on the caffeine-induced Ca released from the sarcoplasmic reticulum. The crude extract was the most effective, followed by the ethanolic extract, the aqueous and the chloroformic extracts (*10*).

Hypertension

The ethanol extract from leaves showed inhibition of the angiotensin converting enzyme (ACE) (*21*). This might explain the antihypertensive effect in traditional medicine but an additional decrease in Ca^{2+} mobilization could also be relevant (*10*). Apart from the hypoglaecemic effect, the renal fluid and electrolyte handling were unchanged after the SBE treatment, whereas the mean arterial pressure was reduced. This suggests that the hypotensive effect of the extract is mediated through influences on components of the cardiovascular system. The hypotensive effects might be from reduction of endothelium-synthesised nitric oxide concentration (*15*).

Anti-Mycobacterial Activity

A few South African medicinal plants were tested for anti-mycobacterial activity of which Marula was included. Marula did not show good activity with the bark extracts with a MIC greater than 100 μg/ml and therefore showed no activity as this was the highest concentration tested (*22*).

Anti-Malarial Activity

Bark decoctions are traditionally used as a prophylactic and cure for malaria (*23*). Marula exhibited high *in vitro* and *in vivo* activity when tested singly to evaluate its anti-malarial properties (*24*). To elaborate on these, further experiments were carried out to determine the activity in combination with other plants and other assays. Marula however was the least potent extract against *Plasmodium berghei* in mice when compared to *Turrea robusta* and *Lannea schweinfurthii* (*24*). It also did not show the best *in vitro* anti-plasmodial results with water and methanol extracts (*24*). In a review paper on South African plants with antimalarial activity Marula is also mentioned to have a low anti-malarial activity with an IC_{50} of 30-40μg/ml against a D10 strain of parasite (*25*).

Toxicity

Several studies have been reported on the toxicity of *S. birrea* extracts. These studies have, however, revealed contrasting results which might be linked to the solvents used and the compounds extracted by each. *Sclerocarya birrea* was tested for its effect in the bacterial AMES test on two cell lines. The leaf, root and bark extracts were all tested for genotoxic effects. The DCM and 90% methanol extracts of *S. birrea* showed no activity on the TA98, and TA 100 cell lines (*26*) . In contrast, *S. birrea* was one of a number of South African medicinal plants that showed genotoxicity in the micronucleus test, in the form of structural and numerical and chromosome aberrations (*12*, *20*, *26*). The MTT assay also showed that high doses of SBE (600-1000 ug/ml) decreased cell viability and that the proximal cells exhibited more sensitivity. The extract was prepared with 95 % methanol (*15*) .

Marula was included in a study to evaluate toxic plants used by Traditional Health Practitioners (THP) and plants that are used when toxic signs surface. *Sclerocarya birrea* was mentioned as being treatment of toxic signs and not as one of the toxic plants. Traditionally, a decoction of the bark of *S. birrea* is taken internally as a purge and a macerate of the twig-bark against snake-bite. The effects could probably be ascribed to the presence of tannins in the plant (*27*). Cold DCM:methanol (1:1) and purified water extracts were also tested for toxic effects (*14*). The organic bark extract showed clear toxicity where the aqueous extracts of the stem and bark and organic extract of the stems showed no toxicity. The organic bark extracts might raise concern for its chronic use. Very low toxicity was observed on VERO cells with methanol and water bark extracts with CC50 values 361.24 and 3375.22 μg/ml respectively (*24*). No toxicity was observed against the brine shrimp larvae (*18*).

Secondary Metabolites and Nutrient Composition

Studies have been conducted that examine the natural plant products that may be associated with the plants medicinal activity. The benefit of this plant also resides in its nutritional content that probably contributes to the beneficial effects of *S. birrea*. When the nutrient content of the seeds of Marula were compared with seeds of four other browse trees from the hardveld region of Botswana namely *Lonchocarpus capassa, Zizyphus mucronata, Kirkia acuminata* and *Rhus lancea,* it was found that the Mg, Ca, P and K levels were within the acceptable ranges for all the seeds except for *S. birrea*. All the seeds are also a rich source of protein except for *S. birrea* which has a very low dry matter digestibility. The Mn, Cu and Zn also measured the lowest in *S. birrea* (*8*). Nevertheless, the seed is known to contain high amounts of Vitamin C, proteins and oils in the kernels (*8*).

During periods of grain shortage, people in Niger increase their reliance on wild plant foods to supplement their diets. The pit of *S. birrea* is one of these supplements and it was found that the pits contain relatively large amounts Cu, Mg and Zn. The Fe, Mn, Na and K compared to other edible supplements were very low. The protein content of the pit was also high, however, the protein fraction contains relatively low proportions of leucine, phenylalananine, lysine and threonine. It therefore falls short of the WHO requirements for essential amino acids. The amino acid composition with exception of lysine is comparable to human milk and whole hens' eggs (*28*).

The pits have been described as a delicacy and yield an oil with a quality (fatty acid composition) comparable to olive oil, but with 10 times greater stability owed to the tocopherol/sterol composition (*28*). The pit contains large amounts of fatty acids which constituted 47 % of the dry weight. The fatty acid, oleic acid constituted two thirds of all the fatty acids in the pit. The essential fatty acid linoleic acid was present, but alpha-linolenic acid was not (*29*). The oil content of *S. birrea* kernels increased with drawn out harvest dates from 23% at the first harvest date to 63% at the last date of harvest (*19*).

The ripe fruit also contains vitamin C with a content of 168 mg/100 g which is approximately three times that of oranges and comparable to the amounts present in guavas (*7*). The ascorbic acid content of the best clones was found to be approximately 10 times higher than orange and pomegranate from the local market (*5*). The ascorbic acid content varies considerable, depending on the area and the genetic variation of the trees (*28*) . The main sugar in the gum is galactose (63%) without any rhamnose (*30*) . The fruits are rich in Vitamin C (194 mg per 100 g at 85 % moisture) (*28*). The ascorbic acid content is also high and ranges between 7 and 21 mg/g dry weight (*5*). Jams retained between 50 -84 % of the ascorbic acid after 45 days and stayed high during the pasteurization process (*5*) .

When the phenolic composition of three fruits were compared, it was found that *S. birrea* had the highest total phenolics, flavonoids and condensed tannins. The other fruits in the study were *Flacourtia indica* (Batoka plum) and *Opuntia megacantha* (prickley pear). *F. indica* produces fruit that is eaten fresh and has a pleasant but rather sour taste whereas *O. megacantha* produces pear-shaped golden green or purple sweet and juicy fruits within a tough skin that is studded with spines. The total phenolic content of *S. birrea* was eight times higher than that of *F. indica* and *O. megacantha. S. birrea* pulp contained twice as many flavonoids in the pulp when compared to the peel (*31*).

The bark has an astringent taste and the antidiarrhoeal effects have been experimentally linked to procyanidins (*9*). Hypoglaecemic effects have been associated with flavonoids (*11*) . It is known that triterpenoid compounds such as oleanic acid, ursolic acid and uvaol possesses hypoglaecemic, anti-inflammatory and hypotensive properties in animal models (*15*). There are a variety of ways to control the hyperglaecemic syndrome. As well as insulin and its substitutes, a number of other substances have an effect, for example somatostatins, pituitary and sex hormones, corticosteroids and prostaglandins. The variety of hypoglaecemic plant constituents and their diversity in action is not surprising as they may act on different aspects of diabetes. A number of plants (e.g. *Allium*, *Corchorus*) only act in the presence of a minimum amount of ß-cells and their action is mediated through insulin (*32*). Plant extracts rich in organic sulphides seem to remove insulin-inactivating compounds through their SH-groups. Similarly nicotinic acid is known to be an insulinase inhibitor (*32*). Anthocyanidins are believed to act by improving the vascularisation of the pancreas. Insulin is fixed on the proteins at the cell surface, thus cellular membranes play an important role in diabetes. Anthocyanidins are said to slow down or inhibit these modifications of the capillary walls, especially in the early stages of the disease. The improvement of diabetes itself obtained from some plants rich in anthocyanosides could possibly be due to recovery of the vascularisation of the pancreas. Other flavonoids which also appear to act on the capillaries could have similar action (*32*). Appreciable high levels of total phenolic compounds (14.15 mg GAE/g), proanthocyanidins, gallotannins (0.24 mg GAE/g) and flavonoids (1.21 mg CE/g) are found in Marula.

The highest concentrations were found in the young stem extracts (*17*). The polyphenol content of the fruit peaked at three weeks post-abscission, but data was not collected subsequent to this (*5*). A quantitative study of the phenolic constituents of wild and cultivated leaves was carried out by HPLC-UV/PDA and LC-MS (*33*). Phytochemical analysis of the wild plants led to the isolation of one new flavonol glycoside, quercetin 3-O-alpha-l-(5' '-galloyl)-arabinofuranoside (*1, 33*) .

Marula also exhibited high antioxidant and free radical scavenging activity, even higher than standard synthetic antioxidants from the stem bark and young stems in the DPPH assay. In fact flavanoids and phenolic derivatives may be responsible for this antioxidant activity (*33*). Marula provides a substantial source of secondary metabolites which acts as natural antioxidants and acetylcholinesterase inhibitors (*17*) .

Marula as an Income Generator for Resource Poor Communities

New initiatives in agroforestry are seeking to promote poverty alleviation and environmental rehabilitation in developing countries through the use of indigenous trees (*28*). Domestication of local tree species is investigated to be explored in local and international markets (*28*). Identification of indigenous trees with potential commercial use in the international food industry especially in the dry regions of southern Africa could be a solution to sustainable income generation for communities (*28*). Marula is one of these indigenous trees with potential to become a sustainable income generator in resource poor communities. Marula has been proposed as a sustainable income generator involving communities and use of communal land in the Bushbuckride area of South Africa (*34*).

Marula trees can yield up to 1.5 tonnes fruit per tree (*28*) . The potential of the Bushbuckridge area which borders the Kruger National Park in South Africa is considerable higher than the current value derived from that area (*34*). The total value of Marula was estimated at 194.12 million US$ with a unit value of 1191.6 US$/ha in the Bushbuckridge area (*34*) .

The urgency to conserve, improve and bring fruit trees into cultivation has been widely recognized in southern Africa (*35*). Due to the lack of cultivation practices and ecological interactions, various agronomic practices revealed the suitability of Marula for resource poor communities (*35*). Marula did not react positively on plant growth with the addition of fertilizer, manure and dry-season irrigation (*35*). Manure seemed to have a positive effect on Marula. Marula is however a slow grower which did not flower or fruit in the 36 months after planting (*35*). Marula shed their leaves in the dry season and form new leaves during the rainy season (*35*). Marula is an alternate bearing species as well as dioecious species which complicates cultivation of these trees. Differential yield from year to year is an important trait from the prospective of commercial planting, especially in the case of a new species intended for the horticultural industry (*5*).

Commercial Use

The Mitsubishi Corporation have brewed a beer named Afreeka which underwent trials in the UK in 1997 (*28*). The internationally popular liqeur Amarula is marketed by the Distillers Corporation. In Zambia a wine Marulam is marketed commercially (*28*). Numerous small enterprises in southern Africa produce jam and jellies made from Marula (*28*). In 1985 it was reported that 600 tonnes of juice was produced in South Africa (*28*). Since then the figure increased dramatically as it was introduced in several alcoholic beverages internationally. The limiting factor is considered to be collection from wild trees (*28*).

The increasing demand for natural ingredients improving health and appearance is also attracting beverages as the fastest growing segment of the functional food market (*36*). African fruits are explored in the beverage industry to aid the super fruits such as berries and grapes as sources vitamins minerals and antioxidants (*36*).

There is a growing trend of replacing synthetics and reverting to the use of natural oils in the cosmetic and pharmaceutical industries (*37*). Moringa oil is stable and is known for its use in cosmoceutical industry. Both Moringa and Marula oil are classified as high oleic (18:1) oils with having about 70% of high oleic (18:1). Marula oil contains very small amounts of long chain fatty acids which impacts on the way the oil feels on the skin in a cosmetic application (*37*). It is significantly less stable than Moringa oil and with a lower slip value than Moringa oil it would not glide as easy on the skin. Marula oil also has a higher spread value (20.6%) than Moringa oil (15.5 %) which would be problematic in highly pigmented formulations such as lipstick and mascara (*37*). It is reported that communities consume the fresh fruit and kernels and produce beer or wine with juice in a lesser extent (*38*).

The wood is used for fuel wood, but not for fencing, poles or utensils as the quality is poor. In South Africa, fallen fruit is collected mostly by women. During a collection trip about 25-45 kg fruit can be collected with a total of between 1200-1500 kg per season. A third of the collected fruit are sold to either Ditell/Mirma in Phalaborwa (Amarula Cream Liquer) or the Mineworkers Development Centre (MDC) Marula project. No fruit is sold to other buyers. An 80 kg bag of fruit is sold to these companies for $3. Beer/wine and selling of kernels are responsible for a smaller portion of the income for community members. Beer/wine relates to $145 per trader per season. About $31 could be earner from selling kernels per season. Cultivation of Marula is prevalent in South Africa in areas which were investigated (*38*).

Process for fruit collection and beer/wine production in Namibia includes processing directly under Marula trees in the owners' fields. The regularity of processing depends more on the availability of unemployed women in the household, on the strength of social ties, and on the abundance of trees in the area, rather than on the number of trees in people's own fields.

Conclusion

There appears to be agreement by the scientific community on the hypoglaecemic effects of *S. birrea* extracts. Several other positive attributes such as anti-inflammatory and antibacterial effects have also been studied. There is however still confusion and much speculation by researchers concerning the secondary metabolite content responsible for the medicinal properties. There is also not clarity on the mode of action on how *S. birrea* extracts are responsible for the attributes ascribed by the researchers. In addition there is no agreement on the potential toxicity of the plant. It seems, however, as if the plant could be relatively safe to use with no acute toxicity (*3*). The effects of the plant in vivo have not been studied and chronic use seems to raise some concern (*14*) .

The anti-bacterial effects described by Eloff (*16*) were found in the non-polar fractions (*10*). Ojewole (*3*) and Dimo et al. (*11*) found activity in the DCM:methanol or aqueous extracts. Van den Venter et al. (*14*) found the DCM:methanol extract of the stem to have the highest activity. The DCM:methanol extract of the bark showed some toxicity and the aqueous extracts of the stem and bark seems to have relatively low toxicity with acceptable activity.

The plant material can be harvested between January and June. The Marula has the potential to become a fruit crop on a commercial scale since the fruits are tasty, rich in antioxidants and particularly suitable for industrial uses (*5*). There seems to be considerable scope for Marula to be used as an income generator in poor communities with options to earn a living. Although the opportunity exists for communities to collect Marula fruit and sell it to companies, the income seems to be very low and not very dependable. It seems to be a contradicting situation as the liquer Amarula is considered to be one of the top ten selling liquers internationally.

As there are various products to be developed from Marula there should be more options for communities to earn a living by selling and processing the fruit (*28*, *38*). Some of these products include juice, wine, beer, cosmetics and oil from the pits (*28*). Current developments in producing health drinks and other beverages might prove to open up more possibilities for Marula fruit collectors (*36*).

Acknowledgments

We thank the National Research Foundation (South Africa), Medical Research Council (South Africa) and the University of KwaZulu-Natal for their financial assistance. Sections of this review were inspired from discussions with traditional medicine bioprospecting platform of NIKSO (DST). Karen de Jager is thanked for the photographs.

References

1. Van wyk, B.-E.; Gericke, N. *People's plants. A guide to useful plants of Southern Africa;* Briza Publications: South Africa, 2000; p 351.
2. Gouwakinnou, G.; et al. Local knowledge, pattern and diversity of use of *Sclerocarya birrea. J. Ethnobiol. Ethnomed.* **2011**, *7* (8), 1746–4269.
3. Ojewole, J. Hypoglaecemic effect of *Sclerocarya birrea* [(A. Rich.) Hochst.] [Anacardiaceae] stem-bark aqueous extract in rats. *Phytomedicine* **2003**, *10*, 675–681.
4. Emanuel, P.; Shackleton, C.; Baxter, J. Modelling the sustainable harvest of *Sclerocary birrea* subsp. caffra fruits in the South African lowveld. *Forest Ecol. Manag.* **2005**, *214*, 91–103.
5. Hillman, Z.; Mizrahi, Y.; Beit-Yannal, E. Evaluation of valuable nutrients in selected genotypes of marula (*Sclerocarya birrea spp. caffra*). *Sci. Hortic. (Amsterdam, Neth.)* **2008**, *117*, 321–328.
6. Mathabe, M.; et al. Antibacterial activities of medicinal plants used in the treatment of diarrhea in Limpopo Province, South Africa. *J. Ethnopharmacol.* **2006**, *105*, 286–293.
7. Viljoen, A.; Kamatou, G.; Baser, K. Head-space volatiles of marula (*Sclerocarya birrea subsp. caffra*). *S. Afr. J. B.* **2008**, *74*, 325–326.
8. Aganga, A.; Mosase, K. Tannin content, nutritive value and dry matter digestibility of *Lonchocarpus capassa*, *Zizyphus mucronata*, *Sclerocarya birrea*, *Kirkia acuminata* and *Rhus lancea* seeds. *Anim. Feed Sci. Technol.* **2001**, *91*, 107–113.
9. Van Wyk, B.-E.; Van Oudtshoorn, B.; Gericke, N. *Medicinal plants of South Africa*; Briza Publications: South Africa, 2000; p 304.
10. Belemtougri, R.; et al. Effects of *Sclerocarya birrea* (A. rich) hochst (anacardiaceae) leaf extracts on calcium signaling in cultured rat skeletal muscle cells. *J. Ethnopharmacol.* **2001**, *76*, 247–252.
11. Dimo, T.; et al. Effect of *Sclerocarya birrea* (Anacardiaceae) stem bark methylene chloride/methanol extract on streptozocin-diabetic rats. *J. Ethnopharmacol.* **2007**, *110*, 434–438.
12. Ojewole, J. Evaluation of the anti-inflammatory properties of *Sclerocarya birrea* (A. Rich.) Hochst. (family: Anacardiaceae) stem-bark extracts in rats. *J. Ethnopharmacol.* **2003**, *85*, 217–220.
13. Iwu, M. *Handbook of African Medicinal Plants*; CRC Press: Boca Raton, FL, 1993.
14. Van de Venter, M.; et al. Antidiabetic screening and scoring of 11 plants traditionally used in South Africa. *J. Ethnopharmacol.* **2008**, *119*, 81–86.
15. Gondwe, M.; et al. *Sclerocarya birrea* [(A. Rich.) Hochst] [Anacardiaceae] stem-bark ethanolic extract (SBE) modulates blood glucose, glomerular filtration rate (GFR) and mean arterial blood pressure (MAP of STZ-induced diabetic rats. *Phytomedicine* **2008**, *15*, 699–709.
16. Eloff, J. Antibacterial activity of Marula (Sclerocarya birrea (A. rich.) Hochst. Subsp. caffra (Sond.) Kokwaro) (Anacardiaceae) bark and leaves. *J. Ethnopharmacol.* **2001**, *76*, 305–308.

17. Moyo, M.; et al. Phenolic composition, antioxidant and acetylcholinesterase inhibitory activites of *Sclerocarya birrea* and *Harpephyllum caffrum* (Anacardiaceae) extracts. *Food Chem.* **2010**, *123*, 69–76.
18. McGaw, L.; Van Der Merwe, D.; Eloff, J. In vitro anthelmintic, antibacterial and cytotoxic effects of extracts from plants used in South African etnoveterinary medicine. *Vet. J.* **2007**, *173*, 366–372.
19. Mariod, A. A.; et al. Fatty Acids, Tocopherols, Phenolics and the Antimicrobial Effect of *Sclerocarya birrea* Kernels with Different Harvesting Dates. *J. Am. Oil Chem. Soc.* **2010**, *87*, 377–384.
20. Fennel, C.; et al. Assessing African medicinal plants for efficacy and safety: pharmacological screening and toxicology. *J. Ethnopharmacol.* **2004**, *94*, 205–217.
21. Galvez, J.; et al. Pharmacological activity of a procayanidin isolated from *Sclerocarya birrea* bark: antidiarhoeal activity and effects on isolated guinea-pig ileum. *Phytother. Res.* **1993**, *7*, 25–28.
22. Green, E.; et al. Inhibitory properties of selected South African medicinal plants against Mycobacterium tuberculosis. *J. Ethnopharmacol.* **2010**, *130*, 151–157.
23. Hutchings, A. *Zulu Medicinal Plants: an inventory*; University of Natal Press: Pietermaritzburg, 1996.
24. Gathriwa, J. W.; et al. The in vitro anti-plasmodial and in vivo anti-malarial efficacy of combinations of some medicinal plants used traditionally for treatment of malaria by the Meru community on Kenya. *J. Ethnopharmacol.* **2008**, *115*, 223–231.
25. Pillay, P.; Maharaj, V. J.; Smith, P. J. Investigating South African plants as a source of new antimalarials. *J. Ethnopharmacol.* **2008**, *119*, 438–454.
26. Elgorashi, E.; et al. Screening of medicinal plants used in South African traditional medicine for genotoxic effects. *Toxicol. Lett.* **2003**, *143*, 195–207.
27. Maiga, A.; et al. A survey of toxic plants in the market in the district of Bamoko, Mali: traditional knowledge compared with a literature search of modern pharmacology and toxicology. *J. Ethnopharmacol.* **2005**, *96*, 183–193.
28. Leaky, R. Potential of novel food products from agroforestry trees: a review. *Food Chem.* **1999**, *66*, 1–14.
29. Glew, R.; et al. Nutritional analysis of the edible pit of Sclerocarya birrea in the Republic of Niger (daniya, Hausa). *J. Food Compos. Anal.* **2004**, *17*, 99–111.
30. Weinert, I.; van Wyk, P. J.; Holtzhausen, L.C. *Fruits of Tropical and Subtropical Origin: Composition, Properties and Uses*; Nagy, S., Shaw, P., Wardowski, W. F., Eds.; Florida Science Source Inc.: Lake Alfred, Florida, 1990; pp 88−115.
31. Ndhlala, A.; et al. Phenolic composition of Flacourtia indica, Opuntia megacantha and Sclerocary birrea. *Food Chem.* **2007**, *103*, 82–87.
32. Bever, B. Oral hypoglaecemic plant in West Africa. *J. Ethnopharmacol.* **1980**, *2*, 119–127.

33. Braca, A.; et al. Chemical composition and antioxidant activity of phenolic compounds from wild and cultivated Sclerocarya birrea(Anacardiaceae) leaves. *Agric. Food Chem.* **2003**, *5*, 6689–95.
34. Blignaut, J.; Moolman, C. Quantifying the potential of restored natural capital to alleviate poverty and help conserve nature: A case study from South Africa. *J. Nat. Conservat.* **2006**, *14*, 237–248.
35. Akinnifesi, F.; et al. Early growth and survival of three miombo woodland indigenous fruit tree species under fertilizer, manure and dry-season irrigation in southern Malawi. *Forest Ecol. Manage.* **2008**, *255*, 546–557.
36. Gruenwald, J. Novel botanical ingredients for beverages. *Clin. Dermatol.* **2009**, *27*, 210–216.
37. Kleiman, R.; Ashley, D. A.; Brown, J. H. Comparison of two seed oils sued in cosmetics, moringa and marula. *Ind. Crops Prod.* **2008**, *28*, 361–364.
38. Shackleton, S.; et al. *Use of marula products for domestic and commercial purposes: Synthesis of key findings from three sites in southern Africa*; Grahamstown Department for International Development: Rhodes University, Grahamstown, 2002.

Chapter 3

Piper guineense (Piperaceae): Chemistry, Traditional Uses, and Functional Properties of West African Black Pepper

H. Rodolfo Juliani,[1] A. R. Koroch,[2] L. Giordano,[1] L. Amekuse,[3] S. Koffa,[4] J. Asante-Dartey,[3] and J. E. Simon*,[1]

[1]New Use Agriculture and Natural Plant Products Program, School of Environmental and Biological Sciences, and the New Jersey Agricultural Experiment Station (NJAES), Rutgers, The State University of New Jersey, 59 Dudley Road, New Brunswick, New Jersey
[2]City University of New York, Borough of Manhattan Community College, 199 Chambers Street, Science Department, New York, New York 10007
[3]Agribusiness in Sustainable Natural African Plant Products (ASNAPP-Ghana), Accra, Ghana
[4]ELWA, Paynesville, Monrovia, Liberia
***E-mail: jimsimon@rci.rutgers.edu.**

West African black pepper (*Piper guineense*) is an important plant used in traditional medicine and as spice. The fruits (the part of the plant traditionally used) are rich in a wide range of natural products including volatiles oils, lignans, amides, alkaloids, flavonoids and polyphenols. The objectives of this paper are to review the chemistry of this unique spice, to develop quality control standards for the commercialization of the fruits and to examine the traditional uses and functional properties of "West African Black Pepper".

Introduction

The tropical family *Piperaceae* has offered in the past and today a rich source of natural products with a wide range of biological activities. The genus *Piper* includes more than 1,000 species being one of the largest genera of basal angiosperms (*1*). *Piper* species are distributed pantropically and the pattern of distribution vary from locally endemic to widespread. Species of this genus can be found in the understory of lowland wet forests as shrubs, herbs, or lianas (*2*). *Piper* species has been long known for their commercial, economical, ecological and also medicinal importance. The economic uses are mainly as spices, herbal medicines, cosmetics and insecticides.

Piper guineense Schum and Thonn is locally known as West African black pepper, Ashanti pepper, Guinea pepper, Bush pepper and Guinea cubebs (*3–5*). West African black pepper is a climbing perennial plant distributed throughout West Africa that can reach up to 12 m high, usually found in high forest areas and has prominent nodes and clasping roots (*3*). The leaves are simple, opposite and ovate, the flowers are small in solitary spikes. The fruits, commonly known as peppercorns, are racemes and usually black or white depending on the time that they have been harvested (*6, 7*). West African black pepper is widely distributed from Guinea to Uganda.

While the plant product of commerce is the fruits, other plant parts such as roots, seeds, stem bark and leaves are used in traditional medicine (*5*). The fruits or berries are usually sold in local markets for food coloring, as a condiment/spice to flavor for example soups or for medicinal uses (*3, 8*). Preparations of leaves, roots and seeds have medicinal properties (*3, 6*). West African pepper has been indicated to treat different medical conditions such as boils, bronchitis, catarrh, chest pains, coughs, dyspepsia, impotence, insect repellant, lumbago, rheumatism, uterine fibroid, wounds, stomach aches and discomforts (*5, 8*). The fruit are used also as tonic, abortifacient, to easy childbirth, oxytocic, for tumors, insecticide and for hemorrhoids (*4*). Traditionally, the herb is prepared in several forms including decoctions, powders or tinctures (*5*).

Roots have been also used as aphrodisiac, colds, respiratory diseases and caries. It was reported that a mixture of leaves, roots and fruits are incorporated in preparations for the treatment of infectious diseases as antibacterial agent (*3*). Leaves are used for abdominal disorders, antihelmintic, chickenpox, bronchitis, cough, headache, lumbar pain, gingivitis, chest complains and diseases, intestinal colic and as antiseptic. In Cameroon, West African pepper leaves are mixed with leaves of *Pentas shimperana* spp. *occidental* to make a yellow soup that is used to treat diarrhea (*9*). In Southeast Nigeria the leaves are used for contraction of the womb, as a pre-labor stimulation (*10*). *Piper* species are also used in folk medicine for the treatment of coughs, intestinal diseases, bronchitis, venereal diseases, colds, rheumatism and diarrhea (*3, 9, 11, 12*).

The objectives of this paper are to review the chemistry, develop quality control standards for the commercialization of the fruits and to review the traditional uses and functional properties of "West African Black Pepper".

Material and Methods

Five samples of fruits of *P. guineense* which were representative of the available West African pepper were purchased at the local market in Sanniquellie (Nimba County, Liberia) in January 2011. With the proper USDA seed importation permit, the fruits were hand carried from Liberia through USDA customs to Rutgers University. The sensory, foreign matter and chemical analyses of the samples were conducted at Rutgers University (USA). Samples of black whole peppercorns from a commercial source of black pepper (*Piper nigrum*) were also analyzed for comparison purposes.

The colors of the peppercorns were determined visually. Each subsample was weighed (2 g) and then gently placed in an oven (85°C) until constant weight was reached for the determination of moisture percent. The dried fruits were then ground (mesh 20) and total ashes, and acid insoluble ashes were determined for each sample using methods described by the Food Chemical Codex. (*13*). A sieve (250 μm) was used to separate the fine particles and then weighed to calculate their percentages in relation to the total mass of dried spice (10 g).

Essential oil extraction and analysis: The volatile oils were isolated from the dried fruits (100 g) by water distillation using a Clevenger-type apparatus, oil yield was calculated as percent on a dry weight basis (mL of oil/100 g dry leaves). The oils were analyzed by gas chromatography (GC) coupled to a mass spectrometry (MSD) and flame ionization (FID) detectors (Agilent GC System 6890 Series, Mass Selective Detector, Agilent 5973 Network, FID detector). Each oil was run in two separate columns (HP-5, 30 m, 0.25 mm ID, 0.25 im Film), attached to the MSD and the FID. The conditions for both inlets and columns were the same: Helium constant flow was set at 1 ml/min. Inlet temperature was 220°C, temperature program, 60°C 1 min, 4°C/min, 200°C 15 min. The temperature for the FID was set at 220°C and for MSD at 280°C. Qualitative analysis was based on a comparison of retention times and indexes on both columns and mass spectra with corresponding data in the literature and mass spectral libraries (Wiley 275) (*14*).

Samples were analyzed using an International Standard HPLC/UV method (ISO 11027: 1993 E) modified by the New Use Agriculture and Natural Plant Products Program, Rutgers University described below. Piperine (Acros organics), reference standard, was used for identification of the compound, which was performed on an Agilent 1100 HPLC equipped with a diode array detector and analyzed using MSD trap software. A Prodigy 5 μ OD3 150 x 3.20 mm (Phenomenex Co.) column was used.

The various pepper seeds were sifted (2.36 mm sieve) then ground to a powder from which 125 mg were extracted in 25 ml of denatured ethanol (HPLC grade – Fischer Scientific Co.) then sonicated for 30 minutes. Approximately 2 ml aliquots per sample were centrifuged (5 min) and supernatants were transferred to amber glass auto-sampler HPLC vials. HPLC separation was performed in an isocratic condition with the mobile phase consisting of an aqueous acetic acid solution (1.0%) and formic acid acetonitrile solution (0.1%) (52 /48 - v/v).

Samples were run with a flow rate of 1.0 ml/min for 30 minutes and detected on UV chromatogram at 343 nm. Elution time for piperine was approximately 4.1 minutes.

Results

The West African black peppercorns were characterized by variations in color and sensory characteristics. Most of Sample 1 peppercorns and to lesser extent Sample 2 showed grayish colors. Samples 3 to 5 showed black/dark brown colors, which were similar in color to the regular commercial black peppercorns (*P. nigrum*) (Table I).

All the West African black peppercorns were characterized by pleasant spicy notes with Sample 1 also having a slight moldy note. The traditional regular black pepper (*P. nigrum*) samples exhibited the typical spicy and woody notes. (Table I).

The West African samples 1 and 2 were characterized by very high levels of foreign materials (1.6 and 3.1%, respectively), while the commercial sample was characterized by very low levels (<0.1%) (Table I). All the West African black peppercorns showed low levels of moisture (<10%). Samples 1 and 3 exhibited high levels of acid insoluble ashes suggesting contamination with sand and earth that was coupled to high amount of foreign matter which indicates the products were not adequately cleaned. The rest of the samples including the commercial black pepper samples showed lower levels (<0.5%).

Piper guineense samples showed varied amounts of the piperine, the component responsible for the spiciness and one of the main characters defining the sensory quality of these *Piper* spices. The content of piperine in *P. guineense* ranged from 0.23 -1.1% while the commercial black pepper accumulated 1.3% of piperine.

The whole peppercorns (*P. guineense*, samples 4-5) showed the highest levels of essential oils (1.2%) while commercial black pepper showed lower levels of volatile oils (0.9%), (Table I). These results showed that West African black pepper contained higher levels of volatiles oils while lower amounts of piperine.

West African black peppers were characterized by distinct pleasant floral aroma that was not present in the regular black peppers (Table I). The volatile components of *P. gineense* contained high levels of linalool (66.7-70.2%) that would be largely responsible for imparting to this spice such pleasant floral notes (Table II), while the black pepper corns showed high levels limonene, β-pinene and (E) caryophyllene, partially responsible for the typical spicy and woody aromas.

To begin to define the quality of *P. guineense* dried fruits, we propose an initial set of standards as presented in Table III, though additional studies are needed to evaluate microbial loads, among others, of this spice.

Table I. Sensory, Foreign Matter and Chemical Analysis of West African Black Pepper (*Piper guineense*) from Liberia and Its Comparison with Commercial Pepper (*Piper nigrum*)

	Piper guineense					*Piper nigrum*
Character	*Sample 1*	*Sample 2*	*Sample 3*	*Sample 4*	*Sample 5*	*Commercial*
Color	Dark brown, grayish	Dark brown with some peppercorns gray	Dark brown	Dark brown	Dark brown	Dark brown
Aroma	Woody and spicy, slightly moldy	Woody and spicy	Woody and spicy	Sweet, floral, less spicy	Sweet, floral, less spicy	Spicy, woody
Moisture (%)	7.1	7.3	9.6	9.1	8.2	10.0
Foreign Matter (%)	1.6	3.1	0.83	0.93	1	< 0.1
Fine Particle (%)	1.07	1.07	0.13	< 0.1	< 0.1	< 0.1
Total Ashes (%)	6.9	5.9	5.5	6.2	6.0	4.93
Acid Insoluble Ashes (%)	0.68	0.26	1.3	0.3	0.4	-
Essential oil (%)	-	0.5	-	1.2	1.2	0.9
Piperine (HPLC, %)	1.1	0.54	0.83	0.24	0.23	1.33

Table II. Comparison of Chemical Composition of the Essential Oils of *Piper guineense* and *P. nigrum*

Components	*RT*	*Piper guineense*	*Piper nigrum*
α-pinene	6.34	1.5-1.7	5.9
unknown (BP 93)	7.37	0.09	14.1
Myrcene	7.48	0.5-0.6	
δ-3-carene			7.9
Limonene			24.2
1,8-cineole	9.06	0.6	6.6
(Z)-b-Ocimene	9.18	3.7-4.1	
(E)-b-Ocimene	9.5	2.9-3.2	
Linalool	11.17	66.7-70.2	-
β-elemene	20.73	0.4	
Cedrene	22.95	2.2- 3.4	
(*E*)-caryophyllene			7
Elemol	25.44	1.0- 1.2	
Elemicin	25.56	0.7	
β-caryophyllene oxide	26.47	1.2-1.3	
Benzoic acid	29.43	5.0-5.6	

NOTE: Retention time (RT) in order of elution in the GC column, phase methyl (95%)/ phenyl (5%) silicone.

Table III. Proposed Quality Control Standards for *Piper guineense* Dried Fruits, Liberia

Characteristic	*Piper guineense dried fruits*
Sensory characters	
Color	Dark brown (or almost black)
Aroma	Sweet and floral, slightly spicy
Taste	Mildly spicy (less than *P. nigrum*)
Macroscopic and sieve analysis	
Foreign parts, % (m/m) max.	0.5
Fine particles (less than 500 mM), % (m/m) max.	1

Continued on next page.

Table III. (Continued). Proposed Quality Control Standards for *Piper guineense* Dried Fruits, Liberia

Characteristic	*Piper guineense dried fruits*
Physical-Chemical parameters	
Moisture, %, (m/m) max.	10
Total ashes, % (m/m) max.	6
Total insoluble ashes, % (m/m) max.	0.5
Essential oils, % (v/m) min.	0.5
Piperine, % (m/m) min.	0.2

Discussion

Phytochemical studies of *Piper* species have led to isolation of a wide range of different natural products including lignans, amides, alkaloids, flavonoids and aromatic compounds. (*1*, *6*, *15–17*). The compound responsible for the pungency is the amide piperine and for the aroma the terpenoids and phenyl propanoids (*1*, *18*).

The quantity and quality of different natural products in *Piper* species depend on the plant tissue (*18*), geographical location of the plant (*19*) and season in which it was harvested (*20*). The fruits have shown to accumulate a large variation of different alkaloids (*18*). Different amides have been isolated from the fruits such as piperine, trichostachine, piperlongumine, guineensine (*21–25*), pipericide (*26*), wisanidine (*25*, *27*), $\Delta^{\alpha\beta}$ hydrowisanidine (*28*), $\Delta^{\alpha\beta}$ hydrowosanine (*29*), Wisanine (*23*, *25*, *30*), *N*-isobutyl-2-*trans*-4-eicosadieneamide (*24*, *31*), $\Delta^{\alpha\beta}$ dihydro-piperlonguminine (*25*, *32*), pellitorine (*25*, *33*) and *N*-isobutyloctadeca-*trans*-2-*trans*-4-dienamide (*29*, *34*).

Also, amide alkaloids have been extracted from other parts of the plant. From the roots it has isolated wisanine and wisanidin (*31*, *35*), 4,5-dihydro-piperine (*19*), from the stems piperidine (*19*) piperine, wisanine, peperlonguminine, trichostachine (*36*) and from leaves $\Delta^{\alpha\beta}$ hydrowisanine, trichostachine, piperine, wisanidine and wisanine (*25*). The major accumulation of piperine throughout the plant was mainly in the fruits, in contrast to the very low amounts detected in young shoots and roots. However the values were highly variable between the samples (*18*).

Piperine has exhibited different physiological effects, including the stimulation of the production of digestive enzymes, protection against oxidative damage, anti-mutagenic and anti-tumor influences (*37*), analgesic and anti-inflammatory activities (*38*). Piperine was also found to depress the central nervous system (*39*) and improve memory impairment and neurodegeneration of the hippocampus (*40*).

The aroma of West African black pepper is due to the presence of essential oils that are a mixture of different combination of different compounds (Table II). The composition and the total production of essential oils depend on the different landraces and populations, growing conditions, geographic area or also of part of the plant that was extracted. At present, there are no cultivars of West African pepper with any being introduced into cultivation coming in from wild plants.

High variation in the composition of essential oils depending on the geographic area of the samples has been reported. The oils obtained from fruits have been dominated the sesquiterpene β-caryophyllene (more than 50%), β-elemene, bicyclogermacrene and humulene in Cameroon (*41*) and monoterpenes b-pinene in Nigeria (*42*, *43*) very rich in phenylpropanoid derivatives being myristicin the most abundant component (16.55%) and minor constituents were predominantly mono- and sesquiterpenes in Nigeria (*8*). It was also reported a chemotype of *P. guineense* rich in monoterpenes (β-pinene, 23.2-27%) and sesquiterpenes, with the absence of the usual myristicin (*42*, *44*).

The samples analyzed in the present study showed the presence of monoterpenes, sesquiterpenes and phenylpropanoids of which linalool was the major component, showing a unique essential oils profile (Table II).

The essential oils from other plant parts have been studied to a limited extent. Essential oils rich phenylpropanoid derivatives dillapiole (44%) and myristicin (9.8%) (*45*), germacrene (25.1%) and limonene (*41*) or E-β-farnesene (20%) and 25.6% myristicin (*44*) were the major components obtained of leaves and z-E-farnesene (28.7%), limonene (19.7%) and myristicine (10.9%) from stems (*41*). The most common technique of extraction of essential oil is by hydrodistillation; however, using Soxflo technique it was possible to identify new components of the oils, a- giaoeme and α and β-selinene from seeds (*46*).

Other biological active chemical compounds found in West African Black pepper are lignans. They are composed of two C_6, C_3 units linked to b-carbons of the side chain (*6*, *47*). Sesamine (*21*), ashatin (+)-yangambin, and dihydrocurbetin have been reported in this species (*15*, *48*).

For the commercialization of black pepper, not only are high yields of peppercorns important, but also the quality of the product including the pungency (largely due to piperine content), aroma and flavor (largely due to the amount and content of essential oil (*49*) and the sensory characteristics (color, aroma and taste).

Functional Properties

Insecticidal

The use of naturally occurring natural plant products (such as oils, powders, extracts) to protect agricultural products against a variety of insect pests is a common practice in many parts of the world. Now natural plant products are gaining much attention because of the demand for more natural, low cost and environmental friendly alternatives. Insect pests cause huge losses of stored food products being one of the major agricultural problems in countries of the third world.

The use of plant products based on different organs of *P. guineense* to protect grains, legumes and stored products against post- harvest pests has been well documented. Ethanolic extracts from fruits of *P. guineense* were reported as very effective grain protecting agent against serious coleopterous infestation in stored grains such as *Sitophilus zeamais, Tribolium castaneum, Callosobruchus maculates and Oryzaephilus mercator* without affecting seed germination (*50–54*). In addition to the ethanolic extracts, seed oil extract and powder were also tested and proved to be very effective in preventing qualitative and quantitative losses in maize grains by maize weevil *S. zeamais* at 3 months after storage (*55–57*). Seed oils were very effective to repel both female and males of *S. zeamais* (*41*, *58*). Moreover, the chemical and proximate composition studies of grains treated with *P. guineense* showed that this spice did not cause any adverse effect in color, taste, texture or nutritional composition (*55*).

Another important constrains in the production of vegetables and fruits are insect pest infestation that causes drops in yield and reduced quality. Aqueous extracts of *P. guineense* has been reported to protect several vegetable crops with high repellency rate or antifeedant properties (*59–61*). For instance, in brassica crops, aqueous extracts caused 100% of mortality of the larvae of *Plutella xylostella* (*62*), in legume crops (*Vigna unguculata*) reduced the egg viability against of the important pests such as *Maruca vitrata* and *Clavigralla tomentosicollis* (*63*, *64*) and in banana and plantain crop exhibited repellent and antifeedant properties against the banana weevil (*Cosmopolites sordidus)* the major constraint in the crop production causing losses between 40 to 50 percent of the plant crop (*65*).

West African pepper has been used as insecticidal to protect not only grains, vegetables or legumes but also fresh fish and fish stored products. Seeds extracts were active against goldfish monogenean parasites which are economically important parasites of cultured fish (*29*). Pulverized leaves of *P. guineense* on smocked catfish during storage inhibited egg hatchability and adult emergence of *Dermestes maculatus*, reducing thus the percentage loss due to insect infestation and ensuring a steady supply of good quality smoked fish throughout the season (*66*). In addition, the treated fish products were accepted by consumers as there was no evidence of taint, smell or change in the texture of flavor (*67*, *68*) and nutritive value (*69*) of the fish.

Also, the efficacy of *Piper* extracts also has been tested against other pests such as garden pests (*70*), termites (*71*, *72*), millipedes (*73*) and mosquitoes (*24*, *74*, *75*). These results indicate that *Piper* extracts can be used to effectively control a wide range of pests. Plant-based pesticides can be recommended as an eco-chemical and sustainable strategy in the management of agricultural pests, because of their biodegradable nature, systemicity after application, capacity to affect the behavior of target pests and favorable safety profile (*76*). West African pepper could be an alternative to reduce the use of conventional synthetic insecticides since it shows the potential use in stored products protection for organic farmers or small scale farmers. Moreover it is an inexpensive, low-risk, eco-friendly; easily biodegradable and readily available plant material that will not contaminate food products (*58*, *77*).

The capability of pepper extracts as natural insecticides was associated with the concentration of piperamides (*74*), chavicine (*52*, *54*), piperine (*6*, *24*, *50*, *52*, *54*), piperidine (*52*, *56*), pipericide, pellitorine, guineensine (*78*) lignans, isobutylamide (*79*) and phenylpropanoids (*78*). The efficacy of the extracts was not only associated with the unique chemical composition but also attributed to the mode of action of these natural products. The efficiency of *Piper* extracts as insecticidal has been attributed to contact toxicity (*26*, *41*, *50*, *54*, *70*, *78*), synergism between the chemical components (*80*, *81*) repellent properties (*58*, *61*) and antifeedant properties (*65*, *70*). Moreover, it has been suggested that piper extracts could replace contact conventional synthetic insecticides for which resistance has been developed (*1*, *70*).

Antifungal and Antimicrobial

Seed extracts and fractions of West African black pepper were reported to possess antifungal properties against *Rhizopus stolonifer* (*82*), *Candida albicans*, *Cryptococcus neoformans* and some other filamentous fungi responsible for most mycoses with practically no toxic effect (*83*). These results explain its use in traditional medicine to treat certain infectious diseases (*83*). However, the efficacy of the extracts is dose dependant, since lower concentrations of extracts seem to have little or no antifungal activity (*84*). Several reports show moderate activity of aqueous extracts prepared from seeds against *Staphylococus aureus, S. faecalis, B. subtilus, Escherichia coli, Pseudomonas aeruginosa*, and *C. albicans* (*85–87*).

The antifungal activity was attributed to the active principles piperine and piperlonguminine (*88*) and essential oils components (*87*) present in the samples. The essential oils from this plant species have shown antifungal and antimicrobial properties against animal, plant pathogens, food poisoning and spoilage bacteria. The chemical composition of the oils, the structure configuration of the constituents, their functional groups and the relative percentage composition in the oil have been associated with oils' antimicrobial activities (*89*).

Essential oils from the fruit have been effective to inhibit growth of *P. aeruginosa* while ampicillin and gentamycin were ineffective, moderate inhibitory activity against *Bacillus cereus, E. coli* and *C. albicans* (*42*). However, Olonisakin and collaborators (*46*) did not observe any antimicrobial activity against *E. coli* and *P. aeruginosa*. This result could be attributed to the different relative composition of the essential oils. The antimicrobial activities of the essential oils and extracts support the ethno-medical use in folk medicine.

Antioxidant Activity

Reactive oxygen species (ROS) are by products of the cellular metabolism and are associated with numerous pathological conditions (*90*). Among the ROS are the superoxide radicals, produced by various metabolic and physiological processes (*91*) and hydroxyl radicals, the most highly reactive oxygen radical known (*90*). In biological systems there are different mechanisms or molecules (antioxidant) that can protect the cells from the attack of these reactive oxygen species. In the last years, special attention has been paid to the identification and isolation of natural products that have the ability to scavenge reactive oxygen species.

In a comparative study of the antioxidant activity between white (fully ripe fruits) and black (unripe fruits) peppers, it was reported that black peppers were more effective in scavenging free radicals and ROS than the white ones (*92*). This was attributed to higher polyphenol content (*11*, *92*). West African pepper has been reported to possess significant antioxidant properties (*11*). The antioxidant activity was partially ascribed to the presence of phenolic amides (*93*). The oil and oleoresin was reported to have good flavoring and antioxidant properties in pork samples (*94*). The antioxidant activity of the extracts has been characterized by a great superoxide radical (*11*, *92*), high hydroxyl radical scavenging activity (*95*).

The use of *P. guineense* as antioxidant is a promising alternative to the use of synthetic antioxidants, it could play a role in the modulation of free radical induced disorders (*11*), possess potential health benefits by inhibiting lipid peroxidation, justifies the traditional uses in medicine, and potential use as a value-added ingredient for stabilizing food matrixes against lipid peroxidation reactions (*96*).

Other Functional Properties

In a study of the lipid composition of West African black pepper, it was reported that seeds contained moderate amounts of linolenic acid, thus the consumption of these seed oils would add a unique variety of good quality dietary oils to the diets of the local people (*97*). Moreover, the seed/fruit contained relatively high crude protein (18.46%) and K and Na (*98*) and high carbohydrate content (*99*). Yet, it must be noted that while the product exhibits promising nutritional content, the quantity consumed/serving is low reflecting its use as a spice and flavoring ingredient.

Water extracts of *P. guineense* has prolonged anticonvulsant activity at doses which do not cause significant CNS depression. These results can explain the traditional use to suppress human epileptic seizures for prolonged periods (*100*).

Other studies reported that aqueous extracts of West African black pepper significantly increased the levels of testosterone in the serum and testes, cholesterol in the testes, α- glucosidase in the epididymis and fructose in seminal vesicles in rats (*101*, *102*) and modified the sexual behavior of male rats (*103*). These findings explain its traditional use in treatment of male infertility.

Conclusion

West African black pepper (*P. guineense*) from Liberia has a great potential to be commercialized as a unique and new spice for the international markets with several studies showing a wide range of functional properties (both from a sensory and biological perspectives). The chemical components and biological activities of the different plant parts support many of their traditional uses. West African black pepper could be a promising biopesticide used against a wide spectrum of pests that affect the production of human foods. Also, natural biocides can be prepared from the plant, in place of using more toxic synthetic pesticides, since they are locally produced, easily harvested and are highly efficient against a wide range of specific target pests. However, efforts are needed in the harvesting and postharvest handling stages to have a consistent supply of high quality products. Most importantly, West African black pepper has been enjoyed and consumed for thousands of years by indigenous populations. The plant is distinct from the more traditional and familiar black pepper of India (*P. nigrum*), now grown and naturalized in many tropical countries around the world. Yet, West African black pepper can also be blended in with *P. nigrum*, or be branded as a unique African spice.

Acknowledgments

This work was done in support of the USAID funded Liberian initiative, **People, Rules and Organizations Supporting the Protection of Ecosystem Resources (PROSPER) Program which is** led by TetraTech ARD, Inc. (Vermont). This project supports livelihood activities related to the environmentally sustainable development of non-timber forest products. We thank Vaneska Litz and Steve Reid of ARD for their support and assistance.

References

1. Scott, I. M.; Jensen, H. R.; Philogene, B. J. R.; Arnason, J. T. *Phytochem. Rev.* **2008**, *7*, 65–75.
2. Jaramillo, M. A.; Manos, P. S. *Am. J. Bot* **2001**, *88*, 706–716.
3. Iwu, M. M. *Handbook of African Medicinal Plants*; CRC Press: Boca Raton, FL, 1993; pp 221−222.
4. Neuwinger, H.D. *African Traditional Medicine. A dictionary of plant use and applications*; Medpharm. Scientific Publisher: Germany, 2000, pp. 556.
5. *Ghana Herbal Pharmacopoeia*; Science and Technology Policy Research Institute (STEPRI) Council for scientific and industrial research (CSIR): Accra, Ghana, 2007.
6. Parmar, V. S.; Jain, S. C.; Bisht, K. S.; Jain, R.; Taneja, P.; Jha, A.; Tyagi, O. D.; Prasad, A. K.; Wengel, J.; Olsen, C. E.; Boll, P. M. *Phytochemistry* **1997**, *46*, 597–673.
7. Omafuvbe, B. O.; Kolawole, D. O. *Pak. J. Nutr.* **2004**, *3*, 244–249.
8. Ekundayo, O.; Laakso, I.; Adegbola, R. M.; Oguntimein, B.; Sofowora, A.; Hiltuen, R. *J. Agric. Food Chem.* **1988**, *36*, 880–882.
9. Focho, D. A.; Ndam, W. T.; Fonge, B. A. *Afr. J. Pharm. Pharmacol.* **2009**, *31*, 1–13.
10. Nwosu, M. O. *Ärztezeitschrift für Naturheilverfahren* **2000**, *41*, 760–767.
11. Agbor, G. A.; Vinson, J. A.; Oben, J. E.; Ngogang, J. Y. *J. Herbal Pharmacother.* **2007**, *7*, 49–64.
12. Sandberg, F.; Perera-Ivarsson, P.; El-Seedi, H. R. *J. Ethnopharmacol.* **2005**, *102*, 336–343.

13. Committee on Food Chemical Codex. *Food Chemical Codex*; National Academy Press: Washington DC, 1996.
14. Juliani, H. R.; Simon, J. E.; Quansah, C.; Asare, E.; Akromah, R.; Acquaye, D.; Asante-Dartey, J.; Mensah, M. L. K.; Fleischer, T. C.; Dickson, R.; Annan, K.; Mensah, A. Y. *J. Essent. Oil Res.* **2008**, *20* (1), 49–55.
15. Jensen, S.; Hansen, J.; Boll, P. M. *Phytochemistry* **1993**, *33*, 523–530.
16. Parmar, V. S.; Jain, S. C.; Gupta, S.; Talwar, S.; Rajwanshi, V. K.; Kumar, R.; Azim, A; Malhotra, S.; Kumar, N; Jain, R.; Sharma, N. K.; Tyagi, O. D.; Lawrie, S. J.; Errington, W.; Howarth, O. W.; Olsen, C. E.; Singh, S. K.; Wengel, J. *Phytochemistry* **1998**, *49*, 1069–1078.
17. *Piper: a model genus for studies of phytochemistry, ecology and evolution*; Dyer, L. A., Palmer, A. D. N., Eds.; Kluwer Academic/Plenum Publishers: New York, 2004; p 228.
18. Semler, U.; Gross, G. G. *Phytochemistry* **1988**, *27*, 1566–1567.
19. Okogun, J. I.; Sondengam, B. L.; Kimbu, S. F. *Phytochemistry* **1977**, *16*, 1295.
20. Sofowora, A. *Med. Plants Tradit. Med. Afr.* **1982**, 256.
21. Okogun, J. I.; Ekung, D. E. U. *J. Chem. Soc., Perkin Trans. 1* **1974**, *19*, 2195–2198.
22. Addae-Mensah, I.; Torto, F. G.; Dimonyeka, C. I.; Baxter, I.; Sanders, J. K. M. *Phytochemistry* **1977**, *16*, 757–759.
23. Okwute, S. K.; Okogun, J. I.; Okorie, D. A. *Tetrahedron* **1984**, *40*, 2541–2545.
24. Addae-Mensah, I.; Achieng, G. *Planta Med.* **1986**, *52*, 432.
25. Adesina, S. K.; Adebayo, A. S.; Adesina, S. K. O.; Groning, R. *Pharmazie* **2003**, *58*, 423–425.
26. Scott, I. M.; Gagnon, N.; Lesage, L.; Philogene, B. J. R.; Arnason, J. T. *J. Econ. Entomol.* **2005**, *98*, 845–855.
27. Sondengam, B. L.; Kimbu, S. F.; Njimi, T.; Okogun, J. I; Ekong, D. E. U. *Tetrahedron Lett.* **1977**, *4*, 367–368.
28. Sondengam, B. L.; Kimbu, S. F.; Connolly, J. D. *Phytochemistry* **1977**, *16*, 1121–1122.
29. Ekanem, A. P.; Wang, M.; Simon, J. E.; Obiekezie, A. I.; Morah, F. *Phytother. Res.* **2004**, *18*, 793–797.
30. Crombie, L.; Pattenden, G.; Stemp, G. *Phytochemistry* **1977**, *16*, 1437–1438.
31. Addae-Mensah, I.; Torto, F. G.; Oppong, I. V.; Baxter, I.; Sanders, J. K. M. *Phytochemistry* **1977**, *16*, 483–485.
32. Dwuma-Badu, D.; Ayim, J. S. K.; Dabra, T. T.; ElSohly, H. N.; ElSohly, M. A.; Knapp, J. E.; Slatkin, D. J.; Schieff, P. L. *Phytochemistry* **1976**, *15*, 822–823.
33. Gbewonyo, W. S. K.; Candy, D. J. *Planta Med.* **1990**, *56*, 525–526.
34. Tackie, A. N.; Dwuma-Badu, D.; Ayim, J. S. K.; ElSohly, H. N.; Knapp, J. E.; Slatkin, D. J.; Schieff, P. L. *Phytochemistry* **1975**, *14*, 1888–1889.
35. Addae-Mensah, I.; Torto, F. G.; Torto, B.; Achenbach, H. *Planta Med.* **1981**, *51*, 200–201.

36. Adesina, S. K.; Adebayo, A. S.; Adesina, S. K. O.; Groning, R. *Pharmazie* **2002**, *57*, 622–627.
37. Srinivasan, K. *Crit. Rev. Food Sci.* **2007**, *47*, 735–748.
38. Gupta, S. K.; Bansal, P.; Bhardwaj, R. K.; Velpandian, T. *Pharmacol. Res.* **2000**, *41*, 657–662.
39. Lee, S. A.; Hong, S. S.; Han, X. H.; Hwang, J. S.; Oh, G. J; Lee, K. S.; Lee, M. K.; Hwang, B. Y.; Ro, J. S. *Chem. Pharm. Bull.* **2005**, *53*, 832–835.
40. Chonpathompikunlert, P.; Wattanathorn, J.; Muchimapura, S. *Food Chem. Toxicol.* **2010**, *48*, 798–802.
41. Tchoumbougnang, F.; Dongmo, J. P. M.; Sameza, M. L.; Fombotioh, N.; Wouatsa, N. A. V.; Zollo, P. H. A.; Menut, C. *Afr. J. Biotechnol.* **2009**, *8*, 424–431.
42. Oyedeji, O. A.; Adeniyi, B. A.; Ajayi, O.; Konig, W. A. *Phytother. Res.* **2005**, *19*, 362–364.
43. Olonisakin, A.; Oladimeji, M. O.; Lajide, L. *Electron. J. Environ. Agric. Food Chem.* **2006**, *5*, 1531–1535.
44. Zollo, P. H. A.; Biyiti, L.; Tchoumbougnang, F.; Menut, C.; Lamaty, G.; Bouchet, Ph. *Flavour Fragrance J.* **1998**, *13*, 107–113.
45. Martins, A. P.; Salgueiro, L.; Vila, R; Tomi, F.; Canigueral, S.; Casanova, J.; Proenca Da Cunha, A.; Adzet, T. *Phytochemistry* **1998**, *49*, 2019–2023.
46. Ntonifor, N. N.; Brown, R. H.; Mueller-Harvey, I. *J. Agric. Food Chem.* **2002**, *50*, 6295–6300.
47. MacRae, W. D.; Towers, N. G. H. *Phytochemisty* **1984**, *23*, 1207–1220.
48. Dwuma-Badu, D.; Ayim, J. S. K.; Dabra, T. T.; ElSohly, H. N.; Knapp, J. E.; Slatkin, D. J.; Schieff, P. L. *Lloydia* **1975**, *38*, 343–345.
49. Ravindran, P. N.; Kalluparackal, J. A. In *Handbook of Herbs and Spices*; Peter, K. V., Ed.; Woodhead Publishing: Cambridge, 2001; Volume 1, Chapter 7, pp 62–110.
50. Okonkwo, E. U.; Okoye, W. I. *Int. J. Pest Manage.* **1996**, *42*, 143–146.
51. Mbata, G. N.; Oji, O. A.; Nwana, I. E. *Discovery Innovation* **1995**, *7*, 139–142.
52. Lale, N. E. S.; Alaga, K. A. *J. Plant Dis. Protect.* **2001**, *108*, 305–313.
53. Lale, N. E. S.; Yusuf, B. A. *J. Stored Prod. Res.* **2001**, *37*, 63–75.
54. Adedire, C. O.; Akinkurolere, R. O. *Zool. Res.* **2005**, *26*, 243–249.
55. Asawalam, E. F. *Electron. J. Environ. Agric. Food Chem.* **2006**, *5*, 1389–1394.
56. Asawalam, E. F.; Emosairue, S. O.; Ekeleme, F.; Wokocha, R. *Int. J. Pest Manage.* **2007**, *53*, 1–6.
57. Ukeh, D. A.; Arong, G. A.; Ogban, E. *J. Entomol.* **2008**, *5*, 295–299.
58. Ukeh, D. A.; Birkett, M. A.; Bruce, T. J. A.; Allan, E. J.; Pickett, J. A.; Mordue, A. J. *Pest Manage. Sci* **2010**, *66*, 44–50.
59. Scott, I. M.; Jensen, H.; Scott, J. G.; Isman, M. B.; Arnason, J. T.; Philogene, B. J. R. *Arch. Insect Biochem.* **2003**, *54*, 212–225.
60. Okunlola, A. I.; Ofuya, T. I.; Aladesanwa, R. D. *Agric. J.* **2008**, *3*, 181–184.
61. Echereobia, C. O.; Okerere, C. S.; Emeaso, K. C. *J. Biopest.* **2010**, *3*, 505–507.

62. Ntonifor, N. N.; Mueller-Harvey, I.; Brown, R. H. *J. Food Agric. Environ.* **2010**, *8*, 498–502.
63. Ekesi, S. *Phytoparasitica* **2000**, *28*, 305–310.
64. Oparaeke, A. M. *Plant Protect. Sci.* **2007**, *43*, 103–108.
65. Inyang, U. E.; Emosairue, S. O. *Trop. Subtrop. Agroecosyst.* **2005**, *5*, 33–44.
66. Fasakin, E. A.; Aberejo, B. A. *Bioresour. Technol.* **2002**, *85*, 173–177.
67. Akinwumi, F. O.; Fasakin, E. A.; Adedire, C. O. *J. Biol. Sci.* **2006**, *6*, 1023–1028.
68. Akinwumi, F. O.; Adedire, C. O.; Fasakin, E. A. *J. Fish. Aquat. Sci.* **2007**, *2*, 403–409.
69. Akinwumi, F. O.; Fesobi, M. E. *Afr. J. Food Sci.* **2010**, *4*, 143–147.
70. Scott, I. M.; Jensen, H.; Nicol, R.; Lesage, L.; Bradbury, R.; Sánchez-Vindas, P.; Poveda, L.; Arnason, J. T.; Philogène, B. J R. *J. Econ. Entomol.* **2004**, *97*, 1390–1403.
71. Umeh, V. C.; Ivbijaro, M. F. *J. Agr. Sci.* **1999**, *133*, 403–407.
72. Verma, M.; Sharma, S.; Prasad, R. *Int. Biodeterior. Biodegrad.* **2009**, *63*, 959–972.
73. Epidi, T.; Njoku, G. *Sci. Res. Essays* **2010**, *5*, 1005–1008.
74. Scott, I. M.; Puniani, E.; Jensen, H.; Livesey, J. F.; Poveda, L.; Sanchez-Vindas, P.; Durst, T.; Arnason, J. T. *J. Agric. Food Chem.* **2005**, *53*, 1907–1913.
75. Adewoyin, F. B.; Odaibo, A. B.; Adewunmi, C. O. *Afr. J. Tradit.* **2006**, *3*, 79–83.
76. Dubey, N. K.; Shukla, R.; Kumar, A.; Singh, P.; Prakash, B. *Curr. Sci.* **2010**, *98*, 479–480.
77. Ntonifor, N. N. *J. Entomol.* **2011**, *8*, 16–26.
78. Gbolade, A. A. *Discov. Innov.* **2006**, *18*, 123–140.
79. Bernard, C. B.; Krishanmurty, H. G.; Chauret, D.; Durst, T.; Philogène, B. J. R.; Sánchez-Vindas, P.; Hasbun, C.; Poveda, L.; San Román, L.; Arnason, J. T. *J. Chem. Ecol.* **1995**, *21*, 801–814.
80. Scott, I. M.; Puniani, E.; Durst, T.; Phelps, D.; Merali, S.; Assabgui, RA.; Sanchez-Vindas, P.; Poveda, L.; Philogene, B. J. R.; Arnason, J. T. *Agric. Entomol.* **2002**, *4*, 137–144.
81. Jensen, H. R.; Scott, I. M.; Sims, S. R.; Trudeau, V. L.; Arnason, J. T. *Insect Mol. Biol.* **2006**, *15*, 329–339.
82. Eruteya, O. C.; Odunfa, S. A. *Afr. J. Biotechnol.* **2009**, *8*, 2316–2320.
83. Ngane, A. N.; Biyiti, L.; Bouchet, Ph.; Nkengfack, A.; Zollo, P. H. A. *Fitoterapia* **2003**, *74*, 464–468.
84. Lee, S. E.; Park, B. S.; Kim, M. K.; Choi, W. S.; Kim, H. T.; Cho, K. Y.; Lee, S. G.; Lee, H. S. *Crop Prot.* **2001**, *20*, 523–528.
85. Konning, G. H.; Agyare, C.; Ennison, B. *Fitoterapia* **2004**, *75*, 65–67.
86. Okigbo, R. N.; Igwe, D. I. *Acta Microbiol. Immunol. Hung.* **2007**, *54*, 353–366.
87. Nwinyi, O. C.; Chinedu, N. S.; Ajani, O. O.; Ikpo, C. O.; Ogunniran, K. O. *Afr. J. Food Sci.* **2009**, *3*, 77–81.
88. Marques, J. V.; de Oliveira, A.; Raggi, L.; Young, M. C. M.; Lato, M. J. *J. Braz. Chem. Soc.* **2010**, *21*, 1807–1813.

89. Dorman, H. J. D.; Deans, S. G. *J. Appl. Microbiol.* **2000**, *88*, 308–316.
90. Halliwell, B.; Cross, C. E. *Environ. Health Perspect.* **1994**, *102*, 5–12.
91. Bedard, L.; Young, M. J.; Hall, D.; Paul, T.; Ingold, K. U. *J. Am. Chem. Soc.* **2001**, *123*, 12439–12448.
92. Agbor, G. A.; Vinson, J. A.; Oben, J. E.; Ngogang, J. Y. *Nutr. Res.* **2006**, *26*, 659–663.
93. Nakatani, N.; Inatani, R.; Ohta, H.; Nishioka, A. *Environ. Health Persp.* **1986**, *67*, 135–142.
94. Ugwuona, F. U. *J. Food Process Preserv.* **2010**, *34*, 235–246.
95. Abdou Bouba, A.; Njintang, Y. N.; Scher, J.; Mbofung, C. M. F. *Agric. Biol. J. North Am.* **2010**, *1*, 213–224.
96. Odukoya, A. O.; Ilori, O. O.; Sofidiya, M. O.; Aniunoh, O. A.; Lawal, B. M.; Tade, I. O. *EJEAFChe, Electron. J. Environ., Agric. Food Chem.* **2005**, *4*, 1086–1093.
97. Essien, E. U.; Esenowo, G. J.; Akpanabiatu, M. I. *Plant Food Hum. Nutr.* **1995**, *48*, 135–140.
98. Effiong, G. S.; Ibia, T. O.; Udofia, U. S. *EJEAFChe, Electron. J. Environ., Agric. Food Chem.* **2009**, *8*, 917–923.
99. Nwinuka, N. M.; Ibeh, G. O.; Ekeke, G. I. *J. Appl. Sci. Environ.* **2005**, *9*, 150–155.
100. Abila, B.; Richens, A.; Davies, J. A. *J. Ethonopharmacol.* **1993**, *39*, 113–117.
101. Nantia, E. A.; Moundipa, P. F.; Monsees, T. K.; Carreau, S. *Andrologie* **2009**, *19*, 148–158.
102. Mbongue, F. G. Y.; Kamtchouing, P.; Essame, O. J. L.; Yewah, P. M.; Dimo, T.; Lontsi, D. *Indian J. Pharmacol.* **2005**, *37*, 30–32.
103. Kamtchouing, P.; Mbonguea, G. Y. F.; Dimoa, T.; Watchob, P.; Jatsac, H. B.; Sokengd, S. D. *Behav. Pharmacol.* **2002**, *13*, 243–247.

Chemistry , Pharmacognosy, and Validation of Traditional Medicines

Chapter 4

Hypoxis hemerocallidea (African potato): A Botanical Whose Time Has Come?

David R. Katerere*

Department of Pharmaceutical Science, Faculty of Science, Tshwane University of Technology, Arcadia campus, Pretoria, 0001, South Africa
***E-mail: david.katerere@mrc.ac.za.**

Hypoxis hemerocallidea is one of Southern Africa's most important and popular medicinal plants. It is used for a wide a range of traditional medical treatments including urinary tract infections, heart disease, infertility and anxiety. It's most popular contemporary use is for prostate disorders for which there is increasingly good evidence about it's efficacy. Phytosterols are thought to be the main bioactive compounds for this indication and also for anti-lipidemic, anti-diabetic and anti-inflammatory properties exhibited by hypoxis extracts.

Despite the popularity of hypoxis as a herbal and botanical medicine, the research into it's pharmacological application has not been of sufficient depth and width to prove it's utility in modern medicine. Recent data using a colitis model presents an exciting new avenue which might finally yield products. However there continues to be a need for sustained in vitro and more importantly in vivo studies if this useful plant is be fully explored.

Introduction

Traditional medicine has experienced a resurgence in Africa in recent times. This is largely to do with a new sense of cultural pride following independence from colonial authorities who sought to suppress such practices during their tenure (*1*). In post-colonial Africa, there has also been limited access to allopathic medical services. Further, the emergence of new diseases such as HIV, and previously unknown chronic diseases e.g. diabetes and cancer, most of which have no known medical cures has accelerated the growth of Complementary and Alternative Medicine (CAM) (*2*, *3*)

WHO estimates that 80% of the world's population still largely relies on traditional medicine to meet it's healthcare needs (*4*). South Africa has 7000 plants which are used in traditional medicine (*5–7*). One of the most famous, and certainly the most controversial is *Hypoxis hemerocallidea*. Commonly known as African potato, this herb was until recently widely promoted by a now deceased South African minister of health as an alternative to treatment of HIV / AIDS (*8*, *9*) this despite little or no supporting scientific evidence (*10*).

Botanical Information

Hypoxis hemerocallidea Fisch. & C.A. Mey, previously known as *H. rooperi* belongs to the family Hypoxidaceae which consists of 8 genera (*11*). Based on similarities with species of the Amaryllidaceae and Liliaceae family, the genera were previously classified in these taxa (*12*).

The hypoxis genus contains 130 species including *H. hemerocallidea*, *H. acuminate*, *H. iridifolia* and *H. rigidula* of which the first is most widely used and is therefore the primary focus of this chapter. The moniker African potato is to some extent a misnomer as the rootstock is a corm rather than a tuber and the name causes confusion with tuberous *Plectranthus esculentus* (*13*). There is some evidence that the common hypoxis species are used interchangeably in traditional medical practice, sometimes inadvertently because of their close similarities (*14*).

Hypoxis grows as a stemless, perennial geophyte with large black fibrous corms which are bright yellow inside when freshly cut (*15*). The leaves are organized typically into three ranks and maybe slightly hairy. The flowers which appear between October and January (*16*) are star-shaped and bright yellow and the number per inflorescence may vary from two to twelve (*15*, *17*). While it is indigenous to southern Africa, hypoxis has a world-wide cosmopolitan distribution. It grows widely in the savanna grasslands of South Africa, Swaziland, Lesotho , Mozambique, Zimbabwe and into some parts of East Africa (*11*, *16*). It has also been reported to grow in meanders, grasslands and mountains of South America, Australia and coastal areas of Asia (*18*).

Traditional and Contemporary Use

The rootstock of Hypoxis is the plant part of choice for medicinal and dietary use. It is the iconic African potato promoted by proponents of African traditional medicine in South Africa as a cure-all and derided by opponents as quackery (*19*) in the country's highly charged political environment at the turn of the century. In Zulu traditional medicine it is widely known as "zifozonke" literally meaning "panacea" (*20*). It is known as African potato, starflower / South African star grass, Ilabatheka, Inkofe (in Zulu), sterretjie (in Afrikaans), moli kharatsa, monna wa maledu, thitidi (in Sotho), hodzori (in Shona) (*21*).

The aqueous infusion is used as tonic in children and to treat dizziness and mental disorders in adults (*16*). Hypoxis is widely used traditionally particularly in Zulu society for heart disease, bad dreams, anxiety, insanity, barrenness, intestinal parasites, urinary tract infections among other indications (*11*, *22*). Related species mainly *H. colchicifolia* (syn. *H. oligotricha*), *H obtuse*, *H. nyasica* and *H. angustifolia* are also used for similar indications (*11*). In addition *H. obtuse*, which more commonly grows north of South Africa's border is used traditionally for infertility in women, abdominal pains, heart pains, bile emesis, gonorrhea and for aphrodisiac effects (*21*).

Hypoxis also has ethnoveterinary uses. In South Africa *Hypoxis hemerocallidea* and *H.rigidula* have been reportedly used for fertility enhancement, general ailments, heartwater, and to prevent abortion in cattle by the Tswana people (*23*) . Kambizi (*24*) has reported that fresh cut corms are applied to the face by Xhosa women as facial treatments or used to treat burns.

In contemporary society, the use of hypoxis encompasses many of the traditional uses and more recent indications. Today it is marketed both in its crude unprocessed form (at traditional medicine (so-called muthi) markets) (*7*) and also in pharmaceutical formulations and packaging as a phytomedicine in retail pharmacies. In both formats the marketing materials carry many unsubstantiated claims viz. HIV / AIDS, arthritis, myalgic encephalomyelitis (ME), hypertension, asthma, diabetes mellitus, cancer, arthritis, psoriasis, tuberculosis and epilepsy (*22*, *25–28*). It is also increasingly being formulated into dermatological products with purported antiviral and antifungal actions (*11*). In Germany, hypoxis extracts have been widely marketed since the late 1970's for the treatment of prostate adenoma and benign prostate hypertrophy (BPH) (*11*, *29*, *30*). This indication is based on the inhibition of 5α-reductase metabolism of testosterone by phytosterols from hypoxis (*11*).

Dold and Cocks (*7*) estimated that trade in hypoxis exceeds 11 metric tons per year in South Africa alone. Because of the high demand, hypoxis populations are declining and under pressure (*31*). The use of corms is particularly destructive and (*32*) investigated the possible use of the aerial parts as substitutes. They found clear differences between the two plant parts both in chemical constitution and biological activity making the substitution of corms for leaves less attractive.

Phytochemistry

Several compounds have been isolated from hypoxis species and some of them have been linked to their putative biological activity. Hypoxoside, a norlignan diglucoside is an aglycone consisting of a diphenyl-1-en-4-yne-pentane unit (*33*) and has been isolated from all the common Hypoxis species (*34*). have suggested that this compound maybe unique to the genus. Hypoxoside is hydrolysed by β-glucosidase to rooperol which appears to be the active bioavailable compound (*26*).

However it is for its high concentration of phytosterols that hypoxis is best known for, especially sterols, sterolins and stanols. The patent by (*29*) claimed that aqueous extracts of *H. hemerollicadea* had phytosterol yield of 9 mg /

100g, however this could not be reproduced in recent studies by (*13*). These workers found less than 0.01% w/w final extract recovery they have dismissed the therapeutic importance of sterolins in hypoxis based on that data. The likelihood is that they were not using the same material and hence the differences are inconclusive.

The plant hormone (cytokinin) zeatin and it's derivatives were isolated from the corms by Van Staden (*35*).

Biological Activity

It is fair to say that because of its high profile status in South African materia medica, hypoxis has been well studied for its biological effects (Table 1). Methanol extracts of *H. hemerocallidea* corms showed no activity against *Staphylococcus aureus*, *Escherichia coli*, and *Enterococcus faecalis* (*14*). Using freshly harvested corms grown in the laboratory (*32*) previously showed that acetone and ethanol extracts of the leaves had relatively good activity against both *E. coli* (MIC < 0.31 mg /ml) and to some extent *Pseudomonas aeruginosa* (MIC < 0.61 mg / ml). The aerial parts of hypoxis have not been previously investigated so this study was important in showing their possible utility.

Steenkamp et al. (*36*) showed inhibition of *E. coli* by both water and ethanolic extracts of *H. hemerocallidea* with MIC of 62.5 ug / ml, though they estimated that it would not translate to a therapeutic dose in humans (based on the dosage). They further suggested that the antimicrobial activity maybe another reason why hypoxis is used in prostatitis and prostate indications.

Hypoxoside and rooperol showed good activity against *E. coli* and *S. aureus*, with the latter being more active than the positive control (neomycin) (*37*). They also possess good antimuagenic and cytotoxic properties (*38*)

Sathegke (*14*) showed that methanol extracts of hypoxis is non-toxic in a cytcoxicity assay done using Vero cells (African green monkey kidney cell line) and no activity was seen in HeLa cancer cell line. *H. iridifolia* and *H. rigidula* showed slightly better activity in the HeLa cytotoxicity assays. Albrecht et al (*26*) have previously reported that rooperol, the hydrolytic product of hypoxoside showed activity against HeLa cells under 10 μg/ml, while the glycoside was inactive in the same cell line.

Cytotoxic activity against HeLa, HT-29 and MCF-7 cancer cells appeared to increase when β-glucosidase was added to *H. hemerocallidea* which seemed to imply that there were certain compounds (other than hypoxoside) which were being hydrolyzed to active metabolites (*39*)

Van der Merwe et al (*40*) demonstrated that rooperol inhibits leucotriene synthesis in leukocytes and prostaglandin synthesis in platelet microsomes, thus showing *in vitro* antioxidant and anti-inflammatory activity. Good free radical scavenging ability was of aqueous and ethanolic extracts of *H. hemerocallidea* was shown by Steenkamp et al (*36*). Antioxidant and free radical scavenging activities of rooperol has been confirmed in various assays (*11*, *18*, *33*) as would be expected of most phenolic compounds.

Ojewole (*25*) found that an aqueous decoction of the corms of hypoxis had significant antinociceptive and anti-inflammatory activity which was dose dependent (50 – 800 mg / kg p.o.). In the same study, hypoxis extract showed hypoglycemia in both normal and streptozotocin (STZ)-induced diabetes mellitus rats. Gaidamashvili and Van Staden (*41*) showed that lectins isolated from *H. hemerocallidea* may partly explain the anti-inflammatory activity as it elicited a 29% inhibition of cyclooxygenase enzymes which catalyze the conversion of arachidonic acid to prostaglandin. In a study by Laporta et al (*37*), using concentrations below 0.5 mg / ml, there was no inhibition of COX-1 and COX-2 by either the crude extract or the compound hypoxoside. Rooperol, however showed strong inhibition against both COX enzymes with a COX-2 / COX-1 IC50 ratio of 2. It is thought that the compound exerts it's anti-inflammatory actions by also inhibiting the production of NF-κB, AP-1 and similar transcriptional factors important in inflammatory processes (*42*)

Ojewole (*43*) has attempted to validate the ethnobotanical use of hypoxis in managing hysterical fits and epilepsy. Using a rat model, he found that aqueous corm extracts (100 – 800 mg / kg i.p.) delayed the onset of, and significantly inhibited, seizures. In other similar studies in rodent models, Ojewole and his colleagues tried to validate the use of hypoxis in gastroinstestinal and uterine – related conditions, specifically, how do aqueous extracts of hypoxis modulate smooth muscle? (*44–46*). They found that hypoxis extracts reduced gastrointestinal motility, the frequency of defaecation and hence the severity of diarrhea in experimental rats (*43*). The uterolytic effects were also confirmed (*46*). In these experiments the muscle relaxant effects were found to be refractory to β-adrenoceptor, cholinergic and histaminergic blockers leading the workers to conclude that the mode of action might be through a non-specific spasmolytic mechanism.

The dichloromethane (DCM) extract of *H. colchicifolia* leaves was the only one with significant activity of 16 plant extracts assayed for anthelmintic activity using an *in vitro* colorimetric assay measuring free-living nematode larvae viability (*47*). *H. hemerocallidea* was inactive in this study.

A systematic review by (*48*) of various phytotherapies used to ameliorate the symptoms of Benign Prostatic Hyperplasia (BPH) *Serenoa repens* (saw palmetto), *Pygeum africanum* (African plum) and *H. hemerocallidea*, showed that the latter two improved urinary flow rates and symptoms compared to placebo. Several clinical studies have demonstrated that the use of hypoxis extracts in BPH patients resulted in significant sustained symptomatic relief (*49*). The effects of hypoxis on the prostate have always been linked to the purported occurrence of high concentrations of phytosterols. These compounds are known to inhibit 5α-reductase, the enzyme catalyzing the conversion of testosterone to dihydroxytestorone (DHT).

BPH is prevalent in 50% of men over 50 years of age and in 90% of men over 80 years old and it is known to depend on androgens, especially DHT (*50*). Inhibitors of 5α-reductase can also be used for male pattern baldness, hirsutism and prostate cancer. Hypoxis has been widely marketed and used for prostate disorders but there is need for more studies especially in light of recent findings that concentration of phytosterols in the species may actually be much lower than previously recorded (*13*). When this is considered in the light of the efficacy demonstrated in the systematic review of Dedhia and McVary (*48*), it may imply that other compounds which are yet to be identified or tested are responsible for the purported activity.

Phytosterols from hypoxis (mainly stigmasterol, β-sitosterol, brassicasterol and ergosterol) are also marketed for their anti-lipidemic, cholesterol - lowering actions. They accomplish this inhibiting Delta24 (δ 24) reductase enzymes in the cholesterol synthetic pathway (*51*). The lipid lowering effects of phytosterols are however equivocal because the bioavailability of these compounds is variable and dependant on formulation and other co-ingested foods (*39*)

The putative anti-inflammatory, antidiabetic and immune modulating properties of hypoxis have been attributed to β-sitosterol (*52*). The amount of sitosterol in hypoxis on average is variable. A study by (*53*) on immune modulation showed that hypoxis extract was inferior to control in the survival rates of FIV+ (Feline immunodeficiency virus infected) laboratory cat model. However, a sitosterol enriched product showed significantly improved survival rates.

Toxicology and Adverse Effects

The African Herbal Pharmacopeia (*54*) suggests that oral ingestion of hypoxoside showed no toxicity, and no lethality was seen in the brine shrimp assays nor cytotoxicity in Vero monkey cells.

However, it would appear that the use of crude herbal extracts of hypoxis has caused some concern. A study on renal function of wistar rats after both acute and chronic exposure to aqueous *H. hemerocallidea* extracts found that there was impaired function with significant decrease in glomerular filtration rate (GFR) and increased plasma creatinine concentration (*55*). A human study on the safety and efficacy of hypoxis extracts in HIV positive patients in South Africa was prematurely terminated by regulatory authorities reportedly because of bone marrow suppression and immune suppression which has been also been reported in a feline study (*56*). Steroids are known to be both immunesuppressive and may cause osteoporosis (*57*) and this maybe evidence of the high concentration of steroidal compounds in hypoxis.

In mice the LD50 of aqueous extracts of the corm of hypoxis was about 2g / kg meaning that the species shows low acute toxicity (*58*). Low doses of alcoholic and aqueous extracts of hypoxis, while causing a significant increase of weight in suckling rats, did not show any morphometric changes in the small intestine of the test animals (*59*). These workers concluded that hypoxis had no adverse effects on suckling rats.

Dichloromethane and 90% methanol aqueous extracts of hypoxis showed no mutagenic or antimutagenic activity in the Ames Salmonella / microsome mutagenicity assay (*60*).

Daily doses of up to 3.2 g of hypoxis extract showed no clinical toxicity or adverse changes in biochemical or haematological measurements in a phase I clinical trial in human lung cancer patients (*61*). While hypoxis has generally been proved to cause little toxicity, it is the potential for drug-herb interactions which still requires more extensive research. This is an increasingly important direction of research as the concomitant ingestion of herbs with pharmaceuticals is common especially in the treatment of HIV / AIDS (*62*). Therefore the potential for drug-herb adverse reactions is high. Nair et al. (*10*) have shown that hypoxis inhibits human drug metabolizing enzymes, CYP3A5, 3A4 and CYP19 by 11%, 33% and 41% respectively while rooperol showed almost total inhibition of CYP3A4 and 3A5. At concentrations of 5 mg / ml, hypoxis extracts showed high induction of the polyglycoprotein (Pg-p) transporter. No effect was shown on Pg-p by sterols or rooperol (*10*).

Future Directions of Research

Liu et al (*42*) have recently for the first time used a colitis model to demonstrate that methanolic extracts of hypoxis ameliorates severe typhlocolitis in mice. The extract reduced weight loss, severity of bacteria – induced colitis, neutrophil infiltration and intestinal epithelial proliferation making hypoxis and it's constituents an interesting lead in the search for prophylactic therapies for inflammatory bowel diseases. However the extract did not show similar efficacy in a non-bacteria-induced, dextran sodium sulfate (DSS) – induced colitis model. This is certainly an interesting and innovative direction in which research can move.

There remain many unanswered questions e.g. environmental variation and how it impacts on the chemistry and biological activity of hypoxis, the apparently contentious issue of the amounts of sterols and sterolins and their putative activity. Despite all the work that has been done so far, there remains a need to understand the phytochemistry of hypoxis in it's entirety and then to test the compounds in various *in vitro* and *in vitro* assays. Finding the bioactive compound (s) may lead to the progression of such a compound in the pharmaceutical drug pipeline and /or to the standardization of hypoxis extracts.

Conclusion

Hypoxis hemerocallidea is an important medicinal and dietary plant in Southern Africa. However it's contemporary use (particularly in HIV / AIDS and so-called immune-boosting products) has been controversial and political. This may have compromised the scientific research into it's use in these areas. There are some important data which have been gathered from in vitro and rodent studies which may form a good basis for the design of further pre-clinical studies. However there is a general lack of extensive phytochemical studies which may be the first step towards developing evidence – based botanical / phytomedicine products.

Acknowledgments

I wish to thank Dr. Learnmore Kambizi (CPUT) for proofreading the final draft and useful comments.

References

1. Chavunduka, G. *Traditional Medicine in Modern Zimbabwe*; University of Zimbabwe Publications: Zimbabwe,1994.
2. Milan, F.; Arnsten, J.; Klein, R.; Schoenbaum, E.; Moskaleva, G.; Buono, D.; et al. Use of complementary and alternative medicine in inner-city persons with or at risk for HIV Infection. *AIDS Patient Care STDs.* **2008**, *22*, 811–6.
3. Monera, T.; Wolfe, A.; Maponga, C.; Benet, L.; Guglielmo, J. Moringa oleifera leaf extracts inhibit 6 beta-hydroxylation of testosterone by CYP3A4. *J. Infect. Dev. Countries* **2008** (2), 379–83.
4. WHO. *Traditional Medicine Strategy 2002-2005*; World Health Organization, 2002.
5. Cunningham, A. Herbal medicine trade: a hidden economy. *Indicator SA.* **1989**, *6* (51-54).
6. Mander, M. *Marketing of indigenous medicinal plants in South Africa - A case study in KwaZulu Natal*; FAO: Rome, 1998.
7. Dold, A.; Cocks, M. The trade in medicinal plants in the Eastern Cape Province, South Africa. *S. Afr. J. Sci.* **2002**, *98*, 589–97.
8. Moodie, G.; Rooyen, K. V. HIV Concoction Slammed As Toxic: Food combination recommended by Aids dissident to SA's health minister has no support from science, local researchers say. *Sunday Times* **2003**March30.
9. Terreblanche, C. *HIV warning on African potato;* http://wwwiolcoza/news/south-africa/hiv-warning-on-african-potato-1109328, July 15 2003.
10. Nair, V.; Foster, B.; Arnason, J.; Mills, E.; Kanfer, I. In vitro evaluation of human cytochrome P450 and P-glycoproten-mediated metabolism of some phytochemicals in extracts and formulations of African potato. *Phytomedicine* **2007a**, *14*, 498–507.
11. Drewes, S. E.; Elliot, E.; Khan, F.; Dhlamini, J. T.; Gcumisa, M. S. Hypoxis hemerocallidea--not merely a cure for benign prostate hyperplasia. *J. Ethnopharmacol..* **2008**, *119* (3), 593–8Epub 2008/07/08.
12. Betto, P.; Gabriele, R.; Galeffi, C. Determination of the norlignan glucosides of Hypoxidaceae by high-performance liquid chromatography. *J. Chromatogr., A* **1992**, *594*, 131–5.
13. Nair, V.; Kanfer, I. Sterols and sterolins in Hypoxis hemerocallidea (African potato). *S. Afr. J. Sci.* **2008**, *104*, 322–3.
14. Sathekge, N. *Comparison of secondary metabolite content and antimicrobial activity of four Hypoxis species used in traditional medicine [MSc]*; University of Pretoria: Pretoria, 2010.
15. Wyk, B.-E. V.; Oudtshoorn, B. V.; Gericke, N. *Medicinal plants of South Africa*; Briza Publications: Pretoria, 1997.
16. Mukinda, J.; Eagles, P. *The South African Pharmacopiea Monograph Project: South African Traditional Medicines Research Group (SATMERG)*; 2010.

17. Singh, Y. Hypoxis: yellow stars of horticulture, folk remedies and conventional medicine. *Veld Flora* **1999**, *85*, 123–5.
18. Laporta, O.; Perez-Fons, L.; Mallvia, R.; Caturla, N.; Micol, V. Isolation, characterization and antioxidant capacity assessment of the bioactive compounds derived from Hypoxis rooperi corm extracts (African potato). *Food Chem.* **2007b**, *101*, 1425–37.
19. Green, J. *Manto's concoction 'could be harmful' 2003 10 May 2011*; Available from http://www.iol.co.za/news/south-africa/manto-s-concoction-could-be-harmful-1.214582.
20. Amusan, O. O. G.; Sukati, N. A.; Dlamini, P. S.; Sibandze, F. G. Some Swazi phytomedicines and their constituents. *Afr. J. Biotechnol.* **2007**, *6*, 267–72.
21. Gelfand, M.; Mavi, S.; Drummond, R.; Ndemera, B. *The Traditional Medical Practitioner in Zimbabwe*, 2nd ed.; Mambo Press:Gweru, 1993.
22. Watt, J.; Breyer-Brandwijk, M. *Medicinal and Poisonous Plants of Southern and Eastern Africa*, 2nd ed.; Livingstone: London, 1962.
23. Merwe, D. V. d.; Swan, G.; Botha, C Use of ethnoveterinary medicinal plants in cattle by Setswana-speaking people in the Madikwe area of the North West Province of South Africa. *J. S. Afr. Vet. Assoc.* **2001**, *72*, 189–96.
24. Kambizi, L. *Herbal remedies for skin care in the Eastern Cape*; Indigenous Plant Use Forum Conference Abstract; July 2012; University of Venda, Thoyohandou.
25. Ojewole, J. A. O. Antinociceptive, anti-inflammatory and antidiabetic properties of Hypoxis hemerocallidea Fisch. & C.A. Mey. (Hypoxidaceae) corm ['African Potato'] aqueous extract in mice and rats. *J. Ethnopharmacol.* **2006**, *103* (1), 126–34.
26. Albrecht, C. Hypoxoside: a putative prodrug for the treatment of malignancies, HIV infections, and inflammatory conditions. *SAMJ* **1995**, *85*, 302–7.
27. Hutchings, A.; Scott, A; Lewis, G.; Cunningham, A. *Zulu Medicinal Plants - An inventory*, 1st ed.; University of Natal Press: Pietermaritzburg, 1996.
28. Van Wyk B-E, G. N. *People's Plants: a guide to useful plants of southern Africa*; Briza Publications: Pretoria, 2000.
29. Pegel, K. H.; Liebenberg, R. W. (inventors) *Extraction of phytosterol glycosides from Hypoxis tubers*; Germany1973.
30. Tyler, V. Plant drugs in the twenty-first century. *Econ. Bot.* **1986**, *40*, 279.
31. Hölscher B.. *Hypoxis hemerocallidea Fisch., C.A.Mey. & Avé-Lall*; 2009. 10 May 2011. Available from: http://www.plantzafrica.com/planthij/hypoxishemero.htm
32. Katerere, D. R.; Eloff, J. N. Anti-bacterial and anti-oxidant activity of Hypoxis hemerocallidea (Hypoxidaceae): Can leaves be substituted for corms as a conservation strategy? *S. Afr. J. Bot.* **2008**, *74* (4), 613–6.
33. Nair, V. D.; Dairam, A.; Agbonon, A.; Arnason, J. T.; Foster, B. C.; Kanfer, I. Investigation of the antioxidant activity of African potato (Hypoxis hemerocallidea). *J. Agric. Food Chem.* **2007b**, *55* (5), 1707–11Epub 2007/02/14.
34. Nair, V. D.; Kanfer, I. High-performance liquid chromatographic method for the quantitative determination of Hypoxoside in African potato (Hypoxis

hemerocallidea) and in commercial products containing the plant material and/or its extracts. *J. Agric. Food Chem.* **2006**, *54* (8), 2816–21.

35. Van Staden, J. Constituents of Hypoxis rooperi, a valuable medicinal plant in South Africa. *Dtsch. Apoth. Ztg.* **1981**, *33*, 460–4.
36. Steenkamp, V.; Gouws, M. C.; Gulumian, M.; Elgorashi, E. E.; Staden, J. v. Studies on antibacterial, anti-inflammatory and antioxidant activity of herbal remedies used in the treatment of benign prostatic hyperplasia and prostatitis. *J. Ethnopharmacol.* **2006**, *103*, 71–5.
37. Laporta, O.; Funes, L.; Garzon, M. T.; Villalain, J.; Micol, V. Role of membranes on the antibacterial and anti-inflammatory activities of the bioactive compounds from Hypoxis rooperi corm extract. *Arch. Biochem. Biophys.* **2007**, *467* (1), 119–31.
38. Albrecht, C. Hypoxoside as a putative non-toxic, multi-functional prodrug for the treatment of cancer, HIV-AIDS and inflammatory conditions. In: *2nd IOCD international Symposium*; Hostettman, K., Chinyanganya, F., Maillard, M., Wolfender, J., Eds.; University of Zimbabwe Press: Victoria Falls, Zimbabwe, 1996.
39. Boukes, G. J. *The in vitro biological activities of three Hypoxis species and their active compounds.* Ph.D. Thesis, NMMU, 2010.
40. Van der Merwe, M. J.; Jenkins, K.; Theron, E.; van der Walt, B. J. Interaction of the di-catechols rooperol and nordihydroguaiaretic acid with oxidative systems in the human blood. *Biochem. Pharmacol.* **1993**, *45* (2), 303–311.
41. Gaidamashvili, M.; Staden, . Jv. Prostaglandin inhibitory activity by lectin-like proteins from South African medicinal plants. *S. Afr. J. Bot.* **2006**, *72*, 661–3.
42. Liu, Z.; Wilson-Welder, J.; Hostetter, J.; Jergens, A.; Wannemuehler, M. Prophylactic treatment with Hypoxis hemerocallidea corm (African potato) methanolic extract ameliorates Brachyspira hyodysenteriae-induced murine typhlocolitis. *Exp. Biol. Med. (Maywood)* **2010**, *235* (2), 222–30.
43. Ojewole, J. A. Anticonvulsant activity of Hypoxis hemerocallidea Fisch. & C. A. Mey. (Hypoxidaceae) corm ('African potato') aqueous extract in mice. *Phytother. Res.* **2008**, *22* (1), 91–6.
44. Ojewole, J. A.; Awe, E. O.; Nyinawumuntu, A. Antidiarrhoeal activity of Hypoxis hemerocallidea Fisch. & C. A. Mey. (Hypoxidaceae) Corm ('African potato') aqueous extract in rodents. *Phytother. Res.* **2009**, *23* (7), 965–71.
45. Ojewole, J. A.; Olayiwola, G.; Nyinawumuntu, A. Bronchorelaxant property of 'African potato' (Hypoxis hemerocallidea corm) aqueous extract in vitro. *J. Smooth Muscle Res.* **2009b**, *45* (5), 241–8.
46. Nyinawumuntu, A.; Awe, E. O.; Ojewole, J. A. Uterolytic effect of Hypoxis hemerocallidea Fisch. & C.A. Mey. (Hypoxidaceae) corm ['African Potato'] aqueous extract. *J. Smooth Muscle Res.* **2008**, *44* (5), 167–76.
47. Aremu, A.; Ndhlala, A.; Fawole, O.; Light, M.; Finnie, J.; Staden, J. V. In vitro pharmacological evaluation and phenolic content of ten South African medicinal plants used as anthelmintics. *S. Afr. J. Bot.* **2010**, *76*, 558–66.
48. Dedhia, R.; McVary, K. Phytotherapy for lower urinary tract symptoms secondary to benign prostatic hyperplasis. *J. Urol.* **2008**, *179*, 2119–25.

49. Lowe, F. C.; Dreikorn, K.; Borkowski, A.; Braeckman, J; Denis, L.; Ferrari, P.; et al. Review of recent placebo-controlled trials utilizing phytotherapeutic agents for treatment of BPH. *Prostate.* **1998**, *37* (3), 187–93.
50. Carson, C.; Rittmaster, R. The role of dihydrotestosterone in benign prostatic hyperplasia. *Urology* **2003**, *61*, 2–7.
51. Fernández, C. S. Y.; Ferruelo, A. J.; Gómez-Coronado, D.; Lasunción, M. A. Inhibition of cholesterol biosynthesis by Delta22-unsaturated phytosterols via competitive inhibition of sterol Delta24-reductase in mammalian cells. *Biochem. J.* **2002**, *366* (Pt 1), 109–19.
52. Bouic, P. J. D. The role of phytosterols and phytosterolins in immune modulation: a review of the past 10 years. *Curr. Opin. Clin. Nutr. Metab. Care* **2001**, *4* (6), 471–5.
53. Lamprecht, J.; Liebenberg, R.; Freestone, M.; Bouic, P.; Clark, A; Jaarsveld, P. V. A comparison of the survival benefit provided by putative immune modulators in the FIV+ (Feline immunodeficiency virus infected) laboratory cat model. *Int. Conf. AIDS* **2000**, *13* abstract no. TuPeA3021.
54. Brendler, T.; Eloff, J.; Gurib-Fakim, A.; Phillips, L. African Herbal. *Pharmacopeia* **2010**, 324.
55. Musabayane, C. T.; Xozwa, K.; Ojewole, J. A. Effects of Hypoxis hemerocallidea (Fisch. & C.A. Mey.) [Hypoxidaceae] corm (African Potato) aqueous extract on renal electrolyte and fluid handling in the rat. *Ren Fail.* **2005**, *27* (6), 763–70.
56. Niekerk, J. d. V. Food for thought. *S. Afr. Med. J.* **2003**, *93*, 297.
57. Brunton, L.; Goodman, L.; Blumenthal, D.; Buxton, I. *Goodman and Gilman's Manual of Pharmacology and Therapeutics*; McGraw Hill: New York, 2008.
58. Ojewole, J. A. Antinociceptive, anti-inflammatory and antidiabetic properties of Hypoxis hemerocallidea Fisch. & C.A. Mey. (Hypoxidaceae) corm ['African Potato'] aqueous extract in mice and rats. *J. Ethnopharmacol.* **2006**, *103* (1), 126–34.
59. Erlwanger, K. H.; Cooper, R. G. The effects of orally administered crude alcohol and aqueous extracts of African potato (Hypoxis hemerocallidea) corm on the morphometry of viscera of suckling rats. *Food Chem. Toxicol.* **2008**, *46* (1), 136–9Epub 2007/08/25.
60. Reid, K.; Maes, J.; Maes, A.; Staden, J. v.; Kimpe, N. D.; Mulholland, D.; et al. *J. Ethnopharmacol* **2006**, *106*, 44–50.
61. Smit, B; Albrecht, C; Liebenberg, R; Kruger, P; Freestone, M; Gouws, L.; et al. A phase I trial of hypoxoside as an oral prodrug for cancer therapy - absence of toxicity. *S. Afr. Med. J.* **1995**, *85*, 865–70.
62. Bica, I.; Tang, A.; Skinner, S.; Spiegelman, D.; Knox, T.; Gorbach, S.; et al. Use of complementary and alternative therapies by patients with Human Immunodeficiency Virus disease in the era of Highly Active Antiretroviral Therapy. *J. Altern. Complementary Med.* **2003**, *9*, 65–76.

Chapter 5

Pycnanthus angolensis: Bioactive Compounds and Medicinal Applications

Kelsey Gustafson,[1,2] Qing-Li Wu,[1,2] Julie Asante-Dartey,[3] and James E. Simon*,[1,2]

[1]New Use Agriculture and Natural Plant Products Program (NUANPP), Department of Plant Biology and Pathology, Rutgers University, Foran Hall, 59 Dudley Road, New Brunswick, New Jersey 08901
[2]Department of Medicinal Chemistry, Ernest Mario School of Pharmacy, Rutgers University, 160 Frelinghuysen Road, Piscataway, New Jersey 08854
[3]ASNAPP-Ghana, Accra, Ghana
*E-mail: jimsimon@rci.rutgers.edu.

Pycnanthus angolensis (Welw.) Warb. or *Pycnanthus kombo* (Baill.) Warb., commonly known as African nutmeg, is a tree species found in West and Central Africa. Different parts of the tree are traditionally used for a wide range of medicinal purposes, including the treatment of pain and fungal infections. Some recent natural products research has involved the seed fat, from which kombo butter is produced. The seed fat contains unusually high levels of myristoleic acid, a precursor of cetyl myristoleate that is of interest for its potential use in the treatment of arthritis. Kombo butter is also a source of phytochemicals that are structurally related to the tocotrienols and have shown significant anti-inflammatory and antioxidant activity. This chapter focuses on the bioactive compounds found in the seed fat of *P. angolensis*.

Ecology and Ethnobotany

Pycnanthus angolensis is native to tropical Africa and is found in damp secondary forests as far northwest as Mauritania and Senegal, extending to Tanzania and Uganda in the eastern region, with Angola and Zambia at the southern periphery (*1*). The tree is a fruit-bearing evergreen species that provides timber for use in furniture (*1*). It is an important source of shade for banana plantations in Uganda, and for coffee and cocoa plantations in Cameroon (*1*). The fruit contains an oil-rich seed encased in an aril, shown in Figure 1. There is an average annual yield of 60 to 100 seeds per tree (*1*).

Figure 1. Pycnanthus angolensis seed with aril.

While *P. angolensis* is commonly called African nutmeg due to a superficial resemblance to the seeds of the popular spice, it does not belong to the same genus as true nutmeg (*Myristica fragrans*). The seeds of *P. angolensis* are not edible as they have a bitter taste, though no adverse effects have been described and they are reportedly used as a spice in central Africa (*1*). Aqueous extraction of the oil from wild-harvested seeds is done in Africa to make kombo butter, however the seeds are often left to waste. Kombo butter is a reddish-brown, semi-solid fat that can be used to make candles and soap (*1*). The initial aroma of the fat is pleasant, often characterized as having a cacao-like aroma, and later develops less desirable odors, such as trimethylamine (*2*).

Traditional medicine utilizes a variety of parts from the kombo tree in a diverse range of applications, with the bark used most frequently. In Cameroon, ground bark is eaten by mothers for several weeks after giving birth and increases lactation when taken in pepper soup (*3*). It is also rubbed on infected skin to treat scabies (*3*). A decoction of the bark is used as a purgative to treat fever, body aches, and stomach pain (*3*). When mixed with *Diodia scandens* leaves, the bark is believed to cure snakebite (*3*). The bark is also used as a remedy for intestinal helminthiasis and hemorroids in Cameroon (*4*) and for malaria in São Tomé and Príncipe (*5*). In Nigeria, the bark is chewed to treat toothache and infections, and is also reported to treat leprosy (*6*). The bark, leaves, and seed fat are used to treat fungal infections such as thrush in Ghana (*7*). The sap is applied topically to decrease swelling (*3*) and to arrest bleeding (*8*), while the arils are used to treat hernia (*3*).

Phytochemistry

Fatty Acids

As a member of the Myristicaceae family, the seed fat of *P. angolensis* is characteristically high in myristic acid. It also is the richest known natural source of (Z)-9-tetradecenoic acid [**1**], the ω-5 fatty acid commonly referred to as myristoleic acid. The seeds are approximately 56-61.6% fat, which is reported to contain 58.1-64.4% myristic acid and 19.4-26.3% myristoleic acid as shown in Table I (*9*, *10*). Small quantities of myristoleic acid are also found in animal depot fats, marine lipids, microalgae, milk fats, nuts, and vegetables (*11*).

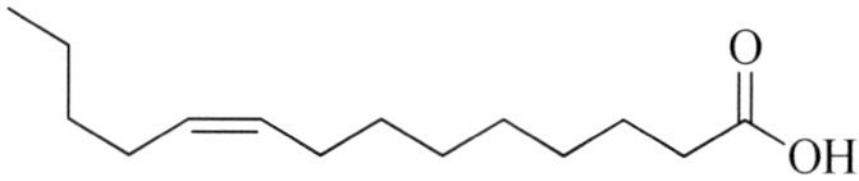

1

Table I. Fatty Acid Composition of *Pycnanthus angolensis* Seeds. SOURCE: Data from reference (*10*).

Fatty Acid	*% Composition*
Lauric 12:0	5.6
Myristic 14:0	64.4
Myristoleic 14:1	19.4
Palmitic 16:0	2.2
Palmitoleic 16:1	2.6
Fatty acids > 18:0	6.0

Myristoleic acid is a precursor of cetyl myristoleate [**2**], a nutraceutical supplement used for treating arthritis and joint pain. These two compounds are sometimes erroneously referred to interchangeably, however cetyl myristoleate is rarely found in nature. It has been reported to occur in spermaceti (*12*), Swiss albino mice, and castoreum oil deposits from male beavers (*13*).

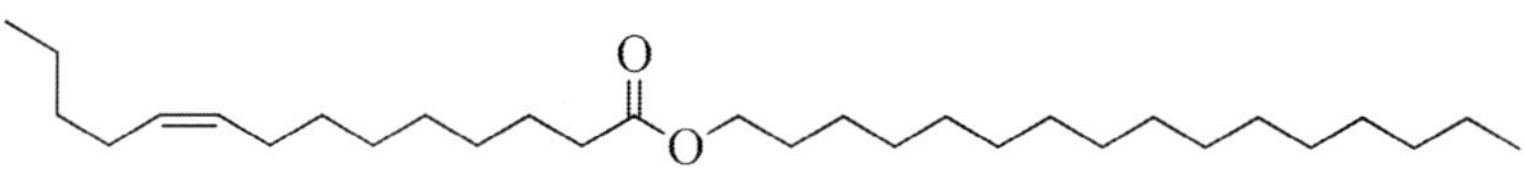

2

Despite the naturally high content of myristoleic acid in *P. angolensis*, enriched beef tallow is the primary commercial source of cetyl myristoleate due to its wide availability and low cost. The amount of myristoleic acid in unprocessed beef tallow is only 0.8-2.5% (*11*), yet with further industrial processing it can be concentrated to 40% (*12*).

Prior to the enrichment process, beef tallow is hydrolyzed and the free fatty acids are extracted. The fatty acid mixture is then enriched in myristoleic acid by fractional distillation and crystallization (*12*) or by fractionation involving urea-adduct formation. Urea complexation is used to isolate polyunsaturated fatty acids from saturated and monounsaturated fatty acids. Addition of urea results in solid-phase complexes with saturated and monounsaturated free fatty acids, allowing separation (*11*). The enriched fraction is subsequently esterified with cetyl alcohol, a derivatized byproduct of palm oil processing (*12*). Esterification occurs without a purification step. This results in a mixture of cetylated fatty acids that includes a large amount of cetyl myristate and other cetyl esters in addition to cetyl myristoleate (*12*).

Cetyl myristoleate produced in this fashion from animal fat appears to be unpalatable to horses and other herbivores. Kombo butter is an alternative source of myristoleic acid for use in the manufacture of cetyl myristoleate supplements, as first described in a US patent (*14*).

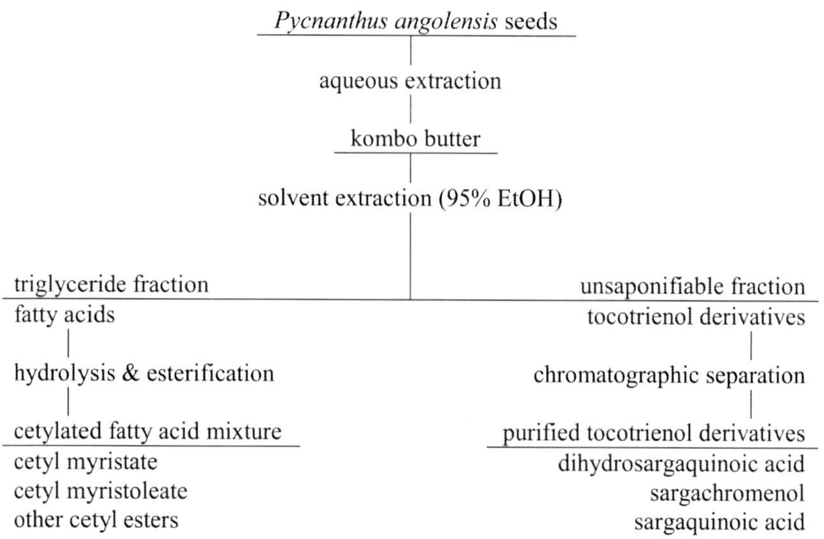

Figure 2. Processing of Pycnanthus angolensis seeds into two distinct products, cetyl myristoleate (at left) and the tocotrienol derivatives (at right).

The seeds of *P. angolensis* are milled into a fine paste and boiled to obtain kombo oil, which is collected, sieved, and solidified into crude kombo butter (Figure 2). Other preparation methods include solvent extraction and mechanical pressing of the seed fat on an industrial scale. The collection of kombo seeds and the production of kombo butter can be an environmentally sustainable practice as the seeds are collected after fruit drops from the wild trees and falls to the forest floor, and the processing effluence can be used as manure. However, the materials remaining from the initial preparation of kombo butter are usually discarded as waste. Agribusiness in Sustainable Natural African Plant Products (ASNAPP) has organized the collection of kombo seeds and the initial processing of kombo butter by local women and two cooperatives in Ghana. This crude kombo butter is then refined and semi-processed prior to export, which has been done in Ghana by the private sector and by the Kwame Nkrumah University of Science and Technology (KNUST). Ghana is the largest source of kombo butter produced for commercial use. Export of kombo butter to date has been largely to the United States to be used as a precursor of cetyl myristoleate in antiarthritic supplements. The ability to partially process the *P. angolensis* seed into kombo butter or an enriched kombo butter product can contribute to added value for those in the source country. Several of the value addition and processing stages can be done in Ghana or in another African country, but all too often only minimal processing occurs in the source country.

Although the myristoleic acid content of kombo butter is naturally high, it is still comparatively lower than the content reported in enriched beef tallow. The New Use Agriculture and Natural Plant Products (NUANPP) lab has investigated dry fractionation as an alternative to fractional distillation and urea complexation for use in developing countries (*15*).

Dry fractionation is the separation of fat by melting point through selective crystallization (*16*). Saturated fatty acids have a higher melting point than unsaturated fatty acids and their crystallization can be guided through temperature and agitation. A controlled reduction in temperature will lead to the precipitation of saturated fatty acids, which can then be separated from the mixture by filtration. Complete separation of fatty acids does not occur due to both the inherent triglyceride structure and to entrainment, which is the retention of uncrystallized oil in the solid phase (*16*). This technique is advantageous due to its low start-up costs.

The initial step in the dry fractionation of kombo butter is to isolate the triglycerides from the unsaponifiable fraction (Figure 3). This can be achieved by addition of 95% ethanol to the melted butter in a 1:1 ratio, which is then allowed to separate at room temperature. The unsaponifiable fraction containing the tocotrienol derivatives is removed as a brown ethanolic supernatant. The waxy pale yellow triglyceride layer formed at the bottom of the mixture is separated into two layers by heating at 60°C and then slowly cooling to 45°C in a water bath until formation of a solid layer occurs. The final temperature is lower than the melting point of myristic acid (55°C) and higher than the melting point of myristoleic acid (-4°C). Thus, a fluid top layer rich in myristoleic acid should be resolved from a solidified saturated fatty acid fraction. The top layer is further fractionated by cooling to room temperature (approximately 23°C) to produce a final supernatant with an increased myristoleic acid concentration. The data showed that the concentration of myristoleic acid could be increased twofold in an enriched fraction (*15*).

Further development of this simple and economical fractionation method has the potential to make kombo butter a competitive plant source of myristoleic acid.

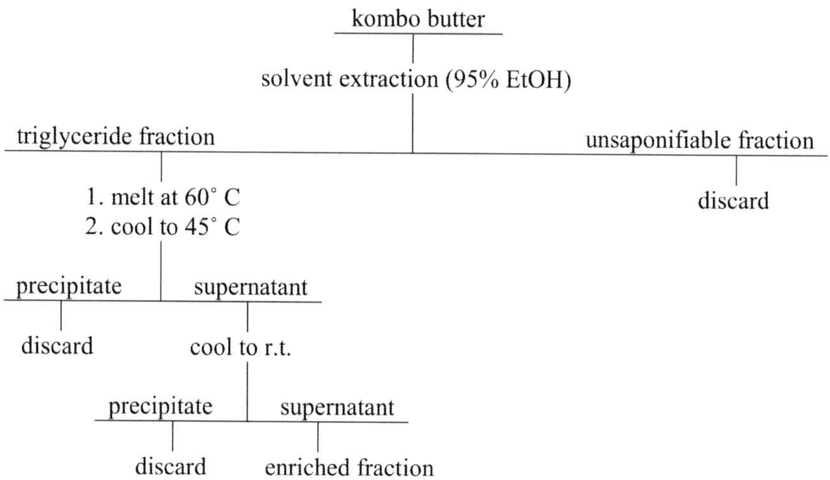

Figure 3. Enrichment of myristoleic acid from kombo butter.

Tocotrienol Derivatives

Tocochromanols, the antioxidants commonly known as vitamin E, are a class of natural products consisting of tocopherols and tocotrienols. Tocotrienols are less prevalent than tocopherols, their saturated analogs, and are primarily found in fruits and seeds to prevent lipid oxidation (*17*). Three compounds closely related in structure to δ-tocotrienol are present in kombo butter. Also described as plastoquinone derivatives or terpenoid quinones, they consist of a carboxylic acid attached to a polyprenyl side chain with either an aromatic or a chromanol ring system. They are present in the unsaponifiable fraction of kombo butter, which can be separated from the fat by solvent extraction (*10, 15, 18*). Fractionation or refining of kombo butter for commercial use in cetyl myristoleate production removes these compounds.

A dihydroplastoquinone derivative originally named kombic acid [**3**] was first isolated from *P. angolensis* as reddish-brown oil in 1983 (*10*). It was later posited that the original structure had been incorrectly characterized and was actually a structural isomer (*18*). The correct structure is believed to be sargahydroquinoic acid [**4**], named in 1987 after it was discovered in the brown alga *Sargassum sagamianum* (*19*). The first report of sargahydroquinoic acid in *P. angolensis* was in a patent application (*20*). Kombic acid has never been described as a natural product from another source nor has its presence in *P. angolensis* been confirmed elsewhere in the scientific literature, although there are exceptions to this in the patent literature.

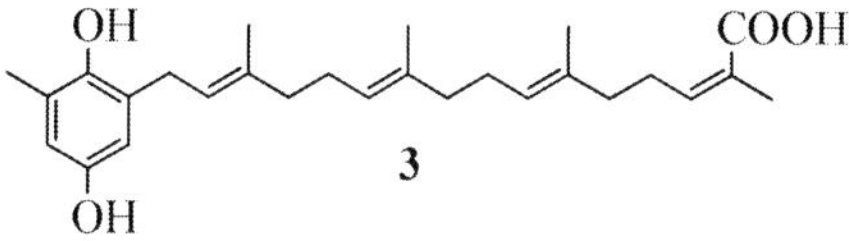

3

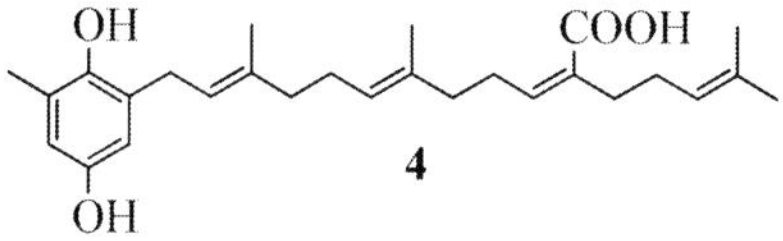

4

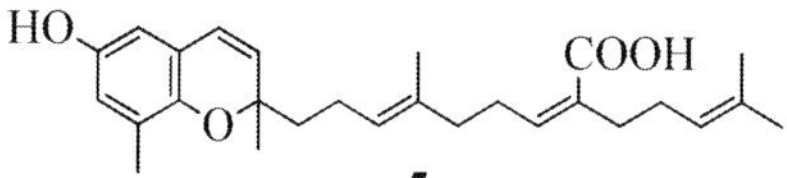

5

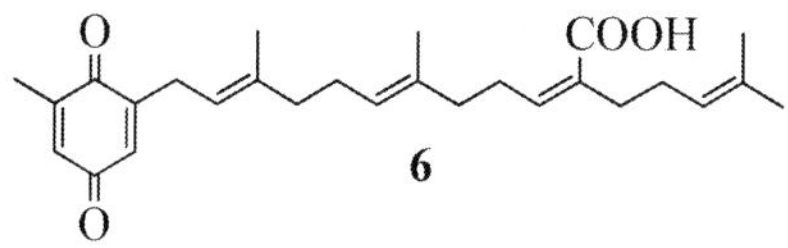

6

The other two tocotrienol derivatives found in kombo butter were initially isolated from the brown alga *Sargassum serratifolium* in 1979 (*21*). Sargachromenol [**5**] and sargaquinoic acid [**6**] weren't discovered in *P. angolensis* until 2000 (*7*) and 2005 (*18*), respectively.

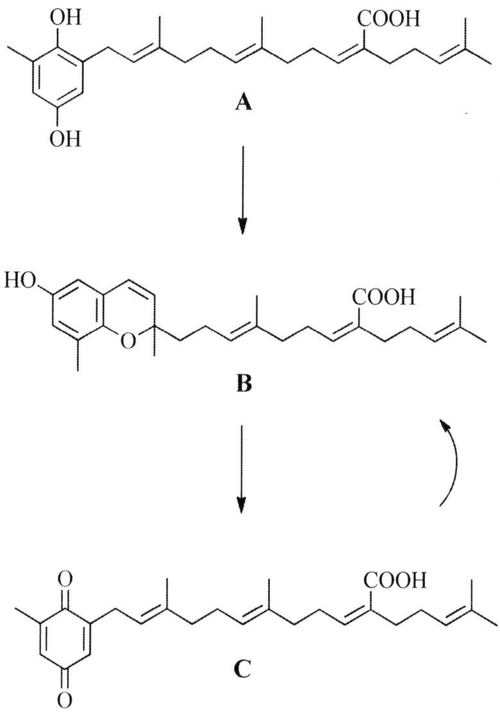

Figure 4. Proposed metabolic scheme of the tocotrienol derivatives in Pycnanthus angolensis seeds.

These compounds are frequently reported in brown algae (*19*, *21*, *22*), but in lower quantities, and have only been identified in three other terrestrial plants. Sargachromenol, sargahydroquinoic acid, and sargaquinoic acid were extracted from the aerial parts of *Roldana barba-johannis,* belonging to the Asteraceae family and native to the highlands of Mexico (*23*). Sargachromenol and sargaquinoic acid have also been found in *Clusia grandiflora* (*24*) and *Iryanthera juruensis* (*25*), both indigenous to the Amazon. *I. juruensis* is similar to *P. angolensis* in that it is also a member of the Myristicaceae family and has oil-rich seeds.

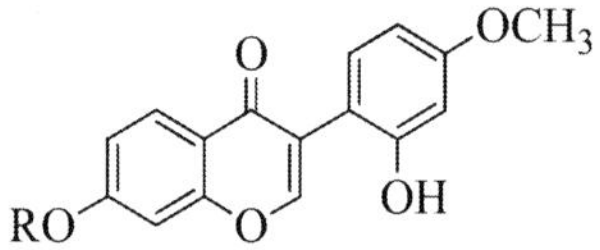

7a: R = H
7b: R = CH_3

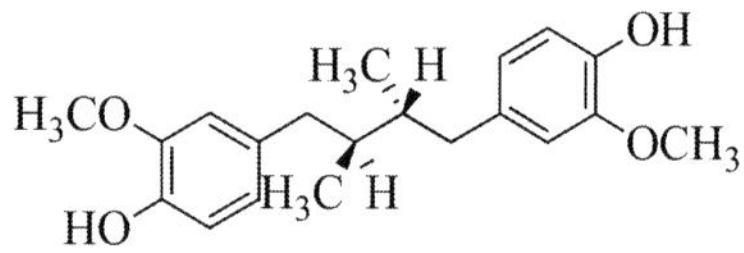

8

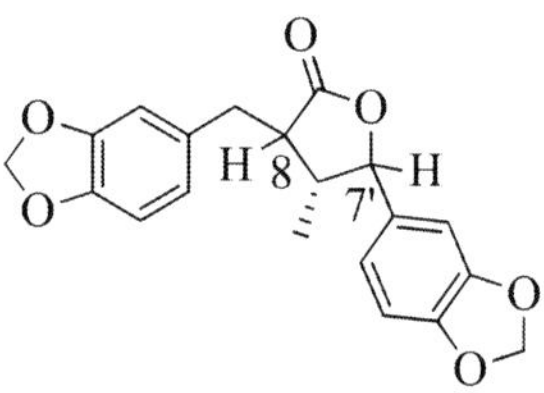

9a: 7' = R, 8 = R
9b: 7' = S, 8 = S

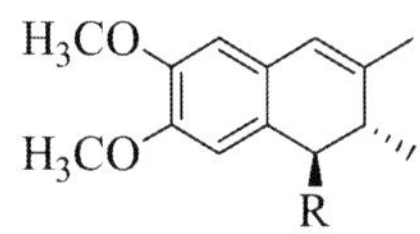

10a: R = 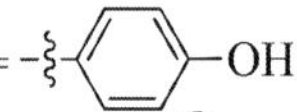

10b: R = 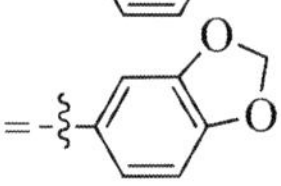

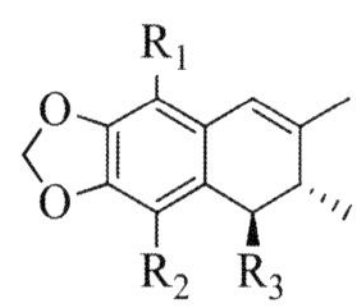

10c: R_1, R_2 = H, R_3 = 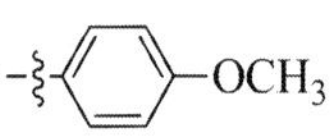

10d: R_1, R_2 = OCH_3, R_3 = 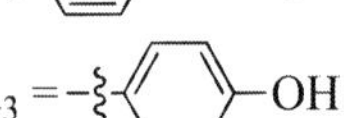

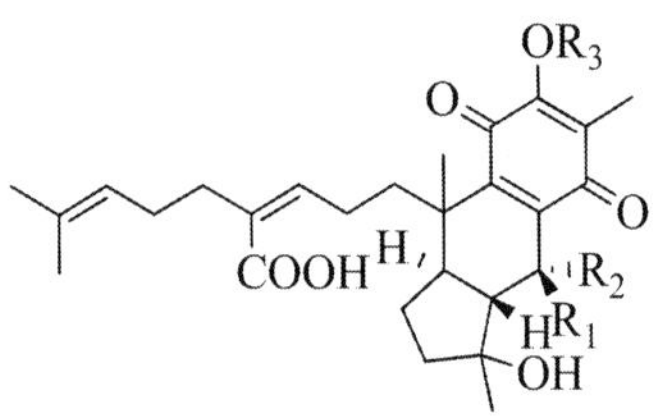

11a: R_1 = H, R_2 = OH, R_3 = H
11b: R_1 = OH, R_2 = H, R_3 = H

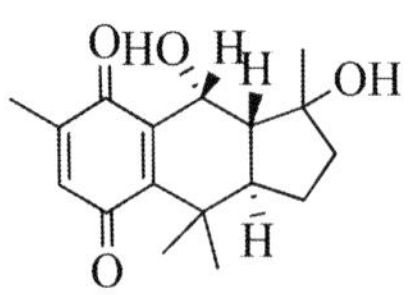

11c

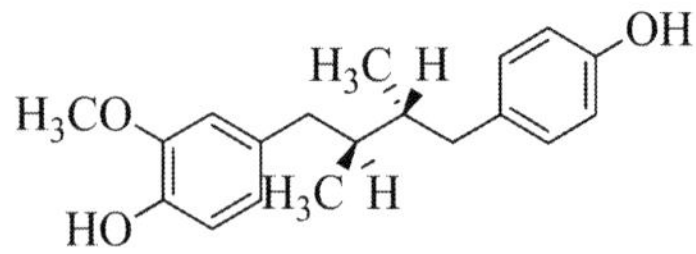

12

The interconversion of these three compounds was demonstrated on the lab bench by oxidation of the methyl ester of sargahydroquinoic acid with manganese dioxide to produce the methyl ester of sargaquinoic acid, which was converted to the methyl ester of sargachromenol by heating in pyridine (*10*). This transformation has also been reported to freely occur in the underivatized compounds on standing (*19*, *21*, *22*). However, the well-established tocochromanol biosynthetic pathway in *Arabidopsis thaliana* suggests a different scheme *in vivo* for these secondary metabolites. The tocochromanol pathway delineates the synthesis of δ-tocopherol from *2-methyl-6*-phytylbenzoquinone (MPBQ), closely related in structure to sargahydroquinoic acid (A in Figure 3) and sargachromenol (B in Figure 4), respectively. The oxidative pathway for tocopherols involves conversion to the quinone (C in Figure 3), producing a sargaquinoic acid analog that can be recycled back to the tocopherol (B in Figure 3) (*17*).

Other Compounds

Additional compounds isolated from other parts of *Pycnanthus angolensis* are listed in Table II.

Table II. Other Compounds in *Pycnanthus angolensis*

Structure	*Compound*	*Location*	*Reference*
7a	2′-hydroxyformononetin	heartwood	(*6*)
7b	7,4′-dimethoxy-2′-hydroxyisoflavone		
8	dihydroguaiaretic acid	bark	(*26*)
9a	pycnanolide A	leaves	(*27*)
9b	pycnanolide B		
10a	pycnanthulignene A	roots	(*28*)
10b	pycnanthulignene B		
10c	pycnanthulignene C		
10d	pycnanthulignene D		
11a	pycnanthuquinone A	leaves, stems	(*7*)
11b	pycnanthuquinone B		
11c	pycnanthuquinone C	bark	(*29*)
12	pycnantolol	bark	(*30*)

Biological Activity

Cetyl Myristoleate

Antiarthritic Activity

Cetyl myristoleate was originally reported to have antiarthritic activity by Harry Diehl (*31*). It was discovered during an investigation of the resistance of Swiss albino mice to adjuvant-induced arthritis. Diehl isolated fractions of compounds present in the mice tissue through homogenization, solvent extraction and separation by column chromatography. He then tested the fractions on rats susceptible to adjuvant-induced arthritis, using thin layer chromatography to identify compounds in active fractions. Diehl isolated cetyl myristoleate as the active compound and subsequently received a patent for its use to treat arthritis in 1977 (*31*). He later published a preliminary study that showed activity for both natural and synthetic cetyl myristoleate (*13*). Several other cetyl esters were also tested, including cetyl oleate and cetyl myristate. Cetyl oleate showed less activity than cetyl myristoleate, and cetyl myristate was found to be inactive, indicating that a degree of unsaturation may be necessary for efficacy. If confirmed, this finding would reinforce the importance of quality control in the production of commercial supplements, as cetyl myristate is frequently reported to be present in higher amounts than cetyl myristoleate in nutraceuticals sold using the name cetyl myristoleate (*12*, *15*).

Very few animal studies have been conducted since and with conflicting results (*32*, *33*), while one clinical trial appears to have inadvertently administered beef tallow with underivatized myristoleic acid instead of cetyl myristoleate (*34*). Despite initially promising results, there is currently a shortage of reliable evidence to support the claims.

If activity can be confirmed, cetyl myristoleate may act as an anti-inflammatory agent in a manner similar to the ω-3 polyunsaturated fatty acids, with either the intact ester or the hydrolysis product as the active compound (*32*).

Tocotrienol Derivatives

Anti-Inflammatory and Antioxidant Activity

Inflammation is an injury-triggered response by the immune system. Nitric oxide and prostaglandins are important mediators of the inflammatory process, but chronic overproduction can lead to oxidative stress and disease. Increased levels of nitric oxide and prostaglandins have been implicated in autoimmune diseases such as rheumatoid arthritis and the progression of neurodegenerative diseases such as Alzheimer's (*35*). They are also believed to contribute to neuronal damage following ischemic stroke (*36*). Nitric oxide is a free radical derived from L-arginine in a reaction catalyzed by nitric oxide synthase (NOS). The synthesis of prostaglandins from arachidonic acid is catalyzed by cyclooxygenase (COX). Chronic inflammation is caused by upregulation of the inducible enzymes iNOS and COX-2 (*35*), potential targets for phytochemicals.

Antioxidants such as the tocochromanols have been shown to modulate the inflammatory process and reduce oxidative damage (*37*). Various tocochromanol derivatives found in plants are being investigated for similar activity. Two of the δ-tocotrienol derivatives from *P. angolensis*, sargachromenol and sargahydroquinoic acid, were assessed for anti-inflammatory activity using the 12-O-tetradecanoylphorbol-13-acetate (TPA)-induced mouse ear edema assay and for antioxidant activity using the 2,2-diphenyl-1-picrylhydrazyl (DPPH) assay (*23*). In the TPA-induced mouse ear edema assay, sargachromenol was shown to have greater activity than α-tocopherol, which was similar to the activity of sargahydroquinoic acid, however both compounds had less activity than the non-steroidal anti-inflammatory drug (NSAID) indomethacin. In the DPPH assay, both compounds were shown to have greater antioxidant activity than α-tocopherol, with sargahydroquinoic acid having more activity than sargachromenol.

Table III. Results of ABTS Free Radical Scavenging Assay. SOURCE: Adapted with permission from reference (*18*). Copyright 2005.

Substance	*TEAC[a] (mM)*
sargahydroquinoic acid	1.33
sargachromenol	0.75
sargaquinoic acid	0.15
kombo butter FFA fraction	400 mM/kg
kombo butter	1120 mM/kg

[a] The TEAC value was reported as the concentration of trolox with the same antioxidant activity as a 1mM concent ration of the substance under experimental conditions

The NUANPP lab investigated the biological activity of all three of the purified δ-tocotrienol derivatives as well as kombo butter and the free fatty acid (FFA) fraction of kombo butter (*18*). Antioxidant activity was evaluated in both the DPPH and 2,2′-azino-bis(3-ethylbenzothiazoline-6-sulphonic acid) (ABTS) free radical scavenging assays. Anti-inflammatory activity was measured *in vitro* by ability to inhibit both nitric oxide production and expression of the iNOS and COX-2 enzymes using lipopolysaccharide (LPS)-induced RAW 264.7 mouse macrophages. LPS is an endotoxin that stimulates an immune response in macrophage cells. Nitric oxide production was measured by colorimetric assay. Inhibition of iNOS and COX-2 protein expression was measured by western blot analysis and inhibition of iNOS and COX-2 mRNA expression was measured by reverse transcriptase-PCR (RT-PCR) with northern blot analysis.

Sargahydroquinoic acid had the highest radical scavenging ability in both the DPPH and ABTS assays (the results of the ABTS assay are shown in Table III), followed by sargachromenol.

Kombo butter, the free fatty acid fraction, and all three pure compounds were found to inhibit both nitric oxide production and iNOS mRNA expression in a dose-dependent manner, with all except for sargachromenol inhibiting iNOS protein expression. The free fatty acid fraction was only found to inhibit COX-2 protein expression, while kombo butter, sargahydroquinoic acid, and sargachromenol demonstrated inhibition of both COX-2 protein and mRNA expression.

Sargahydroquinoic acid and its derivative 1,4-diacetylsargahydroquinoic acid were also tested for neuroprotective activity using the permanent middle cerebral artery occlusion (pMCAO) model (*38*). This model simulates what often occurs in human stroke by producing a long-term vessel blockade. Doses of sargahydroquinoic acid and 1,4-diacetylsargahydroquinoic acid were administered intraperitoneally 2 hours before and 6 hours after pMCAO in adult mice, with euthanization after 24 hours. Both compounds were shown to significantly reduce brain infarct volume and neuronal damage at a dose of 30 mg/kg with a 48% reduction by sargahydroquinoic acid and a 54% reduction by 1,4-diacetylsargahydroquinoic acid as compared with the vehicle. This also corresponded with an improvement in neurological performance as compared with the vehicle-treated group, indicating the potential for use of these compounds as neuroprotective agents following ischemic stroke.

Other Activity

Leonard proposed the use of kombic acid as a stabilizer in edible oils, plastics, and cosmetics, due to its antioxidant properties (*39*). A continuation of this patent considers the potential use of kombic acid and its derivatives as anticancer and cholesterol reducing agents (*40*). Although both patents depict the original structure of kombic acid, a more recent work acknowledges a previous misidentification of sargahydroquinoic acid as kombic acid (*41*).

Sargahydroquinoic acid, sargachromenol, and sargaquinoic acid were also shown to inhibit insect growth, the mechanism of which was correlated with their antioxidant activity (*42*).

Bioactivity of Selected Other Compounds

Pycnanthuquinone A [**11a**] and pycnanthuquinone B [**11b**], isolated from the leaves and stems as the result of a bioassay-guided fractionation, were shown to have antihyperglycemic activity (*7*, *43*). *P. angolensis* was selected for the study due to its traditional use in the treatment of oral thrush, fungal skin infections, and bodyaches, which are symptoms associated with Type 2 diabetes mellitus.

The related structure pycnanthuquinone C [**11c**] from the bark has shown anti-fungal activity (*29*).

Pycnanthulignene A [**10a**] and pycnanthulignene C [**10c**] are cyclolignene derivatives from the roots of *P. angolensis* with demonstrated antimicrobial activity (*28*).

Conclusion

Pycnanthus angolensis is a noteworthy though largely undervalued species with a history of uses in traditional African medicine. Unfortunately, the current body of scientific literature is limited and there is a lack of consistency regarding identification of the tocotrienol derivatives. There also exists a deficiency of reliable data on the activity, metabolism, and toxicity of myristoleic acid and cetyl myristoleate. Future work should address these issues. Beyond the potential shown for medicinal use, the development of *P. angolensis* in plant-based therapies should have a tremendous positive economic impact on local communities in West Africa. These communities would serve as the providers of the raw materials and the value added products of *P. angolensis* to be brought into the global marketplace.

Acknowledgments

We thank Kodzo Gbewonyo of BioResources-Ghana and New Jersey, US and Dan Acquaye of ASNAPP-Ghana for providing kombo butter to us and with whom we began examining this product as a potential source of myristoleic acid for use in the production of cetyl myristoleate, which was in the international marketplace but coming from other sources. Our initial work was in support of developing Ghana as a supplier of kombo butter and focused on its chemistry and quality control. We thank Dr. Cara Welch, a former graduate student who conducted processing research on enriching the butter with myristoleic acid and to Response Products, US for their interest and support. We thank Dr. Mingfu Wang, a former team member at Rutgers whom with Prof. Jim Simon began to study the large amount of waste material that was being discarded in Ghana following the preparation of cetyl myristoleate. Those studies lead to the characterization of sargahydroquinoic acid, sargaquinoic acid, and sargachromenol, and our subsequent investigation of their biological activity. We thank Profs. Chi-Tang Ho, Min-Hsiung Pan and Rafi Mohammed for their involvement in the anti-inflammatory activities and screens; Prof. Wenbin Dang for his work on nervous system disorders; the Ghanaian forest communities; researchers at KNUST involved in the domestication and commercialization of medicinal plants; and our ASNAPP colleagues and friends who provided us kombo butter materials with the hope that something of interest and commercial use could be identified and thus provide a market demand for an indigenous tree that was not being commercialized nor used locally in Ghana. Finally, we recognize and thank the New Jersey Agricultural Experiment Station, Rutgers University.

References

1. Mapongmetsem, P. M. *Pycnanthus angolensis* (Welw.) Warb. In *Plant Resources of Tropical Africa 14: Vegetable Oils*; van der Vossen, H. A. M., Mkamilo, G. S., Eds.; Backhuys Publishers: Leiden, Netherlands, 2007; pp 137−141.
2. Lewkowitsch, J.; Varburton, G. H. *Chemical technology and analysis of oils, fats and waxes*, 5th ed.; Macmillon and Co., Ltd.: London, 1914; Vol. 2, p 565.
3. Thomas, D. W.; Thomas, J. M.; Bromley, W. A.; Mbenkum, F. T. Korup Ethnobotany Survey. In *Final Report to the World Wide Fund for Nature*; World Wide Fund for Nature: Surrey, U.K., 1989.
4. Betti, J. L. Medicinal plants sold in Yaounde markets, Cameroon. *Afr. Study Monogr.* **2002**, *23*, 47–64.
5. Madureira, M. D. C.; Martins, A. P.; Gomes, M.; Paiva, J.; da Cunha, A. P.; do Rosário, V. Antimalarial activity of medicinal plants used in traditional medicine in S. Tomé and Príncipe Islands. *J. Ethnopharmacol.* **2002**, *81*, 23–29.
6. Omobuwaj, O. R.; Adesanya, S. A.; Babalola, G. O. Isoflavonoids from *Pycnanthus angolensis* and *Baphia nitida*. *Phytochemistry* **1992**, *31*, 1013–1014.
7. Fort, D. M.; Ubillas, R. P.; Mendez, C. D.; Jolad, S. D.; Inman, W. D.; Carney, J. R.; Chen, J. L.; Ianiro, T. T.; Hasbun, C.; Bruening, R. C.; Luo, J.; Reed, M. J.; Iwu, M.; Carlson, T. J.; King, S. R.; Bierer, D. E.; Cooper, R. Novel antihyperglycemic terpenoid-quinones from *Pycnanthus angolensis*. *J. Org. Chem.* **2000**, *65*, 6534–6539.
8. Abbiw, D. K. *Useful Plants of Ghana: West African uses of wild and cultivated plants*; Intermediate Technology Publications: London, 1990; p 190.
9. Atherton, D.; Meara, M. L. Component fatty acids and glycerides of some *Myristica* fats. *J. Soc. Chem. Ind.* **1939**, *58*, 353–357.
10. Lok, C. M.; Groenewegen, A.; Stroink, J. B. A.; Ward, J. P. Kombic acid, a hydroquinone polyisoprenoic carboxylic acid from *Pycnanthus kombo* seed fat. *Phytochemistry* **1983**, *22*, 1973–1976.
11. *Bailey's Industrial Oil and Fat Products*, 6th ed.; Shahidi, F., Ed.; John Wiley & Sons, Inc.: Hoboken, NJ, 2005; Vol. 1, pp 124, 163, Vol. 2, pp 127, 182, 264, 268, and Vol. 3, pp 605−606.
12. European Food Safety Authority (EFSA) Panel on Dietetic Products, Nutrition and Allergies (NDA). Scientific opinion on the safety of "Cetyl Myristoleate Complex" as a food ingredient. *EFSA J.* **2010**, *8*, 1686–1699.
13. Diehl, H. W. Cetyl myristoleate isolated from swiss albino mice: an apparent protective agent against adjuvant arthritis in rats. *J. Pharm Sci.* **1994**, *83*, 296–299.
14. Welch, C. R.; Kwansu, K.; Valentina, P.; Wu, Q. L.; Juliani, H. R.; Simon, J. E. Internal Report, New Use Agriculture and Natural Plant Products Program, 2008.
15. Leonard, E. U.S. Patent 7,772,279, 2010.

16. Timms, R. E. Fractional crystallization-the fat modification process for the 21st century. *Eur. J. Lipid Sci. Technol.* **2005**, *107*, 48–57.
17. Mène-Saffrané, L.; DellaPenna, D. Biosynthesis, regulation and functions of tocochromanols in plants. *Plant Physiol. Biochem.* **2010**, *48*, 301–309.
18. Simon, J. E.; Wang M.; Gbewonyo K.; Rafi M. M; Acquaye D. F.; Asianowa Y. U.S. Patent 7,371,413, 2005.
19. Segawa, M.; Shirahama, H. New plastoquinones from the brown alga *Sargassum sagamianum* var. *yezoense*. *Chem. Lett.* **1987**, 1365–1366.
20. Barclay, S. S.; Bosley, J. A.; Cain, F. W.; Koenen, C; Rogers, J. S; Schmid, U. U.S. Patent 7,582,317, 2009.
21. Kusumi, T.; Shibata, Y.; Ishitsuka, M.; Kinoshita, T.; Kakisawa, H. Structures of new plastoquinones from the brown alga *Sargassam serratifolium*. *Chem. Lett.* **1979**, 277–278.
22. Reddy, P.; Urban, S. Meroditerpenoids from the southern Australian marine brown alga *Sargassum fallax*. *Phytochemistry* **2009**, *70*, 250–255.
23. Pérez-Castorena, A. L.; Arciniegas, A.; Ramirez-Apan, M. T.; Villasenor, J. L.; Romo de Vivar, A. Evaluation of the anti-inflammatory and antioxidant activities of the plastoquinone derivatives isolated from *Roldana barba-johannis*. *Planta Med.* **2002**, *68*, 645–647.
24. Monache, F. D.; Marta, M.; Mac-Quhae, M. M.; Nicoletti, M. Two new tocotrienolic acids from fruits of *Clusia grandiflora* Splitg. *Gazz. Chim. Ital.* **1984**, *114*, 135–137.
25. Silva, S.; Pereira, C.; Zanoni, M. V. B.; Yoshida, M. Lipophyllic antioxidants from *Iryanthera juruensis* fruits. *Phytochemistry* **2001**, *57*, 437–442.
26. Njoku, C. J; Hopp, D. C.; Alali, F.; Asuzu, I. U.; McLaughlin, J. L. Dihydroguaiaretic acid: a bioactive component of the stem bark of *Pycnanthus angolensis*. *Planta Med.* **1997**, *63*, 580–581.
27. Onocha, P. A.; Ali, M. S. Pycnanolide A and B: new lignan lactones from the leaves of *Pycnanthus angolensis* (Myristicaceae). *Res. J. Phytochem.* **2011**, *5*, 136–145.
28. Nono, E. C. N.; Mkounga, P.; Kuete, V.; Marat, K.; Hultin, P. G.; Nkengfack, A. E. Pycnanthulignenes A-D, antimicrobial cyclolignene derivatives from the roots of *Pycnanthus angolensis*. *J. Nat. Prod.* **2010**, *73*, 213–216.
29. Wabo, H. K.; Tatsimo, S. N.; Tane, P.; Connolly, J. D. Pycnanthuquinone C: a new terpenoid-quinone from *Pycnanthus angolensis*. *Planta Med.* **2007**, *73*, 187–189.
30. Abrantes, M.; Mil-Homens, T.; Duarte, N.; Lopes, D.; Cravo, P.; Do Madureira, M.; Ferreira, M. J. U. Antiplasmodial activity of lignans and extracts from *Pycnanthus angolensis*. *Planta Med.* **2008**, *74*, 1408–1412.
31. Diehl, H. W. U.S. Patent 4,049,824, 1977.
32. Hunter, K. W.; Gault, R. A; Stehouwer, J.; Tam-Chang, S. Synthesis of cetyl myristoleate and evaluation of its therapeutic efficacy in a murine model of collagen-induced arthritis. *Pharmacol. Res.* **2003**, *47*, 43–47.
33. Whitehouse, M. W. Concerning the anti-arthritic action of cetyl myristoleate in rats: an interim report. *Inflammopharmacology* **1999**, *7*, 303–310.

34. Lee, S. H. The efficacy and safety of beef tallow extract including cis-9-cetylmyristoleate in patients with osteoarthritis. *J. Food Sci. Nutr.* **2005**, *10*, 285–289.
35. Mollace, V.; Muscoli, C.; Masini, E.; Cuzzocrea, S.; Salvemini, D. Modulation of prostaglandin biosynthesis by nitric oxide and nitric oxide donors. *Pharmacol. Rev.* **2005**, *57*, 217–252.
36. Chan, P. H. Reactive oxygen radicals in signaling and damage in the ischemic brain. *J. Cereb. Blood Flow Metab.* **2001**, *21*, 2–14.
37. Zhang, B.; Tanaka, J.; Yang, L.; Sakanaka, M.; Hata, R.; Maeda, N.; Mitsuda, N. Protective effect of vitamin E against focal brain ischemia and neuronal death through induction of target genes of hypoxia-inducible factor-1. *Neuroscience* **2004**, *126*, 433–440.
38. Simon, J. E.; Wu, Q. L.; Deng W. World Patent WO 2010111584, 2010.
39. Leonard, E. U.S. Patent 6,489,494, 2002.
40. Leonard, E. U.S. Patent 6,713,512, 2004.
41. Leonard, E. U.S. Patent 7,534,817, 2009.
42. Céspedes, C. L.; Torres, P.; Marín, J. C.; Arciniegas, A.; Vivar, A.; Pérez-Castorena, A.; Aranda, E. Insect growth inhibition by tocotrienols and hydroquinones from *Roldana barba-johannis*. *Phytochemistry* **2004**, *65*, 1963–1975.
43. Luo, J.; Cheung, J.; Yevich, E. M.; Clark, J. P.; Tsai, J.; Lapresca, P.; Ubillas, R. P.; Fort, D. M.; Carlson, T. J.; Hector, R. F.; King, S. R.; Mendez, C. D.; Jolad, S. D.; Reaven, G. M. Novel terpenoid-type quinones isolated from *Pycnanthus angolensis* of potential utility in the treatment of type 2 diabetes. *J. Pharmacol. Exp. Ther.* **1999**, *288*, 529–534.

Chapter 6

Prospects of *Croton membranaceus* for Prostate Health

A. A. Appiah,[*,1] W. A. Asomaning,[2] I. V. Oppong,[2] G. Duker-Eshun,[1] J. Clement,[3] L. K. N. Okine,[2] G. Achel,[5] O. Gyampo,[5] S. Adjei,[4] A. K. Nyarko,[4] F. A. Aboagye,[1] and D. A. Edoh[1]

[1]Centre for Scientific Research into Plant Medicine, Mampong-Akuapem, Ghana
[2]Faculty of Science, University of Ghana, Legon, Ghana
[3]Western Carolina University, Cullowhee, North Carolina, U.S.A.
[4]Noguchi Memorial Institute of Medical Research, Legon, Ghana
[5]Ghana Atomic Energy Commission, Accra, Ghana
[*]E-mail: alfredampomahappiah@yahoo.com.

Croton membranaceus is used by Ghanaian traditional medical practitioners in the treatment of benign prostatic hyperplasia (BPH). Preparations of the root of this plant have been observed to relieve symptoms of BPH and enhance urination without the accompanying difficulty. There are no observable adverse effects associated with the clinical use of preparations of *C. membranaceus*. While medicinal significance has generated scientific interest in the plant there is real concern relative to the longer tem availability of this indigenous plant in its natural habitat. The objective of this review is to highlight the prospects of *C. membranaceus* for prostate health.

Introduction

Croton membranaceus Mull. Arg. (Euphorbiaceae) grows in West Africa reaching a height of between 1- 2 m. Found mainly in the Krobo-Gyakiti area near the Volta river in Ghana, the Adangme-Krobo people of Ghana refer to it as Buko (*1*). In Nigeria, it is found near Wuru in the region of the confluence of rivers Niger and Benue. The plant appears to prefer hilly habitats close to large rivers that run all year long. Its branches are slender and stellate-pubescent. The leaves are ovate and acutely acuminate, 2 - 8 cm long and 1-5 cm broad. The slender, stellate-pubescent petioles may attain a length of 7 cm and generally have a reddish-brown tinge. The leaves have entire margins and are covered with stellate hairs on both surfaces (Figure 1). *Croton membranaceus* bears only a few monoecious flowers on 5 - 6 cm long racemes. Male flowers are borne on the upper part while female flowers occur at the lower part of the raceme. The flowers are very small and petals may be rudimentary or completely absent. The fruit is an ellipsoid capsule. All parts of the plant including the roots bear a characteristic pleasant odor. This fragrance is a useful characteristic for the purpose of identifying the plant (*2*). This species also produces a clear reddish latex on the surface of the cut fresh stem, a characteristic generally associated with the medicinal properties of *Croton species* (*3*, *4*).

Figure 1. Aerial parts (left), a branch of Croton membranaceus (right) and roots of a three year old C. membranaceus collected in Ghana, (Dry weight = 55g).

Ethnomedical Uses

Croton membranaceus Mull. Arg. (Euphorbiaceae) is a highly valued medicinal plant used by Ghanaian traditional medical practitioners in the treatment of symptoms of benign prostatic hyperplasia (BPH). The root preparations of this plant have been dispensed to BPH patients at the Centre for Scientific Research into Plant Medicine (CSRPM) in Ghana for more than 30 years. Oku Ampofo (*5*) observed that when the tincture of *C. membranaceus* roots is diluted and drunk by BPH patients, urination is enhanced without the accompanying difficulty. Daniel (*6*) observed that relief from the symptoms of the disease usually occurred within the first three weeks of treatment with this *C. membranaceus* root decoction. Adom Winful (*7*) noted that treatment with *C. membranaceus* root decoction relieved two BPH patients of acute dysuria (difficulty to urinate) within six hours, and that the effect of *C. membranaceus* did not necessitate catheterization. In a case study, Dogbatsey (*8*) observed that the administration of *C. membranaceus* to a 60 year old patient with BPH symptoms produced two significant results: the patient was able to pass urine freely after 30 days of administration; and the weight of the prostrate of the patient reduced from 230 g to 115 g (50% reduction) over a period of three months. The root extracts of *C. membranaceus* are also used to treat measles (*5*). Unconfirmed claims by some herbal practitioners suggest that *C. membranaceus* in combination with other medicinal plants is effective in the treatment of prostate cancer and uterine fibroid (*9*).

Chemistry

Compounds that have been isolated from the root of *C. membranaceus* include an alkaloid, a coumarin, diterpenoids and phytosterols (*2*, *10–13*) (Figure 2).

Larixol, phytosterols (β-sitosterol, stigmasterol and campesterol) and a fatty acid have been isolated from the stem (*13*) (Figure 2). Chromatographic separation of the leaf extract led to the identification of only phytosterols (*13*). Aboagye *et. al.* (*10*) first isolated julocrotine from the roots of *C. membranaceus*. Aboagye (*2*, *14*) also reported the isolation of *cis*-terpine and N[N-(2-methyl butanoyl) glutaminoyl]-2-phenylethylamide from the roots of *C. membranaceus* (Figure 3).

The isolation of N[N-(2-methylbutanoyl) glutaminoyl]-2-phenylethylamide from another species, *C. humilis* has been reported (*15*). Yet, biochemically, the detection of this compound from extracts of *C. membranaceus* may result from the reaction of julocrotine and ammonia that could occur in processing/extraction rather than being a natural occurring compound (Figure 4).

This type of reaction could result from the following isolation procedure used by Aboagye (*2*): The dried powdered root of *C. membranaceus* was macerated overnight with 5% HCl. The acid extract was basified with ammonia and extracted with chloroform. The chloroform extract was concentrated under reduced pressure and re-extracted with 10% tartaric acid solution. The acid extract was basified again with ammonia and then extracted once more with chloroform to obtain fraction A. The chloroform solution containing tartaric acid-insoluble matter was dried with anhydrous sodium sulphate to obtain fraction B. Column chromatographic separation of fraction A led to the separation and identifcation of julocrotine whilst that of fraction B yielded N[N-(2-methyl butanoyl) glutaminoyl]-2-phenylethylamide.

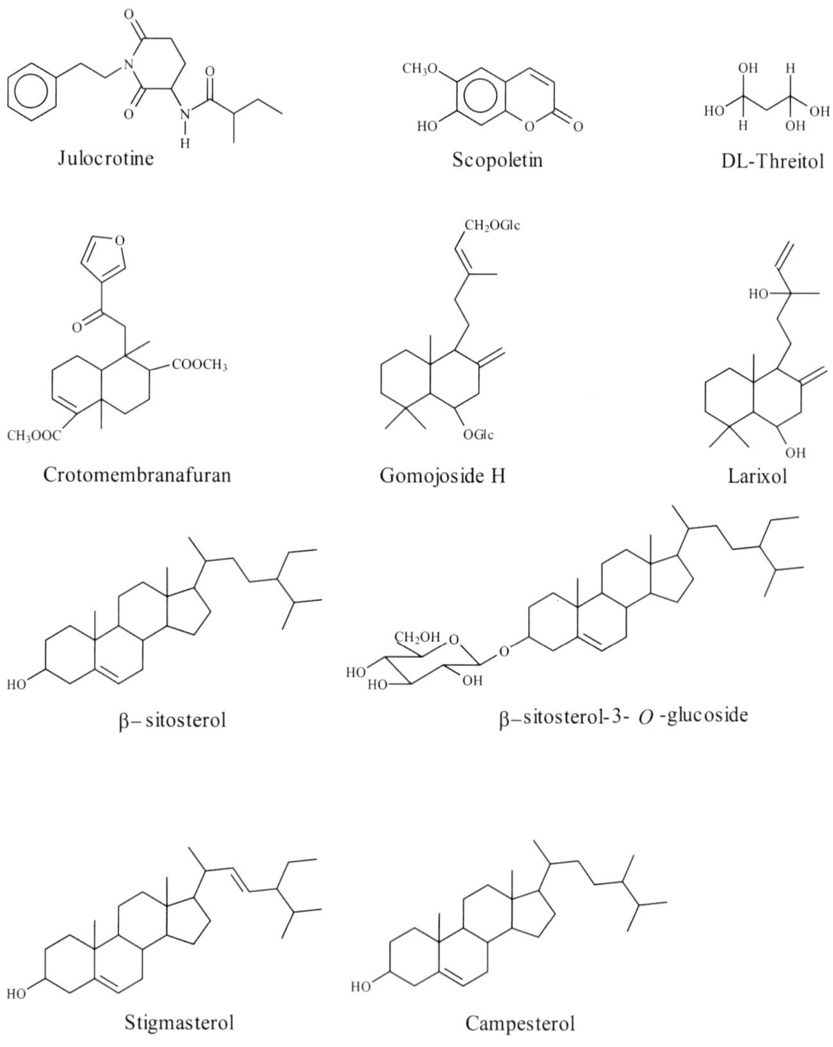

Figure 2. Structures of compounds isolated from the roots of Croton membranaceus.

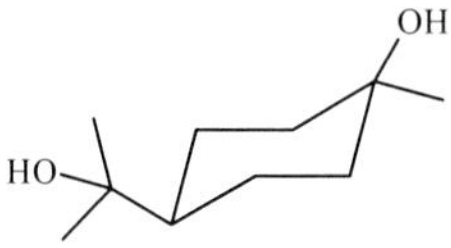

cis-Terpine

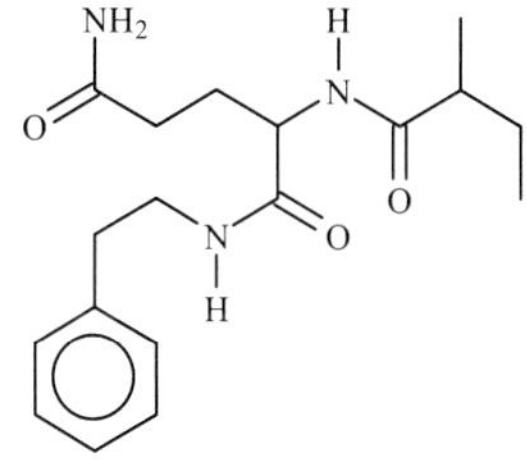

N[N-(2-methylbutanoyl) glutaminoyl]-2-phenylethylamaide

Figure 3. Terpenes and amides isolated from the roots of C. membranaceus.

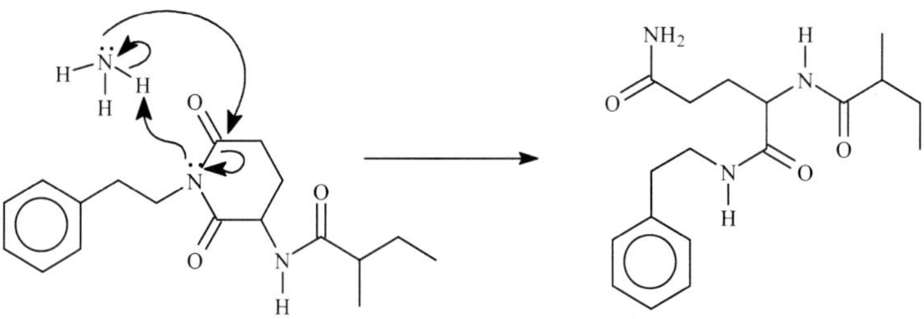

N[N-(2-methylbutanoyl) glutaminoyl]-2-phenylethylamide

Figure 4. Possible artefact in the formation of the amides in C. membranaceus.

Cineole

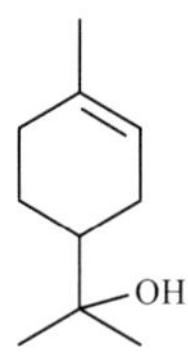

α-Terpineol

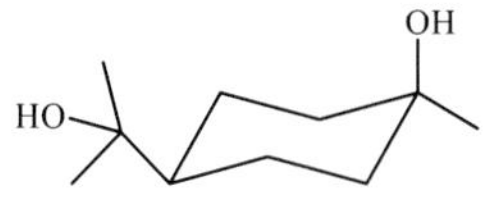

cis-Terpine

Figure 5. Possible hydration of 1,8-cineole or α-terpineol to cis-terpine.

cis-Terpine is not a naturally occurring compound. It is one of the two geometrical isomeric forms of the synthetic compound 1,8-terpine (*16*). Therefore the *cis*-terpine isolated by Aboagye is most likely an artifact which might have been formed as a result of hydration of either 1,8-cineole or α-terpineol during the isolation process (Figure 5).

Trace Elements

In his analysis, Appiah (*13*) found that *C. membranaceus* accumulates significantly high levels of manganese in the root (339 ± 4 mg/kg), stem (252 ± 3 mg/kg) and the leaf (701 ± 5 mg/kg). The author estimated that the amount of manganese in a daily dose of a *C. membranaceus* root preparation was approximately 132.5 % of the Recommended Dietary Allowance (*17*) and 27.7 % of the Tolerable Upper Intake Level (*17*). The amounts of the copper, chromium, iodine, iron and zinc in a daily dose of the *C. membranaceus* root preparation were below 20% of the respective Recommended Dietary Allowance/Adequate Intake levels.

Pharmacological Properties

Inhibition of 5α-reductase and Testosterone-Induced Growth of Rat Prostate

Aboagye (*2*) demonstrated that the crude ethanol extract of *C. membranaceus* as well as one of its isolates, julocrotine (*10*), possessed 5α-reductase inhibitory activity.

Under the working hypothesis that *C. membranaceus* extracts inhibit 5α-reductase which results in the inhibition of prostate growth, Appiah (*13*) applied the Hershberger assay (*18*) as modified by Seidlova-Wuttke *et al.* (*19*) to study the effect of *C. membranaceus* extracts on testosterone induced growth of rat prostate. In this study, intact immature 35-day-old rats were fed with testosterone and orally administered with aqueous extract of *C. membranaceus* root, stem or leaf, or with finasteride as positive or water as negative controls for 5 days. On day 6, the animals were weighed, dissected and the prostates were collected, cleaned from fat tissue and weighed. Means and standard errors of the means (S.E.M.) of the weights of prostate (mg) per weight of rat (g) were calculated. The data was then subjected to *T*-test for multiple comparisons. A P-value $<$ 0.05 was considered statistically significant. In the testosterone treated animals, the average prostate weights per gram body weight of rat was approximately 1.6 times that of the animals not treated (T-free) with testosterone. The testosterone induced increased weight of prostate, expressed as milligram prostate weight per gram body weight of rat was significantly inhibited by *C. membranaceus* root extract as well as the stem extract by approximately 19%. Thus, both the root and stem extracts of the plant inhibited testosterone induced growth of the rat prostate. The leaf extract was found to be ineffective. Julocrotine, which has been found to inhibit the enzyme 5α-reductase was isolated from the root. The chromatographic separation of the stem did not yield julocrotine. This suggests that there might be more than one constituent of *C. membranaceus* that is capable of inhibiting testosterone induced prostate growth in rats. Statistically, the effect of *C. membranaceus* root extract (6 mg) and that of the stem extract (6 mg) were the same as that of the positive control finasteride (0.5 mg).

Anticancer Activity

In their cytotoxic activity studies Bayor *et al.* (*12*) found that the methanolic extract of *C. membranaceus* root exhibited cytotoxic activity against the DLD-1 and MCF-7 cells. They also found that a new furano-clerodane diterpenoid, crotomembranafuran, they had isolated from *C. membranaceus* root, exhibited modest activity against human prostate cancer (PC-3) cells, (IC_{50} = 4.1 ± 0.6 µg/ml; 10.6 µM) but was inactive against both DLD-1 and MCF-7 cells (IC_{50} > 5 µg/ml). They therefore suggested that there are likely to be other cytotoxic compounds present in the plant.

Scopoletin, a constituent of *C. membranaceus*, has also been found to inhibit the proliferation of HL-60 and PC3 (prostate cancer) cells by inducing apoptosis (*20–22*). There is no report on biological studies on larixol, another constituent of *C. membranaceus*, but manool (Figure 6), a compound structurally related to larixol has been found to be the most active isolated chemical constituent from Greek propolis against HT-29 human colon adenocarcinoma cells (*23*). It is therefore possible that larixol may have some medicinal properties which could contribute to the observed anticancer effects (*12*) of *C. membranaceus*.

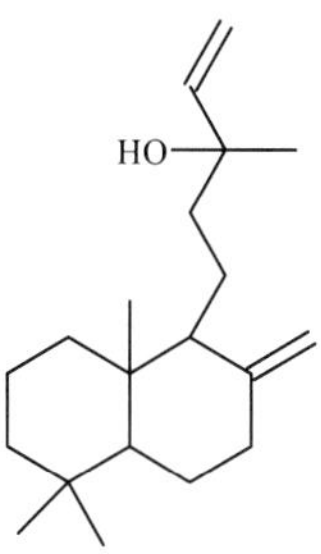

Manool

Figure 6. Chemical structure of manool from Croton membranaceus.

Smooth Muscle Relaxant and Antihypertensive Activity

Scopoletin is known to lower blood pressure in laboratory animals. The possible mechanisms of the hypotensive effect of this compound are believed to be its smooth muscle relaxant activity and its non-specific spasmolytic effect (*24*).

Studies have shown that scopoletin inhibits the indirect electrical stimulation-evoked contractions of the cat nictitating membrane *in vivo*. It also inhibits the contractions of isolated perfused central ear artery of rabbit, induced by electrical stimulation or intraluminal noradrenaline administration. Scopoletin reduces the amplitude and frequency of the spontaneous, myogenic, rhythmic contractions, and exogenous noradrenaline-evoked contractions of the rat isolated portal vein. Scopoletin also inhibits the spontaneous, myogenic, pendular, rhythmic contractions of the rabbit isolated duodenum and attenuates the indirect electrical stimulation-provoked or exogenous noradrenaline-induced relaxations of the muscle preparation. It also depresses the electrical stimulation-evoked contractions of the chick isolated oesophagus (*24*, *25*). Scopoletin has also inhibited the spasmogenic activities of a wide variety of agonists on guinea-pig isolated ileum (*24*). Constituents of *C. membranaceus* with smooth muscle relaxant activity may be very important in the treatment of BPH.

AChE Inhibitory Activity

By a virtual screening procedure, Rollinger *et al.* (*26*) observed that scopoletin and its glucoside scopolin are potential acetylcholinesterase (AChE) inhibitors. These two compounds showed moderate, but significant, dose-dependent and long-lasting inhibitory activities. In an *in vivo* experiment, scopoletin and scopolin increased the extracellular acetylcholine (ACh) concentration in rat brain to about 170% and 300% compared to basal release, respectively. At the same concentration, they observed that the positive control galanthamine increased the ACh concentration to about the same level as scopoletin (*26*). Scopoletin has been patented as a compound that may be used for the prevention or treatment of disorders which involve low levels of acetylcholine in the brain, and therefore can ameliorate disorders connected with deficits in learning and memory functions like Alzheimer's disease, senile dementia, ataxia, myasthenia gravis and Parkinson's disease (*27*).

Antioxidant Activity

The crude methanol extracts of *C. membranaceus* root, stem and leaf exhibited significant concentration-dependent DPPH radical scavenging activity but none of the extracts were more effective at scavenging DPPH radical in this assay than the positive control, gallic acid. The IC_{50} of the methanol extracts of the root, stem and leaf were found to be 100.24, 126.14 and 206.17 mg/l respectively while that of gallic acid was 3.40 mg/l (*13*). The radical scavenging activity of *C. membranaceus* extracts makes it a potential agent for prostate health and other diseases in which oxidants or free radicals are implicated.

In their study of the antioxidant properties of scopoletin, Shaw *et al.* (*28*) observed that scopoletin scavenged superoxide anion in the xanthine/xanthine oxidase reaction system in a concentration-dependent manner, but did not inhibit xanthine oxidase. These workers concluded that scopoletin may be of use in preventing superoxide anion-induced damage *in vivo*. Scopoletin has also been found to inhibit hepatic lipid peroxidation and increased the activity of antioxidants, superoxide dismutase and catalase (*29*).

Manganese is known to play an important role in a number of physiological processes as a constituent of multiple enzymes and an activator of other enzymes (*30*). Manganese superoxide dismutase (MnSOD), for example, is the principal antioxidant enzyme in the mitochondria. Supplementation with manganese has been shown to increase SOD activity indicating increased antioxidant activity (*31*). The nutritionally significant high level of manganese in *C. membranaceus* may therefore contribute significantly to its medicinal values.

Antimicrobial Activity

Using the agar diffusion and broth dilution techniques, Bayor *et al.* (*32*) found that the methanol extract of *C. membranaceus* root showed a significant ($p < 0.01$) antimicrobial activity against *Staphylococcus aureus, Bacillus subtilis, Escherichia coli, Pseudomonas aeruginosa* and fungi *Aspergillus niger* and *Candida albicans* with minimum inhibitory concentrations (MICs) ranging from 0.53 – 1.43 mg/ml. They also found that the effect of gomojoside H (isolated from *C. membranaceus*) on *S. aureus, B. subtilis* and *P. aeruginosa* (MICs < 10 μg/ml) was similar to that of gentamicin (*32*).

Other Pharmacological Properties of Some Constituents of *Croton membranaceus*

Studies by Panda and Kars indicated that scopoletin has the potential to regulate hyperthyroidism and hyperglycemia (*29*). Dietary phytosterols are known to reduce cholesterol levels in laboratory animals (*33*). In their systematic review of the evidence for the efficacy of β-sitosterol in men with symptomatic BPH, Wilt *et al.* concluded that β-sitosterol significantly improves BPH symptoms and urinary flow parameters (*34*).

Toxicity

The oral LD_{50} of the freeze dried aqueous extract of all three plant parts (root, stem and leaf) of *C. membranaceus* was found to be greater than 5000 mg kg^{-1} body weight of rat, indicating a wide margin of safety (*13*).

Possible Mechanism of Action of *C. membranaceus* as a Remedy for BPH

The known chemical constituents and biological activities of *C. membranaceus* extracts and isolates, coupled with the clinical observations made following the administration of preparations of the plant extract, allow some inferences to be made about the plausible mechanism of action of the use of *C. membranaceus* in the treatment of symptomatic BPH. Medicinal preparations of the plant have been found to reduce BPH symptoms within a relatively short period after oral administration (*8*). It has also been reported to reduce prostate size (*9*). These observations suggest that the activity of the plant could be due to more than one mechanism of action and more than one bioactive compound. The plausible mechanisms of action (*35*) may include two or more of the following:

1. 5α-reductase inhibition
2. Alpha adrenergic antagonism
3. Antioxidant action
4. Acetylcholinestrase (AChE) inhibition
5. Inhibition of cholesterol absorption

5α-reductase Inhibition

The reduction in size of the prostate of a BPH patient observed by Dogbatsey (*9*) following the administration of a *C. membranaceus* root preparation might be due to a 5α-reductase inhibitory activity. The effect of 5α-reductase inhibition is lower blood and tissue DHT levels and a reduction in size of the prostate. In their in vitro studies, Aboagye *et al.* (*2*, *10*) demonstrated that the crude ethanol extract of *C. membranaceus* as well as one of its isolates, julocrotine, possess 5α-reductase inhibitory activity. *In vivo* studies have also shown that the root and stem extracts of *C. membranaceus* inhibit testosterone induced prostate growth in rats (*13*), an effect which is most likely to be due to 5α-reductase inhibition.

Julocrotine, a constituent of *C. membranaceus*, appears to have some structural relationship with the synthetic 5α-reductase inhibitor finasteride, i.e., both of these compounds have an alkylamide moiety. This moiety might be relevant in 5α-reductase inhibitory activity studies of similar compounds (Figure 7).

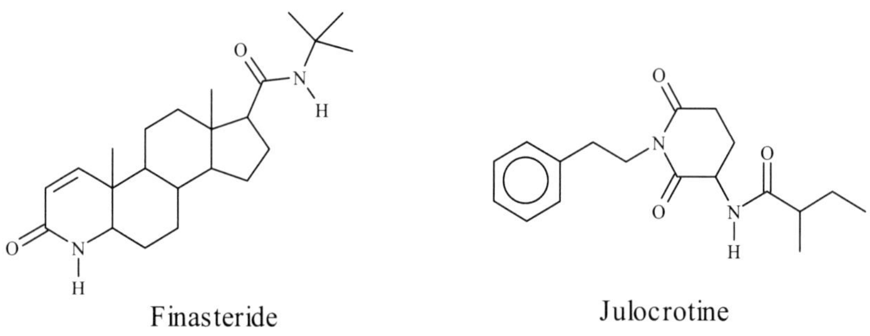

Figure 7. Julocretine a constituent of Croton membranaceus, structuraly related to synthetic 5α-reductase inhibitor finasteride.

Alpha Adrenergic Receptor Antagonism

Adrenergic receptor antagonists (Alpha blockers) are known to start reducing symptoms of BPH within two weeks (*35*) as compared to 5α-reductase inhibitors which might take more than 3 months (*35*). Alpha blockers such as Terazosin are drugs that relax the smooth muscles of the prostate capsule and the bladder neck by blocking the transmission of noradrenalin. Relaxing the muscles around the bladder neck helps relieve urinary obstruction. The quick onset of action of *C. membranaceus* against dysuria in BPH patients observed by Adom Winful (*8*) and Dogbatsey (*9*) might be due to an alpha adrenergic receptor antagonistic effect.

Lambert *et al.* (*11*) isolated scopoletin from *C. membranaceus*. Scopoletin has been found to relax smooth muscles and inhibit the spasmogenic activities of a wide variety of agonists on guinea-pig isolated ileum (*24*). Scopoletin reduces the amplitude and frequency of exogenous noradrenaline-evoked contractions of the rat isolated portal vein. It also attenuates the exogenous noradrenaline-induced relaxations of the rabbit isolated duodenum (*24*, *25*).

Scopoletin seems to have some structural relationship with the synthetic alpha-blocker terazosin which is used in the treatment of BPH. Structurally, the hydroxymethoxybenzo moiety of scopoletin compares with the dimethoxybenzo moiety of terazosin. These moieties may be of interest in the activity studies of these two compounds (Figure 8).

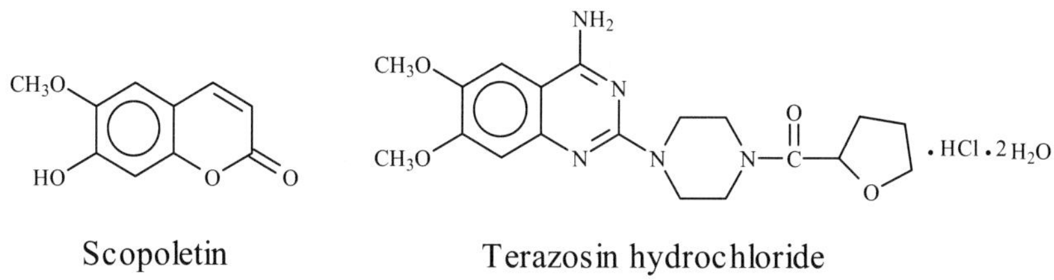

Figure 8. Chemical structures of scopoteletin and synthetic alpha-blocker terazosin.

Antioxidant Action

It has been shown that the crude methanol extracts of *C. membranaceus* root, stem and leaf exhibit significant concentration-dependent DPPH radical scavenging activity (*13*). Scopoletin, a constituent of *C. membranaceus* root (*11*) is known to increase the activity of the antioxidants, superoxide dismutase and catalase (*29*). Scopoletin also scavenges superoxide anion in the xanthine/xanthine oxidase reaction system in a concentration-dependent manner (*28*). *Croton membranaceus* has also been found to contain nutritionally significant levels of manganese, an element that plays an important role in a number of physiological processes as a constituent of multiple enzymes and an activator of other enzymes (*30*). Manganese superoxide dismutase (MnSOD) for instance is the principal antioxidant enzyme in the mitochondria (*31*).

Acetylcholinestrase (AChE) Inhibition

Scopoletin has shown significant, dose-dependent and long-lasting AChE inhibitory activity (*26*). This effect may increase extracellular acetylcholine (ACh) concentration and result in the relaxation of the smooth muscles around the bladder neck and in the prostate.

Inhibition of Cholesterol Absorption

The phytosterols β-sitosterol, stigmasterol and campesterol are constituents of *C. membranaceus*, and have chemical structures similar to that of cholesterol (Figure 9). Dietary phytosterols are known to reduce cholesterol levels in laboratory animals (*33*). Alone and in combination with other phytosterols, β-sitosterol reduces blood levels of cholesterol, and is sometimes used in treating hypercholesterolemia because it inhibits cholesterol absorption in the intestine (*36*). This effect is due to the structural similarity of the phytosterols and the cholesterol (*37*).

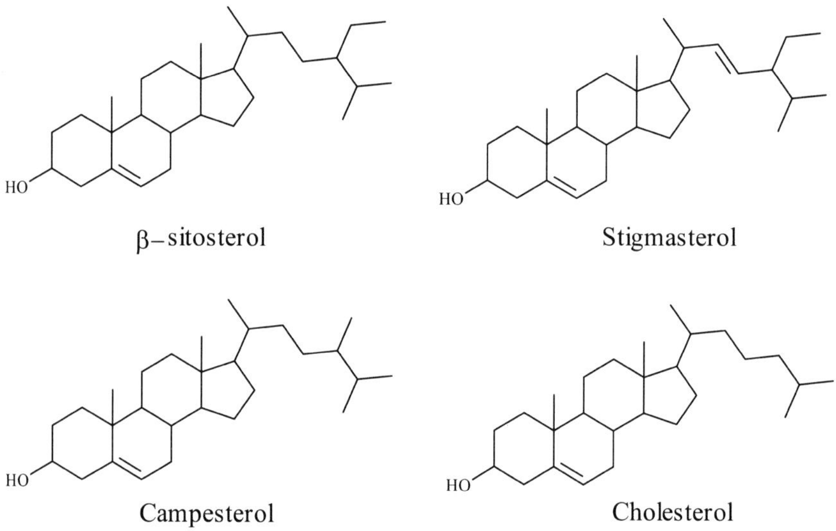

Figure 9. Chemical structures of phytosterols (β-sitosterol, stigmasterol and campesterol) and cholesterol found in Croton membranaceus.

Conclusions

The phytochemical constituents and the pharmacological properties of *C. membranaceus* provide some scientific bases for its use in the treatment of BPH and prostate cancer. The plant may also be useful in the management of hypertension, diseases in which oxidants or free radicals are implicated and in the prevention and/or treatment of disorders which involve low levels of acetylcholine in the brain.

The isolation of most of the compounds of *C. membranaceus* was done using normal phase column chromatography. There is a need for the isolation and characterization of additional compounds using other chromatographic methods such as reversed-phase chromatography and High Performance Liquid Chromatography (HPLC)

One of the chemical constituents of *C. membranaceus* root is scopoletin. This compound has been found to have a number of interesting biological activities which includes smooth muscle relaxant, anti-cancer, antioxidants, and antihypertensive. It has also been found to inhibit acetylcholinestrase (AChE). Because of its AChE inhibitory activity, it has been patented as a compound that may be used for the prevention or treatment of disorders which involve low levels of acetylcholine in the brain, and therefore, can ameliorate disorders connected with deficits in learning and memory functions like Alzheimer's disease, senile dementia, ataxia, myasthenia gravis and Parkinson's disease. More extensive biological and clinical studies on *C. membranaceus* are needed to confirm and validate this African species full medicinal potential.

References

1. Burkill, H. M. *The useful plants of West Tropical Africa*; Royal Botanic Gardens: Kew, London, 1985, Vol. 2.
2. Aboagye, F. A. M.Phil. Thesis, Biochemistry Department, University of Ghana, 1997.
3. Sandoval, M.; Okuhama, N. N.; Clark, M.; Angeles, F. M.; Lao, J.; Bustamante, S.; Miller, M. J. *J. Ethnopharmacol.* **2002**, *80*, 121.
4. Risco, E.; Ghia, F.; Vila, R.; Iglesias, J.; Álvarez, E.; Canigueral, S. *Planta Med.* **2003**, *69*, 785.
5. Mshana, N. R.; Abbiw, D. K.; Addae-Mensah, I.; Agjanouhoun, E.; Ahyi, M. R. A.; Ekpere, J. A.; Enow-Orock, E. G.; Gbile, Z. O.; Noamesi, G. K.; Odei, M. A.; Odunlami, H.; Oteng-Yeboah, A. A.; Sarpong, K.; Sofowora, A.; Tackie, A. N. *Traditional Medicine and Pharmacopoeia. Contribution to the Revision of Ethnobotanical and Floristic Studies in Ghana.*; OAU/STRC: Lagos, 2000.
6. Daniel, H. E. A. Mampong-Akuapem, Ghana. Personal Communication, 1997.
7. Adom-Winful, E. Mampong-Akuapem, Ghana. Personal Communication, 2009.
8. Dogbatsey. Seminar presentation. Centre for Scientific Research into Plant Medicine, Mampong-Akuapem, Ghana, 2006
9. Odei, E. Mampong-Akuapem, Ghana, Personal communication, 2010.
10. Aboagye, F. A.; Sam, G. H.; Massiot, G.; Lavaud, C. *Fitoterapia* **2000**, *71*, 461–462.
11. Lambert, M.; Stærk, D.; Hansen, S. H.; Jaroszewski, J. W. *Magn. Reson. Chem.* **2005**, *43*, 771–775.
12. Bayor, M. T.; Ayim, J. S. K.; Marston, G.; Phillips, R. M.; Shnyder, S. D.; Wheelhouse, R. T.; Wright, C. W. *Nat. Prod. Commun.* **2008**, *3* (11), 1875–1878.
13. Appiah, A. A. Ph.D. Thesis, Chemistry Department, University of Ghana, 2011.
14. Aboagye, F. A. Unpublished. 1997

15. Stuart, K. L.; McNeill, D.; Kutney, J. P.; Eigendorf, G.; Klein, P. K. *Tetrahedron* **1973**, *29* (24), 4071–4075.
16. Finar I. L. *Organic chemistry: stereochemistry and the chemistry of natural products*, 5th ed.; Longman: Singapore, 1989; Vol 2.
17. Food and Nutrition Board, Institute of Medicine. National Academy Press: Washington, DC, 2001.
18. http://epa.gov/endo/pubs/hershberger_fs.pdf. Retrieved 26/5/2011.
19. Seidlova-Wuttke, D.; Pitzel, L.; Thelen, P.; Wuttke, W. *Maturitas* **2006**, *55S*, S75–S82.
20. Liu, X. L.; Zhang, L.; Fu, X. L.; Chen, K.; Qian, B. C. *Acta Pharmacol. Sin.* **2001**, *22* (10), 929–933.
21. Manuele, M. G.; Ferraro, G.; Barreiro, A. M. L.; López, P.; Cremaschi, G.; Anesini, C. *Life Sci.* **2006**, *79* (21), 2043–2048.
22. Kim, E. K.; Kwon, K. B.; Shin, B. C.; Seo, E. A.; Lee, Y. R.; Kim, J. S.; Park, J. W.; Park, B. H.; Ryu, D. G. *Life Sci.* **2005**, *77* (7), 824–36.
23. Pratsinis, H.; Kletsas, D.; Melliou, E.; Chinou, I. *J. Med. Food* **2010**, *13* (2), 286–290.
24. Ojewole, J. A.; Adesina, S. K. *Planta Med.* **1983**, *49* (9), 46–50.
25. Ojewole, J. A. O. *Pharm. Biol.* **1984**, *22* (2), 81–93.
26. Rollinger, J. M.; Hornick, A.; Langer, T.; Stuppner, H.; Prast, H. *J. Med. Chem.* **2004**, *47* (25), 6248–54.
27. European Patent Register, EP1588705, 2005.
28. Shaw, C-Y.; Chen, C-H.; Hsu, C-C.; Chen, C-C.; Tsai, Y-C. *Phytother. Res.* **2003**, *17* (7), 823–825.
29. Panda, S.; Kars, A. *Phytother Res.* **2006**, *12*, 1103–5.
30. Nielsen, F. H. Ultratrace minerals. In *Modern Nutrition in Health and Disease*, 9th ed.; Shils, M., Olson, J. A., Shike, M., Ross, A. C., Eds.; Williams & Wilkins: Baltimore, MD, 1999.
31. Leach, R. M.; Harris, E. D. Manganese. In *Handbook of nutritionally essential minerals*; O'Dell, B. L., Sunde, R. A., Eds.; Marcel Dekker, Inc: New York, 1997.
32. Bayor, M. T.; Gbedema, S. Y.; Annan, K. *J. Pharmacogn. Phytother.* **2009**, *1* (4), 47–51.
33. http://www.livestrong.com/article/230060-what-are-beta-sitosterol-vitamins/#ixzz1LsPVKKTr. Accessed 4/5/2011.
34. Wilt, T. J.; MacDonald, R.; Ishani, A. *BJU Int.* **1999**, *83* (9), 976–983.
35. Barnes, J. *Pharm. J.* **2002**, *269*, 250–252.
36. Matsuoka, K.; Nakazawa, T.; Nakamura, A.; Honda, C.; Endo, K.; Tsukada, M. *Chem. Phys. Lipids* **2008**, *154*, 87–93.
37. Moreau, R. A.; Whitaker, B. D.; Hicks, K. B. *Prog. Lipid Res.* **2002**, *41*, 457–500.

Chapter 7

Utilization of Medicinal Plants and Their Products in the Treatment and Control of Disease in Fish

Albert P. Ekanem* and Austin I. Obiekezie

Institute of Oceanography, University of Calabar, Calabar, Nigeria
***E-mail: albertekanem@yahoo.com.**

African medicinal plants and their products have shown efficacies in the treatment of diseases in farmed fish. Extracts of *Carica papaya* and *Mucuna pruriens* were effective in the treatment of Ichthyophthiriasis in goldfish (*Carassius auratus auratus*) with high host tolerance. Similarly, ectoparasitic monogeneans were effectively dislodged from the gills and skin of goldfish by the application of extracts of *Piper guineense* at therapeutic concentrations. *Artemesia annua* with known antimalarial activities in human was also effective against fish monogenean parasites of *Clarias gariepinus* in a concentration related manner without detriments to the host. Effective antibacterial activities against Aeromonas and Pseudomonas diseases of the African catfish (Heterobranchus longifilis) were demonstrated in some plants such as *Phylanthus amarus*, *Allium sativum*, *A. annua*, Citrus lemon with minimum inhibitory concentration (MIC) between 25 to 100mg/ml. African plants have shown potential promise in the treatment of diseases in fish and other aquatic animals.

Aquaculture Development in Nigeria

The purpose of aquaculture production is to increase the availability of animal protein through fish culture with the aims of making it available to as many people as possible. Presently, the high cost of animal flesh (cow, goat, pork, ram, chicken etc) has made it inaccessible to the common man in some developing countries.

Moreover, some people, because of religious and cultural affiliations, are critical about the consumption of flesh from some animals, but fish is the only animal protein source that has bridged cultural and religious affiliations, being widely consumed. It is the cheapest source of animal protein in Nigeria and other African countries, commonly found in most people's meal. On the estimate, more than 98% of the fish consumed in Nigeria are cropped from the wild and this stock of wild fish is gradually diminishing due to over fishing. Aquaculture production presents the only alternative means to the wild stock of fish to meet the demand of the increasing Nigerian population.

Aquaculture practice in the Sub-Saharan African countries has been crippled by several factors prominent among which are economic losses from fish mortalities due to outbreak of diseases and lack of affordable drug for their controls and management. Presently, and unlike other agricultural practices, most of the available drugs for the treatment of aquaculture diseases are imported from Europe and America, thereby making the practice very expensive for an average African farmer. Most of the farmers who have tried their hands on fish production are making use of natural stream waters with highly degraded quality, thereby predisposing their candidates (fish) to diseases.

The commonest disease problems encountered are parasitic and bacterial and a few cases of fungal related cases as secondary infections. The most frequent parasitic problems associated with tropical aquaculture include trichodiniasis and monogeneasis. The bacteria associated with fish mortalities are mostly gram negative organisms, the Pseudomonas, Aeromonas and Flexibacter species. These organisms poses enormous problem to hatchery raised fish juveniles and brood fish, resulting in economic losses and abandonment of projects by some farmers.

The research on African medicinal plants was informed by their availability and also by the fact that they have been found effective for the treatment of diseases in human and other animals of veterinary importance. It has long been recognized that naturally occurring substances in higher plants have antibacterial activities (*1*). The use of medicinal plants and their products in the treatment of diseases is a widespread practice (*2*). Currently there is an increasing demand for medicinal plants and plant products as alternative to orthodox medicines especially in developing countries (*3*).

It is also important to develop a therapeutic source that will have little or no impact on the environment, being informed by the deleterious impact on the environment by the use of chemical-based substances for the treatment of aquaculture diseases in the developed countries of the world. Moreover, the problems associated with drug accumulation in the tissues of food fish can be solved or reduced by the use of plants and their products in the management of disease problems.

There are however, relatively few publications on the ethno medical and ethno botanical uses of plants or their products in the treatment of diseases in aquaculture. Large numbers of plants are rapidly becoming extinct and need to be urgently investigated for their medicinal properties. Plant-based substances and other natural products have been extensively used for the treatment of human diseases, ailments and partly for the treatment of diseases in other animals excluding fish.

The need for the development of alternative methods of treatment of aquaculture diseases is informed by the fact that some of the effective chemical-based substances which are presently used (Table I) for the treatments of aquaculture diseases are not recommended for food fish. A typical example is malachite green which is effective for the treatment of protozoan infection in cultured fish but a potential mutagen in human consuming such products (fish) and highly restricted from used in some countries (*4*). Moreover, the continuous use of chemical substances in the aquatic environment is likely to impact negatively on the environment.

The review on the utilization of Medicinal plants and their products in the treatment of fish diseases is designed to provide adequate information on their efficacies, toxicities (Tables II, III and IV) and pharmacological active principles as recommended in the African pharmacopoeia by the World Health Organization.

The use of medicinal plants and their products in the treatment of diseases is a widespread practice (*2*). Currently there is an increasing demand for medicinal plants and plant products as alternative to orthodox medicines especially in developing countries (*3*).

Table I. Mortality of Goldfish Heavily Infested with *Ichthyopthirius multifiliis* after Bath Treatment with Petroleum Ether Extract of *Carica papaya* and Methanolic Extract of *Mucuna pruriens* for 96 h

Concentration (mg/l)	*Fish Mortality (%)*	
	Carica papaya	*Mucuna pruriens*
0 (Control)	48	65
100	40	50
150	30	30
200	15	20

Efficacies of Extracts of *Mucuna pruriens* and *Carica papaya* Against *Ichthyophthirius multifiliis* of Goldfish

The ciliate *Ichthyophthirius multifiliis* is among the most pathogenic parasites of fish maintained in captivity. According to Ekanem et al (*5*), the effects of the crude methanolic extract of leaves of *Mucuna pruriens* and the petroleum-ether extract of seeds of *Carica papaya* against *I. multifiliis* were investigated under in vivo and in vitro conditions. Goldfish (*Carassius auratus*) infected with the parasites were immersed for 72 h in baths with M. pruriens extract, and for 96 h in baths with *C. papaya* extract. There was a 90% reduction in numbers of *I. multifiliis* on fish after treatment in baths of each plant extract at 200 mg/l compared to untreated controls. Consequently, parasite-induced fish mortality was reduced significantly. A complete interruption of trophont recruitment was achieved by immersion in the *M. pruriens* extract. In vitro tests led to a 100% mortality of *I. multifiliis* in 150 mg/l *M. pruriens* extract, and in 200 mg/l of *C. papaya* extract after 6 h. Although the active constituents of the medicinal plant extracts are still unknown, it was demonstrated that *M. pruriens* and *C. Papaya* have potentials for effective control of *I. multifiliis*.

Efficacy of Extracts of *Piper Guineense* Schum. and Thonn. against Skin And Gill Monogenean Parasites of Fish under *in Vivo* and *in Vitro* Conditions

Monogenea is one of the most economically important parasites of cultured fish (*6*). The viviparous Gyrodactylids parasitizes the skin, fin and gills of freshwater and marine fishes, whereas the oviparous Dactylogyrids are mainly gill parasites of freshwater fish (*7*). Two of the most important species of fish monogeneans are *Dactylogyrus extensus* and *Dactylogyrus vastator* (*8*), which are endemic in Asia, Central Europe, Middle East and North America (*9*).

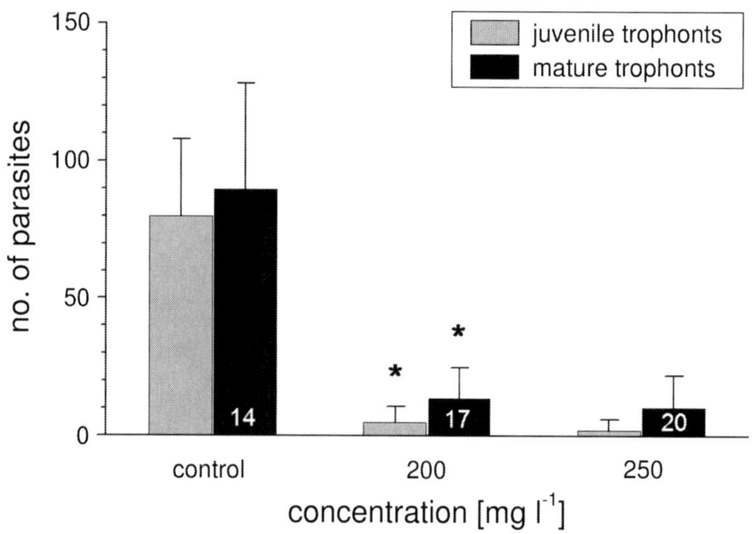

Figure 1. Effect of a 96 h bath treatment with petroleum ether extract of Carica papaya on Ichthyophthirius multifiliis on the skin of goldfish. Given are the mean numbers and standard deviations of juvenile and mature trophonts on the skin (including fins). Asterisk indicate significantly decreased parasite numbers ($p \leq 0.05$) compared to the lower concentrations and control, respectively.

According to Ekanem et al (*10*), methanol extracts of the seeds of *Piper guineense* (Piperaceae) were active against goldfish (*Carassius auratus auratus* L. Pisces Cyprinidae) monogenean parasites. The seed extract of *P. guineense* was administered at different concentrations of 0.5–2.0 mg/L, under *in vivo* and *in vitro* conditions. There was a higher efficacy of the effects of the extracts against fish parasites under *in vitro* situations than under *in vivo*.

According to the report, There was a positive correlation (Fig. 1) between goldfish (*Carassius auratus auratus*) parasite (*Gyradactylus* and *Dactylogyrus*) mortalities and the concentration of methanol extracts of *Piper guineense*. An equal percentage fish mortality was observed in the extract concentrations 0.5, 1.0 and 1.5 mg/L, but later increased with increasing natural product concentration (Fig. 2). There were more parasites on the skin of goldfish than on the gills. Considering the fish and parasite mortalities between the control and the test groups, a marked reduction of fish mortalities was observed in the control group compared with fish mortalities in the treated groups, which increased with increasing extract concentration. There was zero parasite mortality on both the skin and gills of the fish in the control groups.

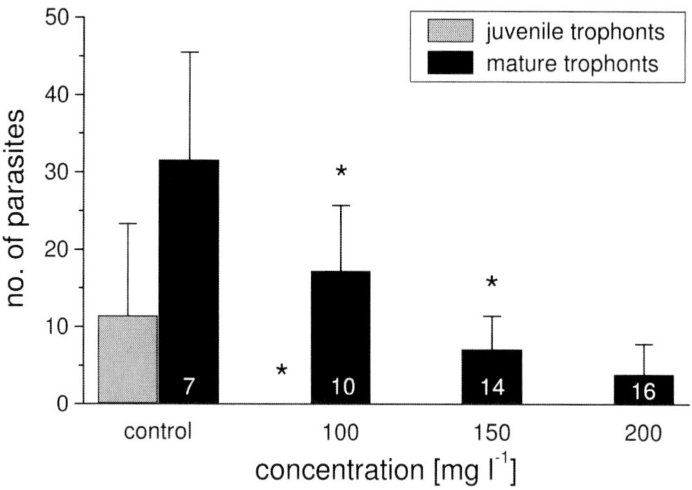

Figure 2. Effect of a 96 h bath treatment with methanolic extract of Mucuna pruriens on Ichthyophthirius multifiliis on the gills of goldfish. Given are the mean numbers and standard deviations of juvenile and mature trophonts on the gills. Asterisk indicate significantly decreased parasite numbers (p≤0.05) compared to the lower concentrations and control, respectively.

A higher efficacy against fish parasites in the *in vitro* tests was observed in the treated groups when the extracts were at a concentration of 1.5 and 2.0 mg/L. There was a 60% and 80% reduction of skin and gill parasites in 1.5 mg/L extract, whereas a 70% and 100% parasite reduction on the skin and gill respectively, were observed in 2.0 mg/L seed extracts of *P. guineense*. Comparing results from the *in vivo* and *in vitro* screens, anthelmintic activity was observed earlier from the extract in the *in vitro* screens than in the *in vivo*.

Efficacy of Ethanol Extracts of *Artemisia Annua* L. Against Monogenean Fish Parasites Of *Heterobranchus Longifilis*

Artemisia annua L., also known as annual wormwood or Sweet Annie, belongs to the family Asteraceae. It is a highly aromatic herbaceous plant of Asiatic and Eastern European origin, widely dispersed throughout the temperate region (*11*, *12*). It was used by Chinese herbalists since A.D. 341 for the treatment of fevers associated with malaria (*13*). Its activity against malarial parasites in primate models was demonstrated in 1971, but the isolation and characterization of the active antimalarial principle, artemisinin, by Chinese scientists were in 1972 (*14*).

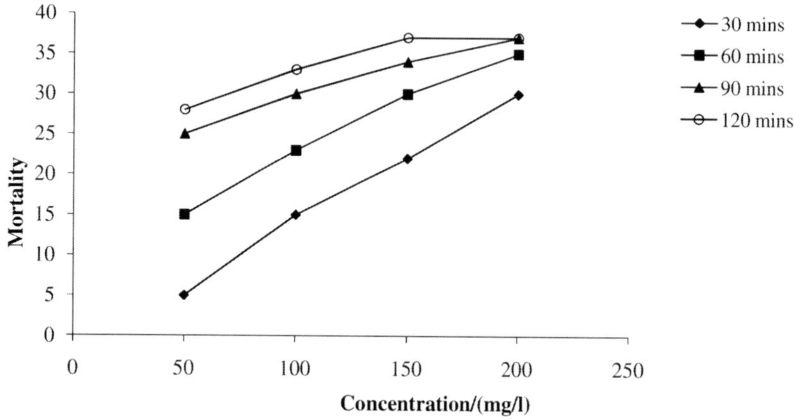

Figure 3. Parasite mortality against concentrations (mg/l) of Artemisia annua at different time intervals.

According to Ekanem and Brisibe (*15*), ethanol extract of *Artemisia annua* was effective in the dislodgement and mortality of monogenean parasites of juvenile *Heterobranchus longifilis* at concentrations ranging from 50 to 200 mg/l. Five hundred 1-week-old juvenile fish were stocked in hapa in earthen pond for 7 days to accumulate parasites. The approximate number of parasites per fish was confirmed by counting the number of parasites attached to body surfaces and the gills with a stereo-microscope before being exposed to the extract under *in vivo* conditions. The bioactivity of the extract was conducted in plastic Petri dishes with three replications and controls. The results obtained from *A. annua* extract (Fig. 3) were matched against those produced by pure artemisinin (Fig. 4) and artesunate powder (Fig. 5), respectively, under similar experimental conditions. There was a faster effect of pure artemisinin crystals on the parasites as compared to *A. annua* extract and artesunate. Coagulation of parasite cells was observed with artemisinin treatment, whereas parasites were merely dislodged from their attachment organs and killed some hours later in the same concentration of *A. annua*. There were positive correlations between the number of parasites dislodged/killed and the concentration of A. annua extract, artemisinin, and artesunate powder, respectively, as well as the duration of exposure of affected fish to the substances. This led to the conclusion that A. annua contains substances that are effective against helminthes parasites of *H. longifilis*.

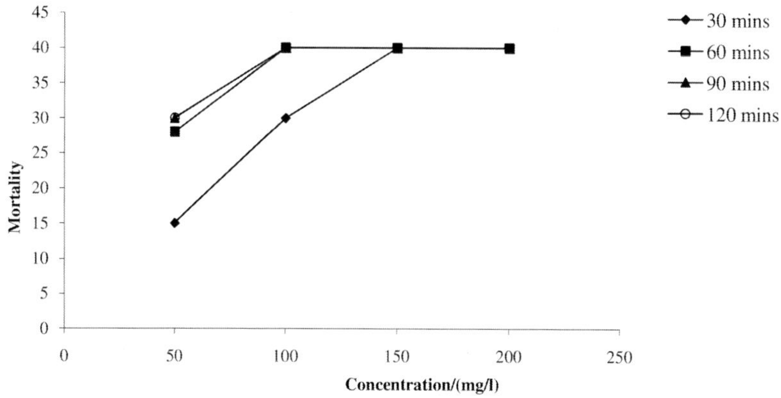

Figure 4. Parasite mortality against concentrations (mg/l) of Artemisin at different time intervals.

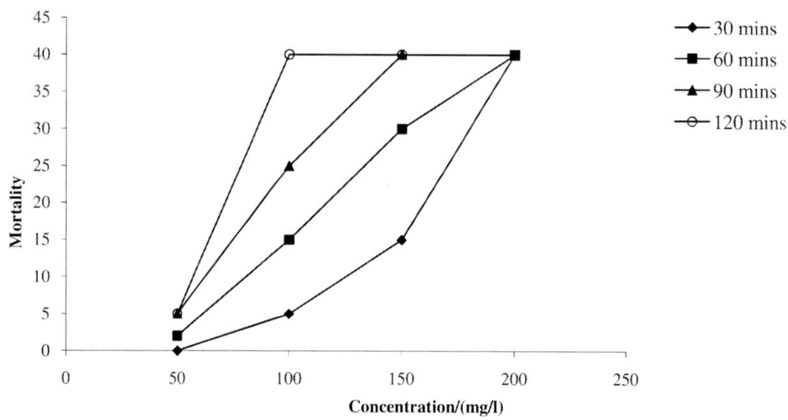

Figure 5. Parasite mortality against concentrations (mg/l) of Artesunate at different time intervals.

Examination of the fish fry at the end of the test periods demonstrated that while some of the fish that previously harbored parasites were found to be free of some of the parasites, others were completely free of the parasites. Comparatively, the number of parasites on the body surfaces of fish in the control was the same throughout the test period.

The concentration of *A. annua* in which 50% of the parasites were killed was 100 mg/l within 60 minutes and a significant number (about 85%) were killed in 200 mg/l. In treatment with artemisinin, 50% of the parasites were killed in 70 mg/l after exposure for 30 minutes while all parasites were killed with their cells coagulated in 100 mg/l within 60 minutes of exposure. Fifty percent parasite mortality was observed in 85 mg/l of Artesunate after exposure for 90 minutes. A mortality rate of about 85% of the parasites occurred when the fish fry were bathed in solution containing 150 mg/l of artesunate for 90 minutes. Interestingly, monogenean parasites were all dislodged from their attachment sites before the occurrence of mortality following treatment with artesunate.

Table II. Toxicity Test of Concentrations (mg/l) of *A. annua* Against Fish Mortality (%)

Conc. Mg/l	*Fish Mortality (%)*			
	24H	*48H*	*72H*	*96H*
250	0	10	0	10
300	0	10	10	10
350	10	20	10	10
400	10	10	10	20
500	20	10	20	20

It was also observed that the parasite loads were reduced with increasing concentrations of *A. annua* extract. There was a positive correlation between the number of parasites dislodged from the body surfaces of fish and the time of exposure of fish to the extract. In addition to all of these, an increased agility in the swimming of fish freed from parasites was also observed when compared to their counterparts in the control with all parasites remaining intact.

Results of the toxicity test showed that extract of *A. annua* was well tolerated by *H. longifilis* juveniles. A minimal mortality observation was made throughout the 96 h period of exposure of fish to the extract. A few fish showed weak swimming activity in 350 to 500 mg/l of the test solutions. The highest percentage mortality observed after 96 h in the highest concentration was 20%.

Table III. Showing Host (Fish) Tolerance to Effective Plants Against Fish Ectoparasites

Plant extract	*Mode of application*	*Curative dose (EC_{50})*	*Threshold conc. (mg/l)*	*Lethal dose (96 h LC_{50})*	*Therapeutic index (TI)= (LC_{50}/EC_{50})*	*Remarks*
Mucuna pruriens	Long bath in water	200 mg/l	1000	1200 mg/l	6	High host tolerance
Carica papaya	Long bath in water	200 mg/l	1200	1600 mg/l	8	High host tolerance
Piper guineense	Long bath in water	1 mg/l	1.5	2 mg/l	2	Indication of host toxicity

Table IV. Standard Threshold Concentrations of Substances and Their Toxicity Gradients to Aquatic Organisms (*16*)

Group	*Threshold concentration (mg/l)*	*Toxicity grade*
A	> 500	Hardly toxic
B	100-500	Weakly toxic
C	10-99	Moderately toxic
D	1-9	Considerably toxic
E	< 1	Highly toxic

References

1. Cha, J. D.; Jeong, M. R.; Choi, H. J.; Jeong, S. I.; Moon, S. E.; Yun, S. I.; Kim, Y. H.; Kil, B. S.; Song, Y. A. *Planta Med.* **2005**, *71*, 575–577.
2. Iwu, M. M. *Ethnomed.* **1982**, *2* (1-4), 45.
3. Murray, N. D. *African Indigenous Medicinal, African Medicinal and Research Function*; Nairobi, Kenya, 1998; pp 37−38.
4. Commiittee on Mutagenic Chemicals in Food, Consumer Products and the Environment. Statement for COT: *Malachite Green and Leucomalachite Green*; Department of Health: London, 1999.
5. Ekanem, A. P.; Obiekezie, A.; Kloas, W.; Knopf, K. *Parasitol Res.* **2004**, *92*, 361–366.

6. Wooten, R. In *Fish Pathology*; Roberts, R. J., Ed.; Bailliere-Tindall: London, 1989; pp 228–242.
7. Yamaguti, S. *Monogenetic Trematodes of Hawaiian Fishes*; University of Hawaii Press: Honolulu, HI, 1968, 287.
8. Musselius, V. A. In *Investigations of Monogeneans in the USSR*; Skarlato, O. A., Ed.; Oxonian: New Delhi, 1987.
9. Paperna, I. *J. Parasitol* **1964**, *50*, 90–93.
10. Ekanem, A. P.; Wang, M.; Simon, J. E.; Obiekezie, A. I.; Morah, F. *Phytother. Res.* **2004**, *18*, 793–797.
11. Bailey, L. H.; Bailey, E. Z. *Hortus Third*; MacMillan Publ. Co.: NewYork, 1976.
12. Simon, J. E.; Chadwick, A. F.; Craker, L. E. *Herbs: An Indexed Bibliography, 1971-1980. The Scientific Literature on Selected Herbs, and Aromatic and Medicinal Plants of the Temperate Zone*; Archon Books: Hamden, 1984.
13. Hien, T. T.; White, N. J. *Lancet* **1993**, *341*, 603–608.
14. Yeung, H.-C. *Handbook of Chinese Herbs and Formulas*; Institute of Chinese Medicine: Los Angeles, 1985.
15. Ekanem, A. P.; Brisibe, E. B. *Parasitol Res.* **2010**, *106*, 1135–1139.
16. Schaperclaus, W. *Fish Diseases*; Oxonian Press, PVT: New Delhi, 1991; Vol. 2.

Chapter 8

Rooibos: Effect on Iron Status in South African Adults at Risk for Coronary Heart Disease

Jeanine L Marnewick,[1] Irma Venter,[2] Fanie Rautenbach,[1] Henry Neethling,[1] and Maritha Kotze[*,3]

[1]Oxidative Stress Research Centre, Cape Peninsula University of Technology, P.O. Box 1906, Bellville 7535, South Africa
[2]Programme: Consumer Science: Food and Nutrition, Faculty of Applied Sciences, Cape Peninsula University of Technology, P.O. Box 1906, Bellville 7535, South Africa
[3]Department of Pathology, Faculty of Health Sciences, University of Stellenbosch, Tygerberg, South Africa
[*]E-mail: MarnewickJ@cput.ac.za.

The effect of rooibos (*Aspalathus linearis)* herbal tea consumption on the iron status of adults at risk for developing coronary heart disease was a secondary objective of a recent clinical study. After consuming six cups of rooibos daily for six weeks, serum iron, ferritin, transferrin, TIBC and Fe saturation were determined to assess the iron status of the participants. Dietary intakes were calculated with the use of dietary records completed during the study periods. No significant changes in these biochemical parameters were recorded after completion of the rooibos and control study periods. Results showed that consumption of rooibos tea did not adversely affect the iron status of this study population when taking into account the habitual dietary composition and genetic variation in the HFE gene affecting iron absorption.

Introduction

Rooibos is an infusion/tisane made from the dried, fermented leaves and stems of *Aspalathus linearis* (Brum.f) Dahlg. (Family Fabaceae; tribe Crotalarieae), a legume indigenous to the Cedarberg region of the Western Cape, South Africa.

Traditional/fermented rooibos is different from black, green and oolong teas made from *Camellia sinensis,* in aroma, flavor, color and phytochemicals as it contains different and unique polyphenolic constituents. Rooibos does not contain caffeine and is considered to have a low tannin content (*1–3*) compared to *C. sinensis* teas. Many studies have reported on the beneficial health effects of green and black teas (*4–6*), however studies have also reported that an increased tea consumption can reduce the iron status in apparently healthy individuals (*7*, *8*). Although evidence is still lacking, it has been advised that "at risk" groups should not consume tea (*Camellia sinensis*) with their meals to minimize the effect of the tea phenolic constituents to inhibit the bioavailability of dietary non-haem iron (*6*, *9*, *10*).

To date very few studies have reported on the effect of rooibos on markers of iron status in humans. The first published study to report on this in 1979 concluded that rooibos (200 mL) with added milk (20 mL) and sugar (20 g) did not significantly affect iron absorption in healthy young men (n=10); an effect also found when consuming water (*8*). The group of young men that consumed black tea showed significant reduced iron uptake (*8*). The authors measured the uptake of 1 μCi ^{59}Fe-citrate/Fe-sulfate in the presence of ascorbic acid after 14 days. The next study reported no effect of rooibos or black tea (200 mL twice per day for 16 weeks) with added milk (40 mL) and sugar (8 g) on iron status parameters in school children (*11*). Iron status markers did show improvement at the end of the study period, and could possibly be as a result of tan antihelminthic treatment the children received (*11*).

Recently, we reported on the modulation of oxidative stress parameters by traditional/fermented rooibos in adults at risk for developing coronary heart disease (*12*). A secondary objective of the study was to monitor the relative influence of traditional/fermented rooibos on various parameters of iron status and two other risk factors for cardiovascular disease in these adults. The present study reports on the influence of rooibos on parameters of iron status, i.e. serum iron, ferritin, transferrin, total iron-binding capacity, on levels of hs-C reactive protein as iron is involved in the normal functioning of the immune system (*13*) and atherosclerosis considered a chronic inflammatory disease (*14*) and on levels of homocysteine as increased levels have been associated with cardiovascular disease (*15*).

Materials and Methods

Subjects

A total of 43 pre-screened volunteers from the City of Cape Town Metropolitan Municipality (South Africa) met the inclusion criteria of at least two or more risk factors for coronary heart disease. None of the volunteers were on any chronic medication nor were they taking any vitamin and/or antioxidant dietary supplements. All participants attended an information session where they were informed about the study details. Written consent was obtained from each participant as well. Research guidelines of the Declaration of Helsinki and Tokyo for human studies were followed, while the study protocol was approved by the Faculty of Health and Wellness Sciences Research Ethics Committee of the Cape Peninsula University of Technology (CPUT).

Study Design

The participants followed a two week wash out period, where after they consumed traditional/fermented rooibos (1200 mL per day) for six weeks followed by a cross over period of four weeks when an equivalent amount of water was consumed (*12*). Each participant prepared the rooibos according to a standard recipe (one tea bag per 200 mL freshly boiled tap water, with a brewing time of 5 min) with the given option to add milk and/or sugar, as this is how it is traditionally consumed. A previous study showed that the addition of milk does not prevent polyphenols from binding with iron (*7*). Participants were instructed to consume the 1200 mL rooibos as 6 cups throughout the day after and in-between meals. During the entire study period, participants were instructed to maintain their habitual lifestyle and dietary intake but avoid the intake of flavonoid–rich beverages, such as tea (except for rooibos in the rooibos study period), cocoa drinks, red wine and fruit juice, and restrict their intake of flavonoid-rich foods, such as apples, citrus fruit, berries and dark chocolate (*16*), along with coffee which also contain iron-binding phenolic compounds (*7*). Participants, after being trained, kept dietary records for three consecutive days per week for two weeks during each study period to monitor and determine their dietary intake. The six dietary records kept per study period were used to determine the participant average daily energy, macronutrient, micronutrient and flavonoid intake per study period. Traditionally fermented rooibos herbal tea bags of superior grade and same batch number, was provided by Rooibos Ltd (Clanwilliam, SA) for consumption during the rooibos study period [would the processing and fermentation process impact the results?]. The total polyphenol content, measured using the Folin-Ciocalteau method (*17*) and iron content of the prepared rooibos tisane are shown in Table I.

HPLC Analysis of Fermented Rooibos Tea Preparations

Chemicals

Rutin, quercetin and luteolin were purchased from Sigma-Aldrich (Johannesburg, South Africa). Orientin, Isoorientin, vitexin, isovitexin, hyperoside and chrysoeriol were purchased from Extrasynthese (Genay, France). Aspalathin was a gift from the South African Medical Research Counsel (Prof WCA Gelderblom). Methanol, dimethylsulfoxide (DMSO) and trifluoroacetic acid were purchased from Merck (Johannesburg, South Africa). Standards were dissolved in methanol and DMSO (1 mg/mL) as per the instructions (provide some detail) of the manufacturer and stored at -40 °C. Aliquots of 20 µg/mL were injected into the HPLC.

Table I. Total Polyphenol, Iron and Flavonoid (mg/L) Contents of Fermented/Traditional Rooibos

Content	*Fermented rooibos plant material*	*Fermented Rooibos infusion*
[$]Total polyphenols	ND	120 ± 4 mg/kg
[*]Iron (Fe) content	318 ± 26 mg/L	0.165 ± 0.002 mg/L
Aspalathin	ND	15.08 ± 0.89 mg/L
Orientin	ND	12.32 ± 0.82 mg/L
Isoorientin	ND	17.92 ± 0.88 mg/L
Vitexin	ND	3.48 ± 0.32 mg/L
Isovitexin	ND	3.85 ± 0.27 mg/L
Hyperoside/Rutin	ND	13.34 ± 0.69 mg/L
Quercetin	ND	0.81 ± 0.06 mg/L
Luteolin	ND	0.22 ± 0.03 mg/L
Chrysoeriol	ND	0.13 ± 0.01 mg/L
Total	ND	67.14 ± 2.21 mg/L

[*] SOURCE: Data from Malik *et al.*,(*18*); [$] Data from Marnewick *et al.*, (*12*); ND = not done.

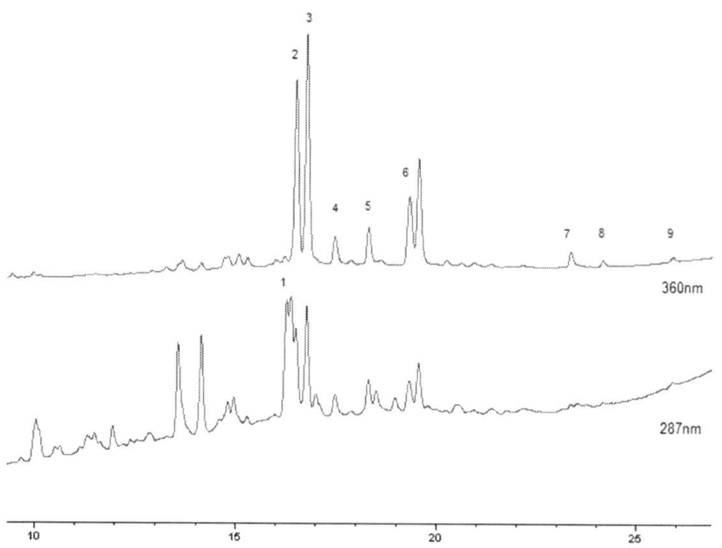

Figure 1. HPLC chromatogram (287 and 360nm) Fermented/Traditional Rooibos extracts, with isoorientin (peak 3) and aspalathin (peak 2) as the main components.

Chromatographic Conditions

The traditionally fermented rooibos tisane was chromatographically separated using the method described by Bramati *et al.* (*19*) with modifications. An Agilent Technologies 1200 Series HPLC system with a diode array detector and a 5μm YMC-PackPro C18 (150 x 4.6mm i.d.) column was used for the separation. Acquisition was set at 287 and 360 nm and the mobile phases consisted of water (A) containing 300 μl/L trifluoroaceticacid and methanol (B) containing 300 μl/L trifluoroaceticacid. The gradient elution started at 95% A changing to 75% A after 5 minutes and to 20% A after 25 minutes. The flow rate was set at 0.8 mL/min, the injection volume was 20 μl and the column temperature was set at 21°C. Peaks were identified based on the retention time of the standards and confirmed by comparison of the wavelength scan spectra (set between 210 nm and 400 nm). Only the major constituents of rooibos were quantified, with aspalathin being one of the major and unique flavonoids in the traditional/fermented rooibos tisane, together with orientin and isoorientin. Trace quantities of quercetin, luteolin and chrysoeriol are also present (Table I, Figure 1).

Dietary Intake Analyses

The daily dietary energy, iron, animal protein, vitamin C, vitamin A, β carotene, calcium, phosphorous, zinc, copper, phytate, oxalic acid and alcohol intakes were estimated using FoodFinder 3, the nutrient analysis software program of the South African Medical Research Council (*20*), and the daily total flavonoid intake using the United States Department of Agriculture (USDA) database for the flavonoid content of selected foods (*21*). Fasting venous blood samples were collected in the mornings (7-9 am) after completion of each study period by two nursing sisters and a phlebotomist.

Sample Analyses

Fasting serum/plasma chemistry analytes were measured at the Oxidative Stress Research Centre (CPUT, South Africa) using automated analyzers (Vitalab SelectraE/Medica EasyRA) with KAT Laboratory and Medical PTY (LTD) reagents and controls (Johannesburg, South Africa). The analytes for iron status included serum iron, ferritin and transferrin, while the total iron binding capacity (TIBC) and transferrin saturation (TS) were calculated (TIBC = transferrin X 22.375; TS = Serum iron/TIBC X 100).

Genetic Analyses

DNA was extracted from whole blood and subjected to mutation analysis of the HFE gene, using polymerase chain reaction (PCR) TaqMan technology (rs1800562).

Table II. Daily Energy, Iron, Animal protein, Vitamin C, Vitamin A, β-carotene, Calcium, Phosphorus, Zinc, Copper, Phytate, Oxalic acid, Alcohol and Total Flavonoid Intakes by the Participants during each Study Period

Daily intake	*Study periods*		
	Washout	*Rooibos*	*Control*
Energy (kJ)	8671 ± 2530	8097 ± 2206	8052 ± 2394
Iron (mg)	12.38 ± 4.46	11.5 ± 4.02	10.4 ± 3.7
Haem iron (mg)	0.67 ± 0.60	0.7 ± 0.52	0.7 ± 0.6
Non-haem iron (mg)	3.93 ± 1.63	4.2 ± 1.4	3.9 ± 1.6
Animal protein (g) [1]	50.8 ± 21.8	55.3 ± 22.8	49.1 ± 20.5
Vitamin C (mg)[2]	112.7 ± 21.0	88.5± 67.1	84.4 ± 67
Vitamin A (μg) [2]	1013 ± 682	903 ± 459	940 ± 610
β-carotene (μg) [2]	2720 ± 2752	2427 ± 2027	2327 ± 2355
Calcium (mg) [3]	773 ± 362	701 ± 245	670 ± 315
Phosphorus (mg) [3]	1244 ± 420	1174 ± 368	1155 ± 397
Zinc (mg) [3]	11.4 ± 4.6	11.1 ± 4.1	10.2 ± 3.8
Copper (mg) [3]	1.6 ± 1.4	1.8 ± 0.5	1.2 ± 0.5
Phytate (mg) [3]	146 ± 121	149 ± 80	154 ± 99
Oxalic acid (mg) [3]	63.4 ± 53.6	73.2 ± 75.1	66.3 ± 96.1
Alcohol (g) [3]	5.2 ± 7.5	5.3 ± 9.0	6.0 ± 7.8
Total flavonoid (mg)*	31 ± 25	343 ± 90	27 ± 24

NOTE: Values in columns indicate means ± standard deviation. * $P<0.05$ when comparing the relevant parameter between the various study periods. [1] SOURCES: Animal tissue protein (meat, fish and poultry) enhances iron absorption and non-tissue protein (milk, cheese and eggs) reduces absorption (*22, 23*) [2] Dietary factors enhancing iron absorption (*24–26*) [3] Dietary factors reducing/inhibiting iron absorption (*10, 26–30*)

Statistical Analyses

The results are presented as means ± SD. Analysis of Variance (ANOVA) was used to determine whether the means of the different study periods differed significantly. When the ANOVA was positive ($P<0.05$), a Student-Newman-Keuls test for pairwise comparison of the different study periods was performed. In all analyses a P value of <0.05 was considered significant.

Results

Forty participants (26 females and 14 males) completed the study. Their average age was 46.8 ± 9.7 years and their body mass index (BMI) (kg/m^2) 28.4 ± 5.5 (*12*). There were no differences in daily energy, iron, haem iron, non-haem iron, animal protein, vitamin C, vitamin A, β-carotene, calcium, phosphorus, zinc, copper, phytate, oxalic acid and alcohol intakes when comparing the various study periods (Table II). The total daily flavonoid intake was significantly (P<0.05) higher during the rooibos intervention period, as expected. This correlated with the significant (P<0.05) increased serum polyphenols in participants during the rooibos intervention period when compared with the control period (*12*).

The iron status indexes, and serum levels of hs-CRP and homocysteine after completion of the washout period, rooibos and control intervention periods of the participants are shown in Table III. No significant differences in the iron status indicators (serum iron, ferritin, transferrin, TIBC and % Fe saturation) were found after the rooibos intervention period when compared with the control period. Chronic rooibos consumption also had no significant (P>0.05) effect on hs-C-reactive protein or on homocysteine levels of the participants when compared with the control period.

Table III. Effect of Traditionally fermented Rooibos on Parameters of Systemic Iron and Levels of hs C-Reactive Protein and Homocysteine After Each Study Period

Analytes	*Study periods*		
	Washout	*Rooibos*	*Control*
Serum iron (μmol/L)	16.23 ± 4.39	15.34 ± 4.71	14.11 ± 3.41
Ferritin (ng/mL)	122.30 ± 92.84	113.3 4± 76.86	101.40 ± 65.03
Transferrin (g/L)	2.49 ± 0.37	2.41 ± 0.37	2.54 ± 0.39
TIBC (μmol/L)	55.69 ± 8.25	53.98 ± 8.38	56.92 ± 8.73
% Fe saturation	30.13 ± 10.75	29.20 ± 10.31	25.54 ± 7.63
hs-CRP (mg/dL)	1.15 ± 0.37	1.19 ± 0.46	1.28 ± 0.40
Homocysteine (μmol/L)	9.63 ± 3.31	10.24 ± 2.97	9.05 ± 4.38

NOTE: Values in columns indicate means ± standard deviation. No statistical significance (P>0.05) was shown when comparing the different parameters of the various study periods.

The effects of rooibos consumption were also considered in the sample of female participants (n=26, average age 48 ± 6 yr) due to the risk women have for iron deficiency during their reproductive years compared to men who meet their lower iron need without much dietary effort owed to higher food intake (*31*, *32*).

When comparing the daily intakes of the female participants for each study period, no significant differences (P>0.05) were noted in the investigated dietary intakes as well as the blood chemistry (Table IV). As previously reported (*10*) the serum ferritin levels of the male participants were significantly (P<0.05) higher than that of the female participants as measured at the end of each study phase (208 ± 146 ng/mL vs 94 ± 42 ng/mL for the washout phase; 159 ± 103 ng/mL vs 86 ± 37 ng/mL for the rooibos phase and 154 ± 78 ng/mL vs 76 ± 38 ng/mL for the control phase).

None of the study participants were homozygous for mutation C282Y, the HFE genotype most commonly with hereditary heamochromatosis (HH, genetic iron overload).

Table IV. Daily Energy, Iron, Animal protein, Vitamin C, Vitamin A, β-carotene, Calcium, Phosphorus, Zinc, Copper, Phytate, Oxalate, Alcohol and Flavonoid intakes by the Female Participants and their Blood Chemistry during the Study Periods

	Study periods		
Dietary intake and blood chemistry	*Washout*	*Rooibos*	*Control*
Energy (kJ)	8008 ± 2180	7147± 1734	7101± 2122
Iron (mg)	12 ± 4.9	10.0 ± 3.2	9.1 ± 3.0
Haem-iron (mg)	0.5 ± 0.3	0.6 ± 0.5	0.5 ± 0.5
Non-heam iron (mg)	3.5 ± 1.2	3.6± 1.2	3.4 ± 1.4
Animal protein (g)	44.7 ± 14.9	46.7 ± 18.2	41.9 ± 16.9
Vitamin C (mg)	115.6 ± 73.0	78.8 ± 59.1	82.2 ± 64.8
Vitamin A (μg)	871 ± 612	808 ± 382	733 ± 421
β-carotene (μg)	2351 ± 2601	2113 ± 1465	1630 ± 1389
Calcium (mg)	731 ± 324	647 ± 221	606 ± 294
Phosphorous (mg)	1136 ± 318	1027 ± 316	1000 ± 319
Zinc (mg)	9.8 ± 3.4	9.3 ± 2.6	8.5 ± 2.5
Copper (mg)	1.6 ± 1.7	1.7 ± 0.5	1.1 ± 0.5
Phytate (mg)	131 ± 99	133 ± 80	140 ± 85
Oxalic acid (mg)	56.8 ± 52.7	59.6 ± 37.4	43.4 ± 40.5
Alcohol (g)	3.6 ± 5.6	2.6 ± 4.8	3.3 ± 4.8
Total flavonoid (mg)	34 ± 25	345 ± 97*	29 ± 26
Serum Iron (μmol/L)	15.2 ± 4.1	14.3 ± 4.2	13.8 ± 3.8

Continued on next page.

Table IV. (Continued). Daily Energy, Iron, Animal protein, Vitamin C, Vitamin A, β- carotene, Calcium, Phosphorus, Zinc, Copper, Phytate, Oxalate, Alcohol and Flavonoid intakes by the Female Participants and their Blood Chemistry during the Study Periods

	Study periods		
Dietary intake and blood chemistry	*Washout*	*Rooibos*	*Control*
Ferritin (ng/mL)	94 ± 42	86 ± 37	76 ± 38
Transferrin (g/L)	2.5 ± 0.4	2.4 ± 0.4	2.6 ± 0.4
TIBC (μmol/L)	55.2 ± 8.2	53.8 ± 9.3	58.8 ± 9.2
% Fe saturation	27.9 ± 8.0	27.5 ± 9.4	24.0 ± 7.6
Hs-CRP (mg/dL)	1.2 ± 0.4	1.2 ± 0.4	1.4 ± 0.5
Homocysteine (μmol/L)	9.0 ± 3.2	9.8 ± 3.0	8.0 ± 3.1

NOTE: Values in columns represent average ± standard deviation. * $P<0.05$ when comparing each parameter for the various study periods.

Discussion

Many dietary and physiological factors have been described to influence the absorption of non-haem iron (*33*). Some dietary factors, e.g. vitamin C and animal protein, enhance iron absorption, while others, i.e. phytate and phenolic compounds, inhibit the uptake of iron (*7*, *34*, *35*). Traditionally fermented rooibos contains a plethora of known as well as unique polyphenolic compounds and it is thus anticipated that they could bind to iron with a resultant influence on iron absorption and ultimately iron status. The structure of polyphenols differ considerably and thus also their effect on iron absorption (*7*, *36*). Many research studies have reported on the possible effects of black tea on the iron status of individuals with consistent evidence still lacking (*6*, *9*, *10*) with far less studies reporting on the effect of herbal teas. Rooibos is considered a healthy beverage not only because of the natural absence of caffeine (*1*) but also because of its low-tannin content especially when compared with *C. sinensis* teas (*2*, *3*) and the presence of highly bioactive polyphenols. The question whether rooibos will influence or play a major role in the iron status of individuals consuming this tea remains a important one to address considering that vulnerable populations and others are limited their available iron content. (or modify as needed). In the current study, the dietary composition, including the iron content and the content of dietary iron absorption enhancing and/or inhibiting factors of the participants remained similar throughout the three study periods with the exception of the flavonoid intake which seems to have had no impact on the participant's iron status, in particular during the rooibos intervention period when the flavonoid intake, predominantly provided by rooibos, increased significantly.

Physiological factors such as the iron status of the individual, i.e. genetic iron-overload disorders such as haemochromatosis, and chronic inflammation also influence iron bioavailability. Hereditary haemochromatosis caused by homozygosity (inheritance of two copies of the faulty gene) for the most common HFE gene mutation C282Y, was also excluded in all the study participants. The two C282Y heterozygotes identified had normal iron stores. None of the study participants furthermore showed any deficient iron status when the study commenced and their iron status was not significantly altered across the study periods; even in the female participants who although due to their age distribution are probably in transition from their reproductive to non-reproductive years may still be at risk for iron deficiency. Levels of Hs C-reactive protein in the study participants were also shown not to differ significantly when comparing each study period, excluding inflammation as a possible factor to have modulated the iron status (*37*) of the participants.

Results from this study also showed regular consumption of rooibos over a period of six weeks did not alter mean plasma homocysteine levels significantly, in line with studies reporting on the effect of black tea (*38*). Previously it was shown that certain dietary polyphenols can increase homocysteine levels (*39*), but the role of homocysteine levels in cardiovascular disease remains controversial (*40*).

When considering the iron status, habitual dietary composition and host factors, we found that the intake of rooibos did not have an adverse effect on the participants' resultant iron status at the end of the intervention study period. Results from this adult study population at risk for developing heart disease agree with previous findings where the study populations differed, i.e. South African learners (6-15 yrs) (*11*) and apparently health young men (21-34 yrs) (*8*) and does therefore not make it unreasonable to suggest that rooibos, in the South African context does not have an adverse effect on markers of iron status. Long-term studies are needed to further investigate the relationship between rooibos consumption and iron status in various South African populations.

Acknowledgments

The authors would like to thank the study nurses, Lorna Thomson and Jane Van Zyl, the participants, Maretha Opperman for assisting with the anthropometry and the energy and nutrient intake analysis, Nelia van Wyk and the Baccalaureus Technologiae students (Hannelise Louwrens and Carla Ullrich) from the Program: Consumer Science: Food and Nutrition, CPUT for their assistance in screening and entering of the dietary data. This study was supported by grants from the South African Rooibos Council, THRIP (National Research Foundation) and CPUT.

References

1. Galasko, G. T. F.; Furman, K. I.; Alberts, E. *Food Chem. Toxic.* **1989**, *27* (1), 49–51.
2. Blommaert, K. L. J.; Steenkamp, J. *Agroplantae.* **1978**, *10*, 49.

3. Reynecke, J.; Coetzee, W. H. K.; Bester, J. J. A. *Farming SA* **1949**, *24*, 397–412.
4. Chacko, S. M.; Thambi, P. T.; Kuttan, R.; Nishigaki, I. *Chinese Med.* **2010**, *5*, 13–21.
5. Gardner, E. J.; Ruxton, C. H. S.; Leeds, A. R. *Eur. J. Clin. Nutr.* **2007**, *61*, 3–18.
6. Ruxton, C. H. S. *Nutr. Bull.* **2008**, *33*, 91–101.
7. Hurrell, R. F.; Reddy, M.; Cook, J. D. *Brit. J. Nutr.* **1999**, *81*, 289–295.
8. Hesseling, P. B.; Joubert, J. R. *SAMJ* **1982**, *62*, 1037–1038.
9. Temme, E. H. M.; Van Hoydonck, P. G. A. *Eur. J. Clin. Nutr.* **2002**, *56*, 379–386.
10. Nelson, M.; Poulter, J. *J. Hum. Nutr. Diet.* **2004**, *17*, 43–54.
11. Breet, P.; Kruger, H. S.; Jerling, J. C.; Oosthuizen, W. *Nutr. Res.* **2005**, *25*, 983–994.
12. Marnewick, J. L.; Rautenbach, F.; Venter, I.; Neethling, H.; Blackhurst, D. M.; Wolmarans, P.; Macharia, M. *J. Ethnopharmacol.* **2011**, *133*, 46–52.
13. Hulsewé, K. W. E.; Van Acker, B. A. C.; Von Meyenfeldt, M. F.; Soeters, P. B. *World J. Surg.* **1999**, *23*, 536–544.
14. Koenig, W. *Cardiol. Rev.* **2001**, *9* (1), 31–35.
15. Hankey, G. J.; Eikelboom, J. W. *Lancet.* **1999**, *354*, 407–413.
16. Weichselbaum, E.; Buttriss, J. L. *Nutr. Bull.* **2010**, *35*, 157–164.
17. Singleton, V. L.; Rossi, J. A. *Am. J. Enol. Vitic.* **1965**, *16*, 144–158.
18. Malik, J.; Szakova, J.; Drabek, O.; Balik, J.; Kokoska, L. *Food Chem.* **2008**, *111*, 520–525.
19. Bramati, L.; Minoggio, M.; Gardana, C.; Simonetti, P.; Mauri, P.; Pietta, P. *J. Agric. Food Chem.* **2002**, *50*, 5513–5519.
20. South African Medical Research Council, Nutritional Intervention Research Unit. *FoodFinder 3 Dietary Analysis Software Programme*; Parow Valley, Cape Town, South Africa, 2002.
21. Department of Agriculture (USDA). *USDA database for the flavonoid content of selected foods*, Release 2.1; Beltsville, MD, 2007.
22. Hallberg, L.; Rossander, L. *Am. J. Clin. Nutr.* **1984**, *39*, 577–583.
23. López, M. A. A.; Martos, F. C. *Int. J. Food Sci. Nutr.* **2004**, *55* (8), 597–606.
24. Hunt, J. R.; Mullen, L. M.; Lykken, G. I.; Gallagher, S. K.; Nielsen, F. H. *Am. J. Clin. Nutr.* **1990**, *51*, 649–655.
25. Layrisse, M.; Garcia-Casal, M. N.; Solano, L.; Barón, M. A.; Arguello, F.; Llovera, D.; Ramírez, J.; Leets, I.; Tropper, E. *Arch. Latinoam. Nutr.* **2000**, *50* (3), 243–248.
26. Hallberg, L.; Hulthén, L. *Am. J. Clin. Nutr.* **2000**, *71*, 1147–1160.
27. Brune, M.; Rossander-Hultén, L.; Hallberg, L.; Gleerup, A.; Sandberg, A-S. *J. Nutr.* **1992**, *122*, 442–449.
28. Arredondo, M.; Núñez, M. T. *Mol. Aspects Med.* **2005**, *26*, 313–327.
29. Solomons, N. W.; Ruz, M. *Nutr. Res.* **1997**, *17* (1), 177–185.
30. Gillooly, M.; Bothwell, T. H.; Torrance, J. D.; MacPhail, A. P.; Derman, D. P.; Bezwoda, W. R.; Mills, W.; Charlton, R. W. *Brit. J. Nutr.* **1983**, *49*, 331–342.
31. Harvey, L. J.; Armah, C. N.; Dainty, J. R.; Foxall, R. J.; Lewis, D. J.; Langford, N. J.; Fairweather-Tait, S. J. *Brit. J. Nutr.* **2005**, *94*, 557–564.

32. Whitney, E. N.; Cataldo, C. B.; Rolfes, S. R. *Understanding Normal and Clinical Nutrition*, 5th ed.; West/Wadsworth Publishing Company, Inc.: Belmont, 1998; pp. xxviii–961, ISBN: 0-534-53334-5.
33. Cockell, K. A. *J. AOAC Int.* **2007**, *90* (5), 1480–1491.
34. Sandstrom, B. *Brit. J. Nutr.* **2001**, *85*, S181–S185.
35. Hallberg, L.; Brune, M.; Rossander, L. *Int. J. Vitam. Nutr. Res., Suppl.* **1989**, *30*, 103–108.
36. Ullmann, U.; Haller, J.; Bakker, G. C. M.; Brink, E. J.; Weber, P. *Phytomedicine.* **2005**, *12*, 410–415.
37. Gisbert, J. P.; Gomollón, F. *World J. Gastroenterol.* **2009**, *15* (37), 4617–4626.
38. Hodgson, J. M.; Burke, V.; Beilin, L. J.; Croft, K. D.; Puddey, I. B. *Am. J. Clin. Nutr.* **2003**, *77*, 907–911.
39. Olthof, M. R.; Hollman, P. C.; Zock, P. L.; Katan, M. B. *Am. J. Clin. Nutr.* **2001**, *73*, 532–538.
40. Mursu, J.; Voutilainen, S.; Nurmi, T.; Alfthan, G.; Virtanen, J. K.; Rissanen, T. H.; Happonen, P.; Nyyssönen, K.; Kaikkonen, J.; Salonen, R.; Salonen, J. K. *Free Rad. Biol. Med.* **2005**, *38*, 527–534.

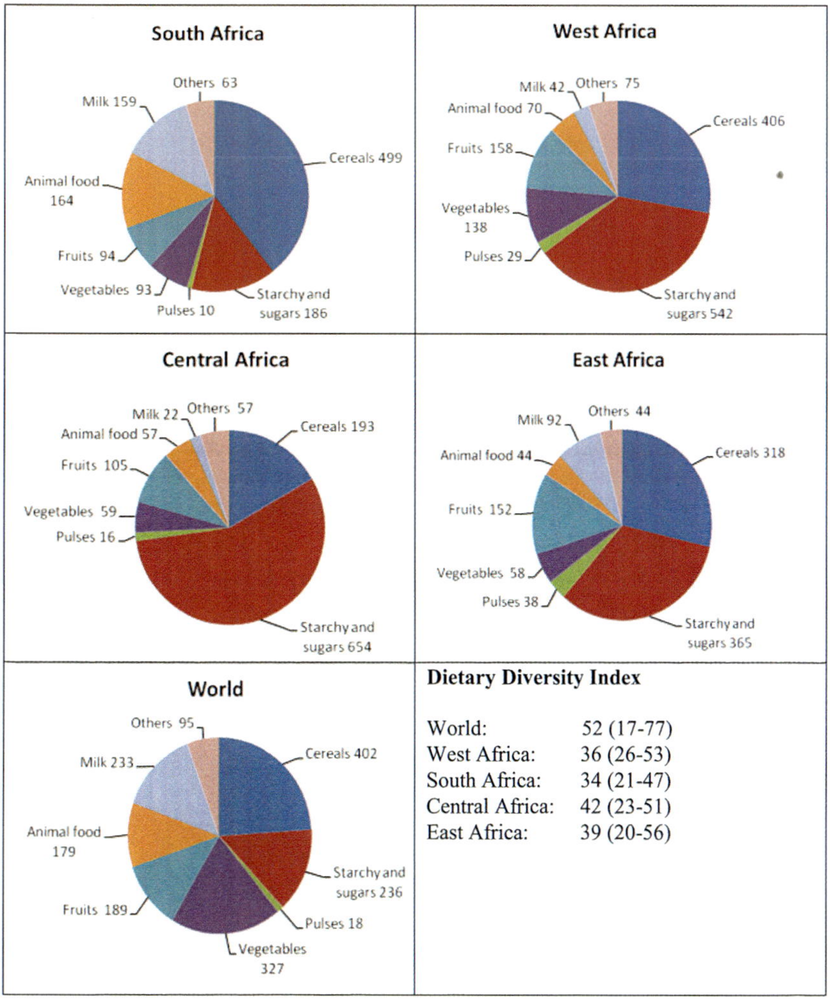

Figure 1. Dietary pattern in the world and regions of Sub-Saharan Africa. Data source: (15). (Chapter 15)

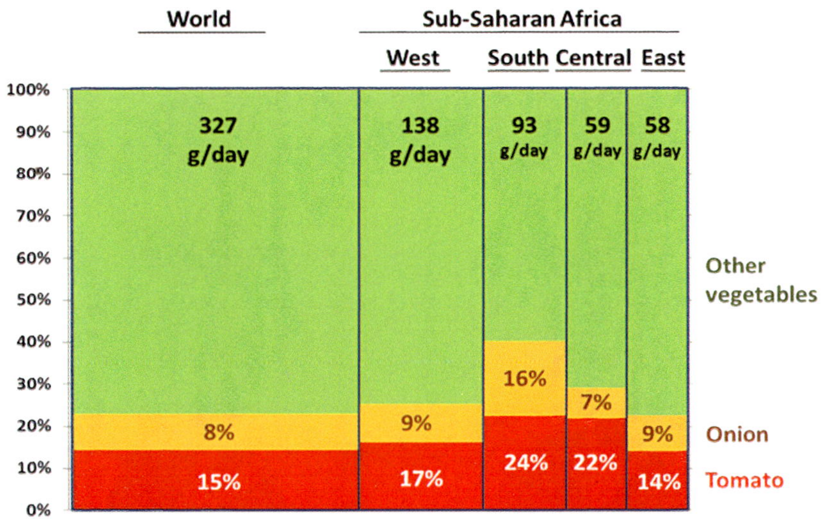

Figure 2. Share of vegetable availability (g/ person/day) in Sub-Saharan Africa. Data source: (1). (Chapter 15)

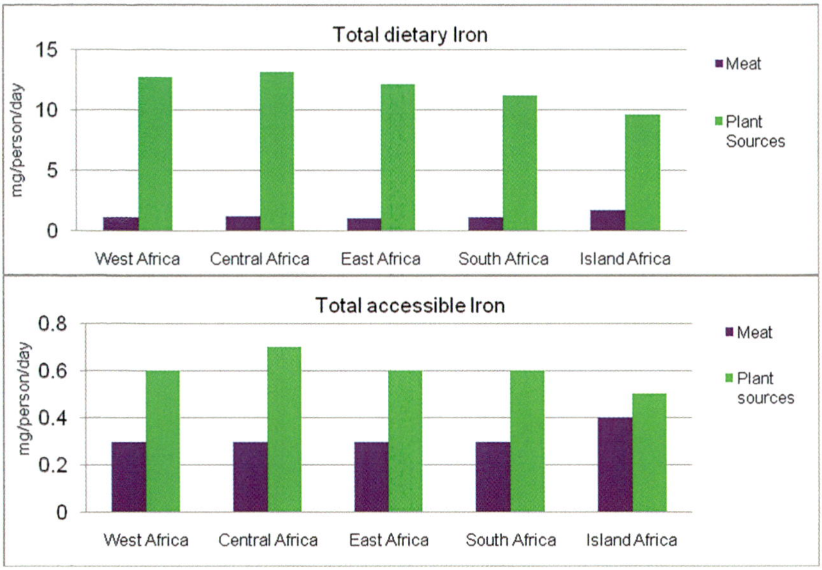

Figure 3. Total and accessible dietary iron from animal and plant sources. Data souce: (15). (Chapter 15)

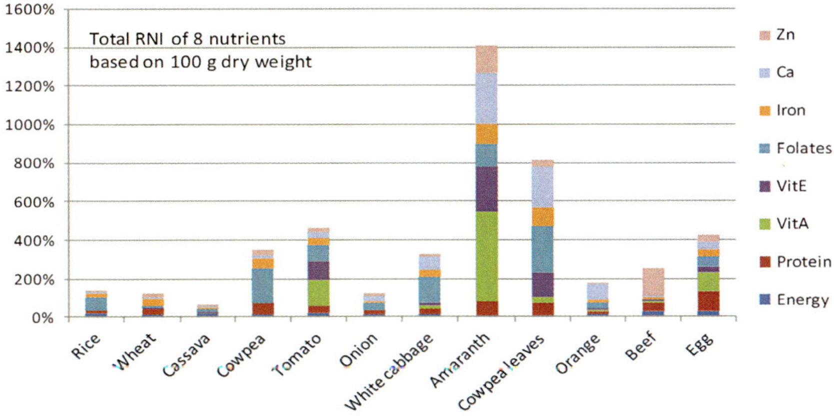

Figure 4. Comparison of total RNI of eight nutrients for staple, animal products, fruits and vegetables. Nutrient data source: AVRDC nutrient data and USDA nutrient database . Note: RNI for vitamin C was not included. (Chapter 15)

Chapter 9

Secoiridoid Glucosides from Fraxinus Excelsior with Effects on LPS-Induced Nitrite Production in RAW 264.7 Macrophages and Human Cancer Cell Lines

Naisheng Bai,[1] Kan He,[1] Marc Roller,[2] Ching-Shu Lai,[3] Xi Shao,[4] Min-Hsiung Pan,[3] and Chi-Tang Ho*,[4]

[1]Naturex, Inc., 375 Huyler Road, South Hackensack, New Jersey 07606
[2]Naturex SA, Site d'Agroparc BP 1218, 84911 Avignon Cedex 9, France
[3]Department of Seafood Science, National Kaohsiung Marine University, Kaohsiung, Taiwan
[4]Department of Food Science, Rutgers University, 65 Dudley Road, New Brunswick, New Jersey 08901
***E-mail: ho@aesop.rutgers.edu.**

Nine secoiridoid glucosides, excelside A (**1**) and excelside B (**2**), nuzhenide (**3**), GI3 (**4**), GI5 (**5**), ligstroside (**6**), oleoside-11-methyl ester (**7**), oleoside dimethyl ester (**8**), 1‴-*O*-*β*-D-glucosylformoside (**9**), and one phenylethanoid, salidroside (**10**), were isolated from the seeds of *Fraxinus excelsior*. The structures were established on the basis of NMR spectroscopic methods supported by MS or HRMS. HPLC method and a calibration curve was constructed for analysis the constituents of the extract. All isolated compounds were tested for cytotoxicity in human cancer cell lines (Hep G2, COLO 205, and HL-60) and anti-inflammatory activities in LPS-treated RAW264.7 macrophage cells. Among them, the water extract **11**, and the compounds **5**, **8**, **9**, **10**, and **4** were modestly active to inhibit nitrite production in macrophages, followed by compounds **3**, **2**, **1**, **6**, and **7**. Compound **4** was slightly effective as an anti-proliferative agent in HL-60 cells with IC_{50} of 82.0 μM. There was no cytotoxicity observed for any compounds except for **4** and the extract in these cells.

Introduction

The plant of *Fraxinus excelsior* L. (Oleaceae) is known as 'common ash' or 'European ash' in temperate Asia and Europe (*1*, *2*). The genus Fraxinus (Oleaceae) is mainly distributed in the temperate and subtropics regions of the Northern hemisphere and the species have economical, commercial and medicinal importance (*3*, *4*). The characteristic chemical feature of Fraxinus species is the presence of coumarins, secoiridoids, and phenylethanoids. The herbs of Fraxinus species have been used in folk medicine in different area of the world for their diuretic and mild purgative effects, as well as for treatment of constipation, dropsy, arthritis, rheumatic pain, cystitis and itching scalp (*4*, *5*). The plant of *F. excelsior* is also widely distributed throughout the South-East of Morocco (Tafilalet), where it is locally known as "Lissan Ettir" and its seeds as "l'ssane l'ousfour". This region is a rich source of ethnobotanicals, and an area in which phytotherapy has been and remains to be well developed (*2*, *6*, *7*). The leaves and bark of *F. excelsior*, native in Europe and Asia, have been used as a diuretic and rheumatic remedy since olden times (*8*). The bark and the leaves of *F. excelsior* are applied in the folk medicine against various diseases, including wound healing, diarrhea and dysentery. Nowadays, the leaves of this species are mainly recommended against fever and rheumatism (*4*). The ethanolic extract of the bark of this plant is a component of the plant drug Phytodolor N with antiinflammatory and antirheumatic properties (*8*). Aqueous seed extract of *F. excelsior* (FE) has been shown to be highly potent in the reduction of blood glucose levels without significantly affecting insulin levels (*6*, *7*, *9*). The phlorizin-like effect of inhibiting renal glucose reabsorption is a potential mechanisms for the hypoglycemic effect of FE (*9*). Previous investigations on the chemical composition of FE led to the characterization of several compound classes including secoiridoid glucosides, coumarins, flavonoids, phenylethanoids, benzoquinones, indole derivatives, and simple phenolic compounds (*4*, *10–12*). Our previous study revealed that inhibition of adipocyte differentiation and PPARα-mediated mechanisms of the isolated secoiridoids might be relevant pathways for the anti-diabetic activity of *F. excelsior* extract (*13*).

It has been demonstrated that nitric oxide (NO) is involved in many inflammatory diseases when NO is produced in large amount. Overproduction of NO and its more reactive N-nitrosating agents such as peroxynitrite, may also represent an essential link between inflammation and carcinogenesis (*14*, *15*). The focus of the present study was to isolate and characterize the potential active principle(s) of *F. excelsior* and evaluate their biological activity in anti-inflammatory and cancer preventive assays. As a result, we reported herein the isolation of nine secoiridoids and one phenylethanoid, as well as the biological testing results on LPS-induced nitrite production in RAW 264.7 macrophage and human cancer cell lines of these compounds.

Materials and Methods

General Experimental Procedures

Optical rotations were measured with a Perkin-Elmer 241 polarimeter. FT-IR was performed on a Perkin-Elmer spectrum BX system (PerkinElmer Instruments, Norwalk, CT). UV spectra were acquired on a Shimadzu, UV-1700 UV-Visible Spectrophotometer. The ^{1}H and ^{13}C NMR spectra were recorded on an Inova-400 (^{1}H at 400 MHz) instrument (Varian Inc., Palo Alto, CA) with CD_3OD (*reference 3.30* ppm) and D_2O as the solvent (Aldrich Chemical Co., Allentown, PA). The 2D correlation spectra were obtained using standard gradient pulse sequences of Varian VNMR software and performed on 4-nuclei PFG AutoSwitchable or PFG Indirect Detection probes. HRFAB-MS was run on a JEOL HX-110 double focusing mass spectrometer. Both negative and positive electrospray ionization-mass spectrometric spectra (ESI-MS) were obtained on an LCQ ion trap (Thermo-Finnigan, San Jose, CA). GC-MS analysis was carried out on an Agilent HP 6890 Series Gas Chromatograph system and *Agilent HP 5973 Mass Spectrometer (*Santa Clara, CA) *with Rxi®-1ms capillary GC column (60 m × 0.25 mm ID × 1.0 μm).*

HPLC Analysis Conditions

HPLC analysis was performed on an Agilent 1100 LC Series using Prodigy ODS3 column (5 micron, 4.6 mm ID × 25 cm) with a flow rate of 1.0 mL/min. Solvent system consisted of 0.1% triflouroacetic acid/water (A) and acetonitrile (B) in the following manner: 0-5 min, 0-20% B; 5-15 min, 20-30% B; 15-25 min, 30-100% B. At the end of the run, 100% of acetonitrile was allowed to flush the column for 10 min, and an additional 10 min of post run time were set to allow for equilibration of the column with the starting eluant. The UV detector was operating at 238 nm, and the column temperature was ambient.

Calibration Curves

Methanol stock solutions containing the two active compounds [nuzhenide (**3**), and GI3 (**4**)] were prepared and diluted to five different final concentrations. A calibration curve was constructed for each of the compounds by plotting peak areas versus compound concentrations.

Plant Material

The seeds of *Fraxinus excelsior* (Oleaceae) were collected in Morocco. A voucher specimen (J02/02/A7) was deposited in the Herbarium of Naturex, Inc.

Extraction and Isolation

A total of 2.5 kg of air-dried and powered seeds of *F. excelsior* was extracted twice with 15 L of water at 95° C for 2 hours. The combined extract was concentrated and dried into powder (**11** or FE994702, 500 g). The powder was re-extracted with MeOH two times (3.5 L each) and the MeOH was evaporated in vacuum. The obtained extract (54 g solid) was reconstituted in 0.5 L of water and was loaded on a C-18 (1 L) (Sigma Chemical Co., St. Louis, MO) column (8.0 cm i.d. × 70 cm) eluted with water (5 L) and 10% MeOH/water (3 L). Fractions with similar HPLC chromatograms were combined and concentrated *in vacuo*. The combined water fractions (21 g solid) were separated over silica gel (Sorbent Technologies, Inc.) by column chromatography (500 g, 3.5 cm × 60 cm), eluting with a step gradient consisting of CH_3Cl-MeOH (10:1, 8:1, 5:1, 3:1, 2:1); In each gradient step, 1.5 L of eluent were used, collecting of 0.5 L . A total of 15 fractions was collected and labeled as W-fractions. These fractions were subjected to column chromatography over MCI GEL CHP-20P (Mitsubishi Kasei Co.) (100 mL, 2.5 cm × 40 cm) and/or Sephadex LH-20 (100 mL, 2.5 cm × 40 cm), eluting with a water-MeOH (4:6) system to yield **3** (210 mg from fraction W-9, t_R = 12.6 min in HPLC), **6** (46 mg from W-2, t_R = 19.3 min), **7** (28 mg from W-11, t_R = 9.4 min), **8** (41 mg from W-10, t_R = 12.7 min), **9** (21 mg from W-8, t_R = 13.4 min), and **10** (22 mg from W-4, t_R = 8.1 min). In a similar manner, the 10% MeOH-water eluates (12 g solid) from the C-18 column were chromatographed over a silica gel column, collecting a total of 15 M-fractions. These fractions were chromatographed over MCI GEL CHP-20P and/or Sephadex LH-20 to yield **1** (16 mg from fraction M-8, t_R = 10.4 min), **2** (33 mg from M-6, t_R = 15.6 min), **4** (238 mg from M-3, t_R = 17.9 min), and **5** (36 mg from M-5, t_R = 20.4 min).

Cell Culture and Chemicals

The COLO 205 cell lines were isolated from human colon adenocarcinoma (ATCC CCL-222); human promyelocytic leukemia (HL-60) cells were obtained from American Type Culture Collection (Rockville, MD). The human HepG2 hepatocellular carcinoma cell lines (BCRC 60025) were obtained from the Food Industry Research and Development Institute (Hsinchu, Taiwan). COLO-205 and HL-60 cell lines were grown at 37 °C in 5 % CO_2 atmosphere in RPMI. Hep G2 cells were grown in Dulbecco's minimal essential medium (DMEM) supplemented with 10% heat-inactivated fetal calf serum (Gibco BRL, Grand Island, NY), 100 units/mL of penicillin and 100 µg/mL of streptomycin, and kept at 37°C in a humidified atmosphere of 5% CO_2 in air. Selected compounds were dissolved in dimethyl sulfoxide (DMSO). Propidium iodide was obtained from Sigma Chemical Co. (St. Louis, MO).

Determination of Cell Viability

Cell viability was assessed by 3-(4,5-dimethylthiazol-2-yl)-2,5-diphenyltetrazolium bromide (MTT) assay (*16*). Briefly, human cancer cells were plated at a density of 1×10^5 cells/mL into 24 well plates. After overnight growth, cells were pretreated with series of concentration of test compounds for 24 h. The final concentration of DMSO in the culture medium was < 0.05%. At the end of treatment, 30 μL of MTT was added, and the cells were incubated for a further 4 h. Cell viability was determined by scanning with an enzyme-linked immunosorbent assay reader with a 570 nm filter.

Nitrite Assay

The RAW264.7 cells were treated with selected compounds and LPS or LPS alone. The supernatants were harvested and the amount of nitrite, an indicator of NO synthesis, was measured by use of the Griess reaction. Briefly, supernatants (100 μL) are mixed with the same volume of Griess reagent (1% sulphanilamide in 5% phosphoric acid and 0.1% naphthylethylenediamine dihydrochloride in water) in duplicate on 96-well plates. After incubation at room temperature for 10 min, absorbance at 570 nm is measured with an ELISA reader (Thermo Labsystems Multiskan Ascent, Finland).

Table I. HPLC Calibration Curve Data for Nuzhenide and GI 3 Compounds

Compounds	*Retention time (min)*	*Linear range (μg/mL)*	*Calibration eq*	*Correlation coefficient (R)*
Nuzhenide	12.6	1.9-1190	y = 9.6728x + 40.647	0.9999
GI 3	17.9	1.2-806	y = 12.29x + 6.1331	1.0

Results and Discussion

Chromatography of the hot water extract of *F. excelsior* as sequential combination of normal, reversed-phase, and gel permeation column chromatography now led to the isolation of nine secoiridoids including the new excelside A (**1**) and B (**2**) (*13*), the known nuzhenide (**3**) (*17*), GI3 (**4**) (*18*), GI5 (**5**) (*11*, *19*), ligstroside (**6**) (*20*), oleoside-11-methyl ester (**7**) (*21*), oleoside dimethyl ester (**8**) (*22*), 1‴-*O*-*β*-D-glucopyranosylformoside (**9**) (*23*), and one phenylethanoid, salidroside (**10**) (*24*, *25*) (Figure 1). The chemical structures were established by spectroscopic methods and compared with literature and the ^{13}C NMR data (Table II & III).

Table II. ^{13}C NMR Data for Compounds 1-3 and 6-10 (CD_3OD)

	1	*2*	*3*	*6*	*7*	*8*	*9*	*10*
No.	δ_C	δ_C	δ_C	δ_C	δ_C	δ_C	δ_C	δ_C
1	94.8	94.7	95.0	95.0	95.3	95.0	95.2	71.3
3	155.2	155.2	155.1	155.1	154.5	155.1	155.3	36.2
4	109.3	109.3	109.2	109.2	110.7	109.2	109.2	129.3
5	31.9	32.0	31.6	31.7	32.9	31.8	31.7	130.7
6	41.1	41.3	41.2	41.1	45.0	40.9	41.0	116.7
7	173.7	173.4	172.9	173.2	179.6	173.5	171.6	158.2
8	124.7	124.8	124.9	124.8	123.8	124.8	125.1	116.7
9	130.4	130.1	130.2	129.9	131.1	130.3	130.5	
10	13.6	13.6	13.7	13.5	13.7	13.5	13.8	
11	168.7	168.7	168.5	168.6	169.0	168.6	168.6	
11-OCH_3	52.3	51.9	52.0	51.9	51.7	51.9	52.0	
7-OCH_3	51.9					52.1		
1′	100.6	100.4	100.6	100.7	100.7	100.8	100.9	104.1
2′	77.6	77.5	74.5	74.6	74.8	74.7	74.7	74.8
3′	77.8	77.8	78.1	78.2	78.4	78.3	77.9	77.7
4′	71.6	71.5	71.3	71.3	71.5	71.3	71.3	72.3
5′	75.2	75.1	77.7	77.7	77.8	77.8	78.3	77.6
6′	70.1	70.1	62.5	62.6	62.7	62.6	62.5	62.4
1″	105.2	105.2	104.2				104.3	
2″	74.7	74.7	74.8				75.0	
3″	77.7	77.6	77.7				78.0	
4″	71.5	71.4	71.2				71.5	
5″	77.8	77.6	74.9				77.8	
6″	62.7	62.7	64.9				62.7	
1‴		67.0	130.5	66.8			71.4	
2‴		35.2	130.8	35.0			36.5	
3‴		130.3	116.0	130.2			137.9	
4‴		131.1	156.6	130.9			131.0	
5″		116.4	116.0	116.2			122.5	
6‴		157.0	130.8	156.9			150.5	

Continued on next page.

Table II. (Continued). ^{13}C NMR Data for Compounds 1-3 and 6-10 (CD_3OD)

	1	*2*	*3*	*6*	*7*	*8*	*9*	*10*
No.	δ_C	δ_C	δ_C	δ_C	δ_C	δ_C	δ_C	δ_C
7‴		116.4	36.2	116.2			122.5	
8‴		131.1	72.1	130.9			131.0	

Table III. ^{13}C NMR Data for Compounds 4 and 5 (CD_3OD)

	4	*5*		*4*	*5*
No.	δ_C	δ_C	*No.*	δ_C	δ_C
A			1‴	71.5	
1	95.2	95.2	2‴	36.4	
3	155.2	155.2	3‴	137.8	
4	109.2	109.2	4‴	130.9	
5	31.6	31.6	5‴	122.4	
6	41.1	41.1	6‴	150.4	
7	172.9	172.9	7‴	122.4	
8	125.1	125.1	8‴	130.9	
9	130.4	130.4	**B**		
10	13.8	13.8	1	95.0	95.0
11	168.5	168.5	3	155.1	155.1
OCH_3	52.0	52.0	4	109.1	109.1
1′	100.8	100.8	5	31.6	31.6
2′	74.6	74.6	6	40.9	40.9
3′	78.7	78.7	7	171.5	171.5
4′	71.3	71.3	8	124.9	124.9
5′	77.7	77.7	9	130.3	130.3
6′	62.6	62.6	10	13.7	13.7
1″	104.3		11	168.5	168.5
2″	74.8		OCH_3	52.0	52.0
3″	77.7		1′	100.7	100.7
4″	71.4		2′	74.6	74.6
5″	75.0		3′	78.7	78.7

Continued on next page.

Table III. (Continued). ^{13}C NMR Data for Compounds 4 and 5 (CD_3OD)

	4	*5*		*4*	*5*
No.	δ_C	δ_C	*No.*	δ_C	δ_C
6″	64.9		4′	71.2	71.2
			5′	77.7	77.7
			6′	62.5	62.5

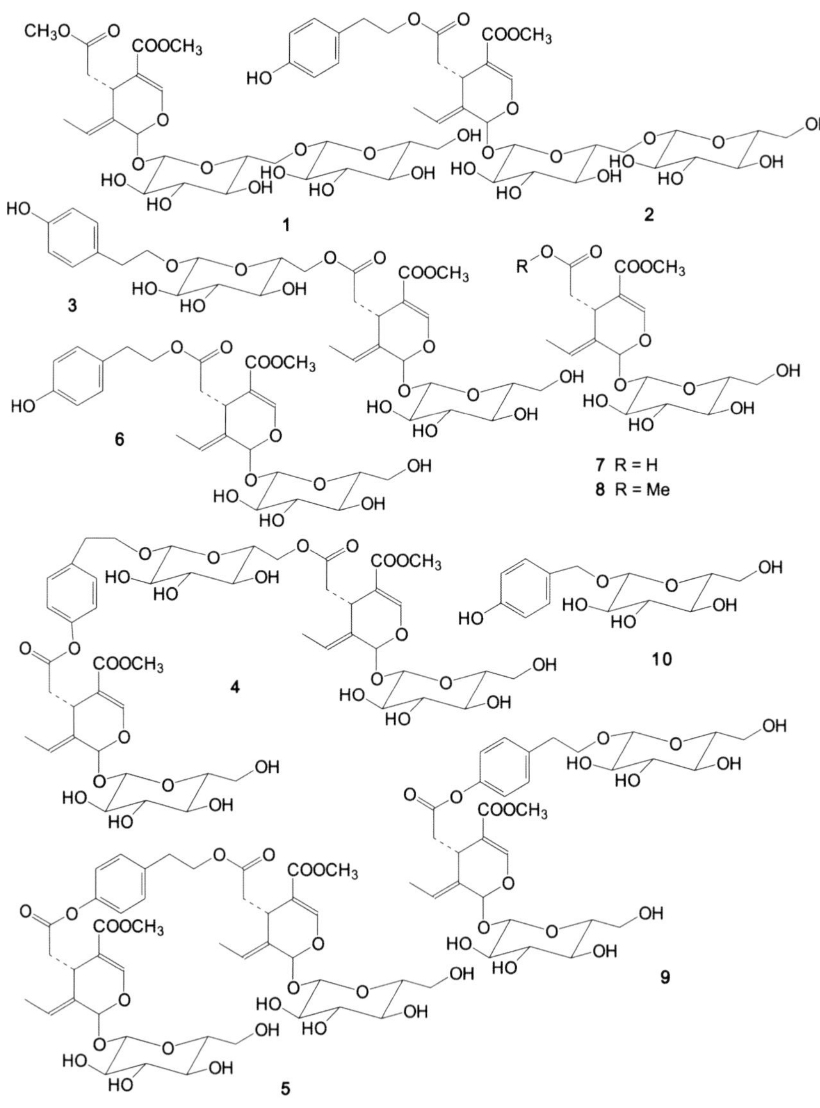

Figure 1. The compounds structure.

A calibration curve was constructed for nuzhenide (**3**), and GI3 (**4**) (Table I). Based on HPLC analysis (Fig. 2), the compounds content in the extract 11 or FE994702 is listed as **2**, 0.41%; **3**, 11.42%; **4**, 6.15%; **5**, 0.63%; **7**, 0.19%; **9**, 1.35%; **10**, 0.20%; and total content of the identified compounds, 17.57%. The compounds were picked up and calculated by UV spectrum analogue comparing with that of the isolated compounds.

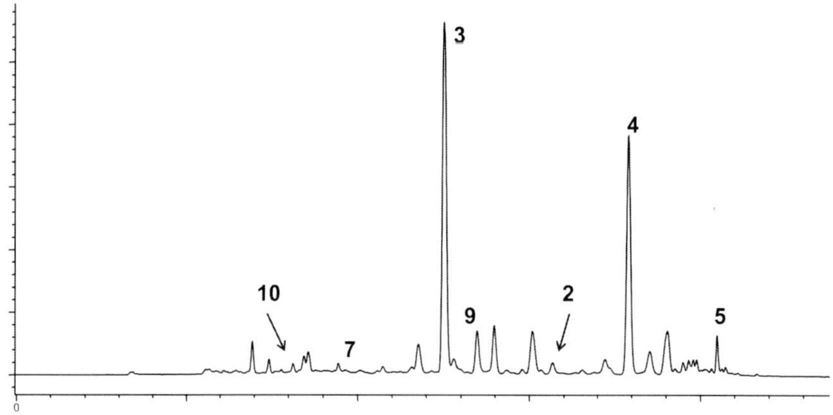

*Figure 2. HPLC chromatogram of the hot water extract **11** from the seeds of F. excelsior.*

The isolated compounds and the extract **1-11** were tested with regard to their effect on nitrite production in LPS-activated macrophages for anti-inflammatory screening. When RAW264.7 cells were treated with test compounds at 40 μg/mL and LPS (100 ng/mL), respectively, the potency of the inhibitory effects on nitrite production showed sequence as **5** > **8** > **9** > **10** > **4** > **3** > **2** > **1** > **6** > **7** (Table IV). Among them, compound **5**, **8**, **9**, and **10** were the strong inhibitors to nitrite production in macrophages and the extract **11** was active on this bioassay. The cytotxicity of compounds and the extract **1-11** was evaluated *in vitro* against HL-60, Hep G2, and COLO 205 cell lines. These cell lines were treated with different concentrations (5-100 μM) of selected compounds for 24 h, and the viability of the cells was determined by MTT assay. As shown in Table V, Compound **4** was slightly effective as an anti-proliferative agent in HL-60 cells with IC_{50} of 82.0 μM. However, there was no cytotoxicity observed for any compounds except for **4** and the extract in these cells. Therefore, the current study confirms that the secoiridoids are the major components responsible for the activities of anti-inflammatory and there are no cytotoxicities for the isolated secoiridoids. Thus, the seeds and the extract of *F. excelsior* are largely safe for normal use.

Table IV. Effect of 1-11 on LPS-Induced Nitrite Production in RAW 264.7 Macrophages

	1	*2*	*3*	*4*	*5*	*6*
Control	0.0 ± 0.3	0.0 ± 0.3	0.0 ± 0.3	0.0 ± 0.3	0.0 ± 0.3	0.0 ± 0.3
LPS	22.8 ± 1.5	22.8 ± 1.5	22.8 ± 1.5	22.8 ± 1.5	22.8 ± 1.5	22.8 ± 1.5
20 μg/mL	23.2 ± 1.5	22.3 ± 0.8	19.4 ± 1.2	20.2 ± 2.3	13.1 ± 0.5	19.6 ± 0.8
40 μg/mL	19.7 ± 0.1	19.1 ± 1.1	17.0 ± 2.0	16.6 ± 1.3	10.3 ± 1.1	20.0 ± 1.3
	7	*8*	*9*	*10*	*11*	
Control	0.0 ± 0.3	0.0 ± 0.3	0.0 ± 0.3	0.0 ± 0.3	0.0 ± 0.3	
LPS	22.8 ± 1.5	22.8 ± 1.5	22.8 ± 1.5	22.8 ± 1.5	22.8 ± 1.5	
20 μg/mL	20.2 ±0.3	18.0 ± 0.6	16.4 ± 0.8	18.0 ± 0.1	16.4 ± 0.2	
40 μg/mL	19.9 ± 0.1	15.9 ±0.3	16.0 ± 0.7	15.6 ± 0.8	16.8 ± 0.8	

Table V. Effect of 1-11 on the Growth of Various Human Cancer Cells

Compound		*Cell line*	
IC_{50} (μM)	*HL-60*	*Hep G2*	*COLO 205*
1	> 100	> 100	> 100
2	> 100	> 100	> 100
3	> 100	> 100	> 100
4	82.0 ± 3.6 [b]	> 100	> 100
5	> 100	> 100	> 100
6	> 100	> 100	> 100
7	> 100	> 100	> 100
8	> 100	> 100	> 100
9	> 100	> 100	> 100
10	> 100	> 100	> 100
11	> 100	> 100	> 100
Doxorubicin[a]	5.0 ± 0.6	10.0 ± 4.3	11.9 ± 4.9

[a] Positive control. [b] Each experiment was independently performed three times and expressed as mean ± SE.

References

1. Hemmer, W.; Focke, M.; Wantke, F.; Götz, M.; Jarisch, R.; Jäger, S. *Allergy* **2000**, *55*, 923–30.
2. Eddouks, M.; Maghrani, M.; Lemhadri, A.; Ouahidi, M. L.; Jouad, H. *J. Ethnopharmacol.* **2002**, *82*, 97–103.
3. Wallander, E.; Albert, V. A. *Am. J. Bot.* **2000**, *87*, 1827–1841.
4. Kostova, I.; Iossifova, T. *Fitoterapia* **2007**, *78*, 85–106.
5. Çalis, I.; Hosny, M.; Khalifa, T.; Nishibe, S. *Phytochemistry* **1993**, *33*, 1453–1456.
6. Eddouks, M.; Maghrani, M.; Zeggwagh, N. A.; Haloui, M.; Michel, J. B. *J. Ethnopharmacol.* **2005**, *99*, 49–54.
7. Maghrani, M.; Zeggwagh, N. A.; Lemhadri, A.; El Amraoui, M.; Michel, J. B.; Eddouks, M. *J. Ethnopharmacol.* **2004**, *91*, 309–316.
8. Kruedener, S.; Schneider, W.; Elsner, E. F. *Drug Res.* **1995**, *45*, 169–171.
9. Eddouks, M.; Maghrani, M. *J. Ethnopharmacol.* **2004**, *94*, 149–154.
10. Iossifova, T.; Kostova, I.; Evstatieva, L. N. *Biochem. Syst. Ecol.* **1997**, *25*, 271–274.
11. Egan, P.; Middleton, P.; Shoeb, M.; Byres, M.; Kumarasamy, Y.; Middleton, M.; Nahar, L.; Delazar, A.; Sarker, S. D. *Biochem. Syst. Ecol.* **2004**, *32*, 1069–1071.
12. Damtoft, S.; Franzyk, H.; Jensen, S. R. *Phytochemistry* **1992**, *31*, 4197–4201.
13. Bai, N. S.; He, K.; Ibarra, A.; Bily, A.; Roller, M.; Chen, X.; Ruhl, R. *J. Nat. Prod.* **2010**, *73*, 2–6.
14. Pacher, P.; Beckman, J. S.; Liaudet, L. Nitric oxide and peroxynitrite in health and disease. *Physiol. Rev.* **2007**, *87*, 315–424.
15. Mantovani, A.; Allavena, P.; Sica, A.; Balkwill, F. Cancer-related inflammation. *Nature* **2008**, *454*, 436–444.
16. Mosmann, J. Rapid colorimetric assay for cellular growth and survival: application to proliferation and cytotoxicity assays. *J. Immunol. Methods.* **1983**, *65*, 55–63.
17. Servili, M.; Baldioli, M.; Selvaggini, R.; Macchioni, A.; Montedoro, G. *J. Agric. Food Chem.* **1999**, *47*, 12–18.
18. LaLonde, R. T.; Wong, C.; Tsai, A. I. M. *J. Am. Chem. Soc.* **1976**, *98*, 3007–3013.
19. Tanahashi, T.; Takenaka, Y.; Nagakura, N. *Phytochemistry* **1996**, *41*, 1341–1345.
20. Takenaka, Y.; Tanahashi, T.; Shintaku, M.; Sakai, T.; Nagakura, N.; Parida *Phytochemistry* **2000**, *55*, 275–284.
21. Sugiyama, M.; Machida, K.; Matsuda, N.; Kikuchi, M. *Phytochemistry* **1993**, *34*, 1169–1170.
22. Shen, Y. C.; Lin, S. L.; Chein, C. C. *Phytochemistry* **1996**, *42*, 1629–1631.
23. Tanahashi, T.; Watanabe, H.; Itoh, A.; Nagakura, N.; Inoue, K.; Ono, M.; Fujita, T.; Morita, M.; Chen, C. C. *Phytochemistry* **1992**, *32*, 133–136.
24. Tolonen, A.; Pakonen, M.; Hohtola, A.; Jalonen, J. *Chem. Pharm. Bull. (Tokyo)* **2003**, *51*, 467–470.
25. Nishimura, H.; Nonaka, G.; Nishioka, I. *Chem. Pharm. Bull. (Tokyo)* **1984**, *32*, 1735–1740.

Chapter 10

Nutritional Value of Fonio (*Digitaria exilis*) from Senegal

Adolfina R. Koroch,[*,1] H. Rodolfo Juliani,[2] and James E. Simon[2]

[1]Science Department, City University of New York, Borough of Manhattan Community College, 199 Chambers Street, New York, New York 10007
[2]New Use Agriculture and Natural Plant Products Program, School of Biological and Natural Resources, and the New Jersey Agricultural Experiment Station (NJAES), Rutgers, The State University of New Jersey, 59 Dudley Road, New Brunswick, New Jersey 08901-8520
[*]E-mail: akoroch@bmcc.cuny.edu.

As food production is facing many challenges due to the increased population growth and climate change, indigenous and new plants and plant products that can provide nutritionally rich foods can contribute to food security, income generating activity and the preservation of genetic materials that otherwise could be marginalized and/or displaced by other crop commodities. Fonio (*Digitaria exilis*) is a West African annual millet characterized by tiny seeds that is prized by local communities and enjoyed regionally. The grass is well adapted to hot, dry climates and poor soils, and thus is able to grow in areas where many other cereals are not suited. In West Africa, fonio is used in traditional medicine but more as a source offood, which is used to prepare porridges, flours, alcoholic and non-alcoholic beverages. The objective of this work is to review the literature on fonio research and provide initial nutritional information of fonio from Senegal.

Introduction: Biology, Cultivation, and Uses

In recent years, considerable attention has been given to crops that not only are healthy and inexpensive to produce but also that provides income-generating opportunities for local communities and small-scale/resource limited farmers.

Fonio (*Digitaria exilis* Staf, Poaceae or Gramineae) is also known as white fonio, acha, hungry rice, fonyo, fundi millet, fundi and hungry millet. The genus *Digitaria* originated in West Africa and comprises about 230 species in tropical, subtropical and warm-temperate regions (*1*, *2*).

For more than thousands of years West Africans have been cultivating and consuming fonio, a source of food for several millions of people. It has been estimated that about 300,000 ha are cultivated each year (*1*, *3*). Fonio is widely cultivated and used from Senegal to Chad, in Fouta Djallon (Guinea), the Bauchi-Plateau in Nigeria and in northwest Benin (*2*, *4*).

Fonio is an annual grass that can reach up to 80 cm height. Its inflorescence is a finger shaped panicle with racemes of 5-12 cm length. The fruits are caryopsis (grain), tiny about 0.5-1 mm diameter, 0.75 – 2 mm length (*5*). Fonio *species* can tolerate a wide range of extremes conditions. For example, it is adapted to hot and dry climates (C4 annual herbaceous plants) and tolerates well poor, shallow, sandy, rocky or acidic soils which are unsuitable for other cereals (*2*, *3*). It is highly adapted to drought and its cultivation is concentrated in areas with an average annual rainfall between 900 and 1000 mm (*2*). Moreover, fonio is not attacked by nematodes, even in soils where other plants are infected with *Meloidogyne sp.* No pathogens have yet been reported to attack fonio species (*6*).

There is a large diversity within the species *D. exilis* and based on the morphology several varieties have been identified (*2*, *7*). For instance, certain fonio varieties have short life cycle and can mature very fast producing grains 6 to 8 weeks after planted, while, other fonio species require over 23 weeks to mature. The growing of different varieties of fonio with different maturation time increases the chance of harvesting enough food for longer periods of time, particularly important in those areas where the growing conditions are unreliable (*3*).

Within the *Digitaria* genus, fonio millets (small grain cereals) are the most important economic crop. One of the advantages of this small grain is the minimal processing, high nutritional value and less labor intensive (*8*). Fonio has been used for different purposes as religious and cultural crop, in traditional medicine and as a source of food. Fonio has an important religious and cultural value in weddings, baptism of newborn child (*1*). In traditional medicine fonio has been used for meteorism, constipation, as a diuretic, treat or eliminate blood clots, treat diarrhea, loss of appetite, dysentery, chickenpox, stomach ache and asthma (*1*, *9*). In traditional foods, this grainis consumed as thin and thick porridges steam cooked with fish, meat legumes or vegetable and used to prepare snacks, alcoholic and non-alcoholic beverages (*10*, *11*).

More recently, research has been conducted to identify local alternatives to wheat flour, while seeking a highly nutritious substitute. Fonio grains can be ground into flour to produce flour bread (*12*), biscuits (*13*), sourdough bread (*14*, *15*) and as a good substitute for semolina (*3*). Biscuits made of fonio supplemented with a mixture otherflours Fonio (wheat- soybean or cowpea) have high organoleptic characteristics, good digestibility and could serve as a vehicle to increase the intake of proteins and calories (*16*, *17*). This small grain could also be used in multi-grain breads, as an ingredient in a mixed seed product.

Fonio grains are digested efficiently by different ruminant livestock (*1*, *3*) and because of the high amounts of phosphorus, potassium and trace minerals fonio grains are reliable materials for formulating poultry and pig feed to meet the nutrient specifications of such feed (*18*).

According to the dietary guidance around the world the inclusion of whole grains in the diet is recommended because of their associations with increased health and reduced risk of chronic diseases. The consumption of whole grains has been linked to reduced risk of obesity, cardiovascular disease, hypertension, stroke and decreased risk of cancers of the upper gut (*19*). Currently, nutrition and lifestyle are main components to diabetes prevention, thus, choosing foods with low glycemic index has a small but clinically useful effect on medium term glycemic control in patients with diabetes (*20*). Fonio has such a low glycemic index and the recommended serving size would need to be carefully adhered to be effective (*21*).

The objective of this work is to review the uses and nutritional value of fonio and report the proximal and elemental analysis of fonio grains from Senegal to identify new uses and applications.

Materials and Methods

Fonio samples from Senegal (prepared and given by Peter Trenchard, USAID-Senegal to James E. Simon, Rutgers) were brought back into the USA, declared to APHIS and US Customs upon entry and then brought to Rutgers where the seeds were cleaned and visually inspected to remove any non-fonio seeds and debris. A sub-sample was prepared and submitted to the Agricultural Analytical Services Lab (Pennsylvania State University Soil and Plant Testing Laboratory) to determine the concentration 11 mineral elements including: phosphorus, potassium, calcium, magnesium, manganese, iron, copper, boron, aluminum, zinc and sodium using the dry ash method (*22*). The results were expressed for the macronutrients (phosphorus, potassium, calcium, magnesium) as g of element per 100 grams on a dry weight basis (g element/100 g DW). For the micronutrient (manganese, iron, copper, boron, aluminum, zinc and sodium), the results were expressed as mg of element per 100 grams on a dry weight basis (mg element/100 g DW). The analysis of moisture, ashes, carbohydrates, crude fat, protein, fiber and, calories were conducted at Miller Laboratories (1675 West 2750 South, Ogden UT 84401) following AOAC procedures (*23*).

Results and Discussions

The moisture content of fonio was 9.5% (Table 1).This result agrees with previous findings of low seed moisture. The low moisture suggests that fonio loses water during storage and explains the longer shelf life of this grain (*11*). Moreover, because of the loss of moisture during storage, is less susceptible to attack by pests which could be particularly advantageous in tropical and subtropical regions (*24*).

Total mineral content (ashes) was 4.3%, suggesting that the total mineral content is low. Fonio does not provide a significant amount of fats and carbohydrates though appears to be a very good source of proteins and excellent source of fiber (Table 1). Fonio showed higher levels of protein as compared with other grains (wheat, 13.6% and rye, 8%), the amount of fiber is much higher in fonio as compared with corn (7%), wheat (2%) and rye (1.6%), the mineral content of fonio was higher (6%) as compared with wheat and rye (less than 2%) (Table 1). Thus, fonio seems to be a rich source of protein and fiber, with low carbohydrate contents, 100g of fonio would provide a total of 347 calories, 21.5g are proteins, and 59.4g of fibers. The results showed that fonio is high in proteins, for example, 100 g could provide the daily dietary allowances for children 1-8 years.

Starch obtained from fonio has been reported having similar structure and physicochemical properties to other conventional grains (*8, 25*). Fonio grains have proteins similar to white rice (*26*) but they are rich in the essential amino acids methionine and cysteine. High protein and fiber fonio bread could serve as a good alternative to wheat bread not only for diabetic patients but also for patients that are allergic to gluten (*12*).

Table 1. Proximate Analysis of Fonio (*Digitaria exilis*) from Senegal

	Fonio	*Corn(27, 28)*	*Wheat(29)*	*Rye(29)*
Moisture (%, m/m)	9.5	10.5(*27*)	-	
Ashes (%, m/m)	4.3	1.4(*28*)	1.5	1.72
Carbohydrates (%, m/m)	0.6	64(*27*)*	63.0	69
Crude fat (%, m/m)	0.34	4.3(*28*)	2.5	2.0
Protein (%, m/m)	21.5	7.7(*27*), 9.5(*28*)	13.6	7.98
Fiber (%, m/m)	59.4	9.5(*2*)	2.15	1.56
Calories (in 100g)	347	-	328	

* as total starch.

Because of the small size, fonio is minimally processed and thus consumed as a whole grain, which potentially could be an important source of nutraceuticals with different (more) heath benefits. Larger-scale processing that would be needed in any scale-up operation for fonio to become an internationally traded food grain may be challengingbecause of the small seed size and the difficulty in cleaning and separating seeds from the same/similar sized rocks and pebbles (Peter Trenchard, personal commun.). No reports have been published which answer this constraint.

Table 2. Elemental Composition of Fonio Seeds (*Digitaria exilis*)

Element	*Fonio*	*Corn(30)*	*Wheat(30)*	*Rye(30)*
Phosphorous (%)	0.06	0.31	0.4	0.38
Potassium (%)	0.04	0.33	0.58	0.52
Calcium (%)	0.01	0.03	0.06	0.07
Magnesium (%)	0.02	0.14	0.18	0.13
Manganese (mg/100 g)	0.3	0.6	5.5	7.5
Iron (mg/100 g)	2.1	2	6	9
Copper (mg/100 g)	0.1	0.2	0.8	0.9
Boro (mg/100 g)	0.1	-	-	-
Aluminum (mg/100 g)	1.7	-	-	-
Zinc (mg/100 g)	1.6	-	-	-
Sodium (mg/100 g)	1.4	-	-	-

In addition, compared to other African grains, fonio was found to contain high levels of calcium and phosphorous (*24*). Thus, this grain has been considered an excellent dietary complement to legumes, which are widely eaten in Africa (*3*, *31*). However, the present study showed that the elemental composition of macroelements were low in Fonio (<0.06%) for the macronutrients phosphorous, potassium, calcium, magnesium.While other cereals showed higher of these elements (0.13-0.58%) (Table 2), fonio can provide modest though yet significant amounts of iron and zinc to the diet. One hundred grams would provide 26% and 12% of daily dietary allowance of iron for men and women, respectively. The same amount would provide 18% of zinc for adults.

Future Prospects

Because so little attention has been given to fonio production and, processing technologies crop yields are low and the production systems far from being optimized. Fonio as a 'new crop' despite its long and rich history presents many challenges including lodging, lower yields than other cereals and grain shattering. In short, it has many characteristics of a nondomesticated crop. Thus, a major challenge is to develop improved varieties with high grain yields with high nutritional value. The cultivation of fonio could be advantageous in areas which are not suitable for other grains. Given that the most vulnerable and economically limited farming communities are often located areas with in marginal agricultural soils, fonio could be an excellent grain crop to grow given it is adapted to poorer soil conditions. With an increasing world population, fonio could play an important role to provide food security given its nutritional value including being a rich source of protein. The cultivation and incorporation of fonio in the diet could also improve the economic status of West Africans farmers by providing additional income generating activities and an additional seed crop. Processing of the harvested seed and developing effective seed cleaning systems to meet food company requirements are areas that also warrant attention.

Acknowledgments

We thank Peter Trenchard, USAID-Senegal for providing the fonio seeds to us to analysis and in serving a crop champion for this undervalued grain crop in Senegal. We also thank ASNAPP-Senegal, Malainy Diatta, Babou Diou for their enthusiasm about fonio and in introducing us to this unique indigenous crop that has so many uses, and positive texture and flavor. This work was partially supported by the USAID initiative, Partnership in Food Industry Development for Natural Products (www.pfidnp.org), and the New Use Agriculture and Natural Plant Products Program.

References

1. Adoukonou-Sagbadja, H.; Dansi, A.; Vodouhe, R.; Akpagana, K. *Biodiversity Conserv.* **2006**, *15*, 2379–2395.
2. Vodouhe, R. S.; Dako, G. E. A.; Dansi, A.; Adoukonou-Sagbadja H. *Fonio: A treasure for West Africa*; Briefing by the International Plant Genetic Resources Institute (West and Central Africa Sub-Office) and colleague organisations; 2007, www.underutilized-species.org/documents/PUBLICATIONS/rv.pdf.
3. Vietmeyer, N.; Borlaugh, N.; Axtell, J.; Burton, G.; Harlan, J.; Rachie, K. Fonio (Acha). In *Lost crops of Africa*; National Academy Press: Washington DC, 1996; Vol 1, pp 59−76.
4. Garí, J. A. *International workshop on fonio, food security and livelihood among the rural poor in West Africa*; IPGRI's Programme for Neglected and Underutilised Species; IPGRI/IFAD: Bamako, Mali, 2001,
5. Vodouhe, S. R.; Achigan-Dako, E. G. Digitaria exilis (Kippist) Stapf. In *Plant Resources of Tropical Africa*; Brink, M., Belay, G. , Eds.; PROTAFoundation and CTA Wageningen; Backhuys Publishers: The Netherlands, 2006;p Vol 1, pp 59−63.
6. Morales-Payán, J. P; Ortiz; J. R.; Cicero, J.; Taveras, F. *Digitaria exilis* as a Crop in the Dominican Republic. Supplement to *Trends in new crops and new uses*; Janick, J., Whipkey, A., Eds.; ASHS Press: Alexandria, VA, 2002.
7. Dansi, A; Adoukonou-Sagbadja, H; Vodouhe, R. *Genet. Resour. Crop Evol.* **2010**, *57*, 827–839.
8. Irving, D. W; Jideani, I. A. *Cereal Chem.* **1997**, *74*, 224–228.
9. Neuwinger, H. D. *African Traditional Medicine. A dictionary of plant use and applications*; MedPharm Scientific Publisher: Stuttgart, Germany, 2000; pp 556.
10. Fanou-Fogny, N.; Koreissi; Romain, Y.; Dossa, A. M.; Brouwer, I. D. *Afr. J. Food Agric. Nutr. Dev.* **2009**, *9*, 1927–1944.

11. Chukwu, O.; Abdul-kadir, A. J. *J. Food Technol.* **2008**, *6*, 214–216.
12. Jideani, V. A.; Salami, R. A.; Jideam, I. A. *Br. Food J.* **2008**, *110*, 271–282.
13. Ayo, J. A; Ikuomola, D. S; Esan, Y. O; Onuoha, O. G; Ayo, V. A; Ekele, V. *Cont. J. Food Sci. Technol.* **2010**, *4*, 7–13.
14. Coda, R.; Di Cagno, R.; Edema, M. O.; Nionelli, L; Gobbetti, M. *Food Microbiol.* **2010**, *27*, 1043–1050.
15. Coda, R.; Di Cagno, R.; Rizzello, C. G.; Nionelli, L; Edema, M. O.; Gobbetti, M. *J. Food Sci.* **2011**, *76*, M329–M335.
16. McWatters, K. H.; Ouedraogo, J. B.; Resurreccion, A. V. A.; Hung, Y. C.; Phillips, R. D. *Int. J. Food Sci. Technol.* **2003**, *38*, 403–410.
17. Ayo, J. A.; Ayo, V. A.; Nkama, I.; Adewori, R. *Niger. Food J.* **2007**, *25*, 77–89.
18. Clottey, V. A.; Avornyo, F.; Addo-Kwafo, A.; Agyare, W. A. *Res. Rural Dev.* **2006**, *18*, Article #95.
19. Jones, J. M.; Engleson, J. Whole Grains: Benefits and Challenges. *Annu. Rev. Food Sci. Technol.* **2010**, *1*, 19–40.
20. Brand-Miller, J.; Hayne, S.; Petocz, P; Colagiuri, S. *Diabetes Care* **2003**, *26*, 2261–2267.
21. Alegbejo, JO; Ameh, DA; Ogala, WN; Ibrahim, S. *J. Pure Appl. Microbiol.* **2011**, *5*, 117–122.
22. Miller, R. O. *Handbook and Reference Methods for Plant Analysis*; CRC Press: New York, 1998.
23. Helrich, K. C. *Official methods of Analysis of the AOAC*; Association of Official Analytical Chemists Inc.: Arlington, VA, 1990.
24. Jideani, I. A. *Plant Foods Hum. Nutr.* **1999**, *54*, 363–374.
25. Jideani, I. A.; Takeda, Y.; Hizukuri, S. *Cereal Chem.* **1996**, *73*, 677–685.
26. Temple, V. J.; Bassa, J. D. *J. Sci Food Agric.* **1991**, *56*, 561–563.
27. D'Alfonso, TH. *Global Corn Quality Variability*; Finnfeeds International Limited: Marlborough, Wiltshire, England, 2002.
28. White, P. J.; Johnson, L. A. *Corn: Chemistry and Technology*; American Association of Cereal Chemists: St. Paul, MN, 2003.
29. McCance, R. A.; Widdowson, E. M.; Moran, T.; Pringle, W. J. S.; Macrae, T. F. *Biochem. J.* **1945**, *39*, 214–222.
30. Hoseney, R. C. *Principles of Cereal Science and Technology*; American Association of Cereal Chemists: St. Paul. MN, 1994.
31. deLumen, B. O.; Thompson, S.; Odegard, W. J. *J. Agric. Food Chem.* **1993**, *41*, 1045–1047.

Quality Control of African Natural Plant Products

Chapter 11

African Nightshades and African Eggplants: Taxonomy, Crop Management, Utilization, and Phytonutrients

Ray-Yu Yang[*,1] **and Chris Ojiewo**[2]

[1]**Nutritionist, AVRDC-The World Vegetable Center, Tainan, Taiwan, ROC**
[2]**Vegetable Breeder, AVRDC-The World Vegetable Center, Regional Center for Africa, Arusha, Tanzania**
[*]**E-mail: ray-yu.yang@worldveg.org.**

The Solanaceae plant family contains many food crops important to agriculture, food security, human nutrition and health. These include globally-consumed peppers (*Capsicum* sp), potato (*Solanum tuberosum*), tomato (*S. lycopersicum*), eggplant (*S. melongena*) and regionally consumed, such as, African nightshades (*Solanum* section *Solanum, S. scabrum* and *S. villosum*) and African eggplants (*S. aethiopicum, S. macropcarpon* and *S. anguivi*). The taxonomy of Solanum is complex, and the "deadly nightshade" reputation of *S. nigrum* and other Solanum weeds or medicinal plants has extended to the edible African nightshades and African eggplants; this has confused people unfamiliar with these plants. The attention of scientists and policymakers has been focused on potato, tomato and peppers. Information on other regionally important *Solanum* crops is lacking. This paper reviews the taxonomy, crop management and the uses of African *Solanum* food crops as well as their nutritional values and glycoalkaloid contents.

Introduction

Vegetable species are diverse in tropical Africa (*1*). Of the 275 species vegetable crops grown in Africa, 207 are indigenous to Africa, 45 were introduced long ago but became adapted, and 23 vegetables are recently introduced (*1*). African indigenous vegetables (AIVs) play important roles in nutrition, food security, food diversity, rural development and sustainable land care (*2*). They may serve as income sources and can be marketed or traded locally, regionally and internationally (*3*). Many traditional vegetables are not only consumed as food but are used for their preventive and curative medicinal properties as well (*4*). Popular AIVs in urban markets and rural settings include okra, amaranth, spiderplant, jute mallow, celosia, jew's mellow, roselle, cowpea leaf, African nightshade, and African eggplant (garden egg and gboma) (*1*, *4*). These vegetables have fed Africans for centuries. Many of them are exceptionally nutritious (*5*) yet most have not received sufficient scientific attention despite their significance for food and nutritional security in Africa.

African nightshade (*Solanum* spp) and African eggplant (*S. aethiopicum, S. macropcarpon* and *S. anguivi*) are important traditional leafy and fruit crops in many parts of East, West, Central and Southern Africa (*3*, *4*, *6*). African nightshade and African eggplant are members of genus Solanum which includes economically important crops such as potato (*S. tuberosum*), tomato (*S. lycopersicum)* and eggplant (*S. melongena*). Some species in the nightshade group such as *S. nigrum* are high in glycoalkaloids and are not suitable for human consumption; however, the "poisonous" label has been blanketedly applied to many edible nightshade species that are important foods in tropical Africa and Asia, thus reinforcing their neglect and underutilization (*1*, *7*). African nightshades and African eggplants have much potential for expanded production and use in agriculture and human nutrition. In addition to their value in human nutrition (*5*), the leaves and fruits can be used as a natural dye (*8*, *9*); their resistance to late blight disease could be exploited as a genetic source for varietal improvement of other solanaceous vegetables (*10*, *11*).

Taxonomy

The Solanaceae family comprises 90 genera and about 2300 species. The genus Solanum is the largest and most diverse, containing many essential food plants such as potato (*S. tuberosum* L.), eggplant (*S. melongena* L.) and naranjilla (*S. quitoense* Lam.); horticulturally useful plants such as winter cherry (*S. pseudocapsicum* L.) and jasmine nightshade (*S. jasminoides* Paxt.); medicinal plants such as bittersweet (*S. dulcamara* L.) and *S. viarum* Dun., both used as sources of corticosteroids. The main vegetable Solanum species cultivated in Africa include African eggplant (*S. aethiopicum* L, *S. macrocarpon* (Gboma) and *S. anguivi*) mainly cultivated for fruit and African nightshade (the *Solanum* section *Solanum*, with more than 30 species), mainly cultivated for their leaves.

African eggplant is also known as garden eggs, mock tomato, and ngogweor nyanyachungu. The crop thrives in the warm, dry environments found in the savannah belt of West and East Africa. Optimal temperature ranges for the crop are 23–35 °C (day) and 18–25 °C (night) and it grow on a wide range of well-drained soils with an optimal soil pH range between 5.5 and 6.8.

Figure 1. Major types of African eggplants.

Solanum aethiopicum L

This phenotypically diverse species is grown for its fruit and leaves. Stems and leaves can be hairy or glabrous leaves, the flowers are bisexual and mostly self-pollinate; Fruits are produced singly or in groups from trusses or short cymes, depending on subspecies and varieties (*1*). The fruits vary in color (green, white, striped, multicolored), shape (round to long, smooth, grooved or ribbed) and size, (small to very big). Taste ranges from bitter to sweet, depending on saponin content, and fruit may be consumed cooked or raw. At full maturity, fruit turn red to orange due to their carotene content. The *S. aethiopicum* is divided into 4 groups (Figure 1):

The Aculeatum Group

Leaves and fruits of this group are inedible and members of this group are mostly used as ornamentals. They have hairy and prickly leaves and stems. The fruits are of various sizes, shapes and colors. This group is thought to have resulted from natural crosses between *S. anguivi* Lam and the *Kumba* group of *S. aethiopicum* (*14*). Members of this group are possible sources of genes for disease resistance breeding (*15*).

The Gilo Group

This group is very common in the humid tropics. It has hairy, inedible leaves, variable fruit shape (round, elongated, egg-shaped or spindly, ribbed or smooth), color (dark and light green, white or striped) and size (from a few to over 100 grams). This group is thought to have resulted from evolution of *S. anguivi* (*14*), and is sometimes (wrongly) referred to as *S. gilo*.

The Kumba Group

This group is most commonly grown in the arid areas of tropical Africa. Also called red African eggplant, orange African eggplant, scarlet African eggplant, or mini pumpkin tree (*12*). The plants are glabrous and have large leaves and medium to big ribbed fruits (5-10 cm in diameter) that are eaten cooked or raw. The hairless leaves are sometimes consumed as green vegetables. The Kumba group probably evolved from the Gilo group (*14*).

The Shum Group

This group is most generally grown for its glabrous leaves that are eaten as a green vegetable; its very small, slightly flattened, round or elongated fruits, though edible, are rarely consumed. This group probably evolved from the Kumba group through genome reduction (*14*).

***Solanum anguivi* Lam**

S. anguivi plants have small fruits which mainly differ in shape, color and taste although most are bitter. The pea-size bitter fruits are either collected from the wild or grown and consumed fresh or dried, mainly for medicinal purposes.

Solanum macrocarpon L

S. macrocarpon can easily be distinguished from *S. aethiopicum* by its large hairless leaves and stems and its long calyx that persists after formation of its smooth green or yellow fruits. The leafy varieties are common throughout West and Central Africa, while the fruit types are mainly restricted to the humid coastal areas of West Africa. Some varieties have both edible leaves and fruits. This species is a popular traditional vegetable in West and Central Africa and also very common in Asia and tropical America. Spiny wild forms are found throughout the humid parts of tropical Africa, and occasionally collected as vegetables. It is sometimes called the Gboma eggplant, named after a village in Liberia.

Figure 2. Major types of African nightshades (S. scabrum, S. tarderemotum, S. florulentum, S. americanum and S. villosum). (Photos by Chris Ojiewo).

African Nightshades

African nightshades are variously referred to as common nightshades, garden nightshades, nightshades, '*S. nigrum* and related species', or '*S. nigrum* complex', members of the genus *Solanum* L. subgenus *Solanum* L. section *Solanum* L. The group is comprised of approximately 30 species (*16*). Although they perform well in a wide range of climatic conditions, nightshades grow best at medium to high altitudes in moist conditions at optimal growth temperatures between 15°C and 35°C. They tolerate shade but grow best under full sunlight. African nightshades are generally intolerant of water deficit, and thrive in tropical rainy seasons and in areas with an annual precipitation of 500-1200 mm. They grow in various soil types, but are best adapted to sandy loams to friable clay soils with a pH range of 6.0-6.5, and high in nitrogen, phosphorus and organic matter (*16*). Some African nightshade species are widely consumed (Figure 2).

S. americanum

These are diploid species with simple umbellate cymes with round sepals, reflexes away from mature berries. Its pedicels are erect, radiating from the peduncle. The fruits are shiny dark purple, small (≤12 mm) and globose (fruit height t= fruit width). The plants produce approximately 4500 berries per plant, but many berries drop off at maturity. It produces very small seeds with 1000 seeds weighing approx. 0.3-0.4 g (*9*).

S. florulentum

This is rarely cultivated tetraploid species produces a bitter-tasting plant that that emits an unpleasant smell. It is used as a medicinal plant and is particularly recommended for malaria patients among the Kisii tribe in Kenya. The corolla of *S. florulentum* often have a purple strip along the mid-veins, usually being more conspicuous on the underside. The styles are clearly exerted beyond the anthers and often geniculate. The plants have triangular-ovate shaped sepals which adhere to mature berries. The plants produce 3000 - 4000 berries, dull dark purple, slightly flattened, and small (≤10 mm diameter) which drop off with pedicels still attached when ripe, making harvest difficult. The 1000 seed weight is 0.4 - 0.6 g.

S. grossidentatum

Recent cytological data shows that *S. grossidentatum* is a tetraploid (*9*) although previously it had been erroneously identified as the morphologically similar diploid *S. sarrachoides*. *S. grossidentatum* has simple extended/lax cymes, with fruits ranging in color from greenish-yellow to yellow, and globose in shape. Plants produce 3000 - 4000 berries per plant that drop with pedicels when ripe. The 1000 seed weight is 0.4-0.6 g. Plants develop a dense, glandular pubescence that give the leaves and stems of *S. grossidentatum* have a grayish color.

S. hirsutum

It is probable that *S. grossidentatum* (collected in western Kenya by Mwai) (*9*), and *S. hirsutum* (collected in Uganda by Olet) (*18*) are conspecific. Morphological descriptions of the latter species by Olet (*18*) approximates very closely the morphological description of *S. grossidentatum* (*9*), the most notable difference being the extended cymes in the description of *S. grossidentatum* (*9*), and umbellate cymes in the description of *S. hirsutum* by Olet (*18*).

S. scabrum

This hexaploid species produces simple umbellate cymes, light-purple and purple-colored corollas and brown to yellowish brown anthers. The sepals are round and reflexed away from mature berries. Except for a few green-stemmed accessions, seedlings have purple hypocotyls. Ripe berries are shiny black and larger ($\geq$14 mm), slightly flattened (fruit height<fruit width) and the vertical section slightly lobed rather than globose. Of the African nightshades species, *S. scabrum* produces the highest leaf, berry and seed yields of up to 75 t/ha, 150 t/ha and 9 t/ha, respectively. Average 1000 seed weight is between 1.0 and 1.3 g and berry number per plant ranges from 1000 - 2000. The berries remain intact on the plant even after ripening.

S. tarderemotum

S. tarderemotum is poorly understood even though it is a relatively important vegetable species in Eastern Africa; it is often referred to by synonyms such as *Solanum eldorettii, S. florulentum* and *S. nigrum* (*19, 20*). It is a tetraploid species with lanceolate sepals and long extended cymes. The corolla often have a purple strip along the mid-veins, usually being more conspicuous on the underside. The plants produce 3000-4000 berries per plant with a 1000 seed weight of 0.4-0.6 g. The berries are similar in size to those of *S. florulentum*, and either light-green or dull dark purple, with some accessions having light green berries with purple patches and slightly flattened. Reasons for preference of this species are the tenderness of the leaves and young shoots which are not as bitter as those of other species.

S. villosum

Plants of these are tetraploid species are short and compact, and produce white campanulate flowers with rare purple coloration on the flowers and calyces. The round yellow berries (3000 - 4000 berries per plant) usually drop from the pedicels when ripe. The seeds are slightly bigger than those of *S. americanum* with a 1000 seed weight of 0.4-0.6 g (*9*). Octoploids have been induced from the tetraploid *S. villosum* plants through seed treatment with colchicine (*6*); the octoploids produce 30-50% more leaf area and up to 35-50% more leaf dry weight and yield 1.3-1.6 times higher leaf fresh weight than their tetraploid progenitors (*17*).

Crop Management

African Eggplants

For the three species, the majority of cultivars grown mainly originate from natural and farmers' selection, and more rarely from conventional breeding. Creative breeding based on hybridization is far less developed on these crops than on the European/exotic vegetables. However, those varieties, mainly local ones, which have survived from any of the selection processes imposed upon them are most appreciated by consumers and generally endemic in production areas. Cultivar selection criteria include high productivity, resistance to diseases, early maturity, strong growth habit, tolerance to heat and market preferences.

Seed Germination and Nursery Establishment

Transplants are usually used to establish a uniform and complete stand of plants. Transplants grown in cells or containers are ideal because they allow field planting without disturbing the root system. The optimum temperature for germination is at 24 to 29 °C. At this temperature, seedlings should emerge in six to eight days. Embryo dormancy is a key issue in most varieties of the Kumba group, consisting of absence of (or limited) germination of new seeds. *Ngalam* of Technisem Seed Company is a variety selected for absence of embryo dormancy (*21*); most of the other recommended varieties need artificial dormancy breaking by soaking seeds in a solution of gibberellic acid (GA_3, 500 ppm; 24 hours). Another solution is natural breaking through storage for 4-5 months. At local level, small scale farmers have their own way of storing seeds consisting of leaving them within the berries scattered on their house roofs, for extraction a few months later. Seeds older than two years may not germinate.

Transplanting and Crop Management

Transplanting should not be done where other Solanaceous crops (tomato, pepper, potato) since these crops share many of the same disease and insect pests. The incidence of bacterial wilt and nematodes can be reduced if eggplant is planted after paddy rice. Sowing and transplanting of African eggplants can be done at any time of the year for full and off-season production where irrigation facilities are available, although in many parts of Africa, crops are mostly rain-fed. To reduce transplanting shock, hardening should be done by reduced watering and slow introduction to direct sunlight. Hardening should start six to nine days before transplanting to reduce transplanting shock.

After about five weeks when the plants are 15 cm high, they are ready for transplanting. Seedlings should be watered well before uprooting and again immediately after transplanting. Mulching is recommended during the dry season to keep the soil moist. Do not place the mulch too close to the stem as it may cause collar rot. The field should be irrigated immediately after transplanting to establish a good root-to soil contact. Irrigation is most critical during the flowering and fruit setting stages. Sandy soils are generally irrigated three times per week, whereas for loamy soils with organic matter, twice a week is generally adequate.

Plant spacing is variable, but single row planting with 50 cm between plants (within the row) and 75 cm between rows/ridges within the bed are common. In small-scale farmers' extensive traditional cropping systems, with poor soils, fertilizer application is not systematically done given the lack of availability and high costs. African eggplant is a heavy feeder requiring up to 100 (kg of N) – 100 (P_2O_5) – 200 (K_2O) that can be applied as 400kg/ha of the NPK (20-10-10) in two splits as a basal application during transplanting and the second one 6 weeks after transplanting. It also requires top dressing with 3 split applications of 120 kg N/ha of Nitrogen fertilizer, the first one together with first NPK application, the second application three weeks after the first or when fruit formation begins and the third application another three weeks after the second application. A balanced ratio between nitrogen and potassium supplies (close to 1:1) is needed to maximize yield. Flowering, pollination and fruit formation generally take place about a month and half after transplanting.

Organic production mainly focused on the use of large quantities of farm manure (50 to 80 tons /m^2) associated to biopesticides and botanicals, is practiced by a limited number of producers mainly for export purposes. This obviously leads to improved fruit quality, though yields are often lower as compared to conventional intensive systems.

Mulching is recommended to reduce weeds, prevent soil compaction, and conserve soil moisture. Plastic mulch must be laid before transplanting and holes are made through which the plants can be transplanted. Organic mulch is usually laid after transplanting. Plastic mulching is outstanding for preventing weeds, while organic mulching cools soil temperatures. For that reason, the combination of these mulches is often used during the hot season.

African Nightshades

Seed Germination and Nursery Establishment

Seeds may be sown directly in the field or raised in a seedbed prior to transplanting. Direct sowing during the rainy season results in strong, large plants which yield more than transplants. Seeds may be mixed with fine textured sand to achieve uniform distribution during sowing, by broadcasting or in rows 15-20 cm apart and not more than 2 cm deep, covered with a thin layer of soil. Seedling emergence takes 5-9 days under favorable conditions. Shoot cuttings can be used as planting material, but the plants tend to branch and spread with reduced yields; they also produce more glycoalkaloids compared to seed-propagated plants (*22*).

Transplanting and Crop Management

Seedlings are ready to transplant to fine-tilled and well-manured fields 3 to 4 weeks after plant are at least 8 cm tall, with 5-6 leaves. Seedlings should be hardened as needed and transplanted in late afternoon followed by sufficient watering to minimize transplanting shock. Recommended spacing is 12 x 15 cm for *S. villosum*, *S. americanum* and *S. tarderemotum*, and 50 x 60 cm for *S scabrum* (*13*, *16*, *22*). Wider spacing (up to 50 x 100 cm) is advisable for seed production crops. This encourages stronger branching, making up for the fewer plants.

Well decomposed poultry, farm yard or compost manure at the rate 15-20 t/ha should be worked into the soil prior to planting. A compound fertilizer such as NPK (20-10-10) should be incorporated at the rate of 120-150 kg/ha during sowing/transplanting. N fertilizers such as urea and ammonium sulphate at the rate of 60kgN/Ha are used as top/side dressings after every 2-3 harvests. Over application of N-fertilizers may reduce dry matter content, make the plants more susceptible to diseases and cause build-up of nitrates in leaves to toxic levels (*22*).

In the absence of rain, mulching is advisable, otherwise daily irrigation is needed for the 1st week after transplanting, but irrigation interval can be reduced after seedling establishment to 2-3 times a week. Overhead irrigation should be avoided due to potential to spread diseases. Frequent weeding is required in the early stages, but once the plants form a canopy, weeds are suppressed (*22*).

Harvesting and Post-Harvest Handling

African Eggplant

Fruits of most varieties are ready for commercial harvest and consumption about 2.5-3 months after transplanting, while physiological maturity occurs about one month later. Farmers should harvest the fruit with a sharp knife or pruning scissors before the skin becomes tough and changes color. Regular harvest encourages proper development of young fruit. Harvest early morning or evening and store produce in a cool, shaded place. Fruits can be packed in onion bags for transport to the marketplace. Leaves are preferably harvested after copious watering of the crop to maximize and maintain freshness. Packaging is done in bundles further to water spraying in case of long distance transport from production areas to marketplace. Preservation of fruits can be done in baskets under ambient conditions for a few days; longer-term storage requires cold temperatures..

African Nightshades

The crop will be ready for the first harvest about 5 weeks after transplanting and can be harvested continuously for up to nine months. The length of the harvested shoot varies from 10-50 cm. Harvest frequency ranges between 7-14 days. Harvested shoots may be kept fresh by wrapping them in small bundles using plastic sheeting or banana leaves, and watered sparingly to avoid rotting. Harvested shoots can also be sundried, wrapped in plastic and used during the dry season when production is limited.

Little or no leaf harvest should be performed on plants meant for seed collection (*22*). Fruits are harvested when fully ripe, and crushed to separate seeds from the pulp. The crushed berries may be washed immediately or left for a few days to ferment for easier separation of seeds and pulp. Extracted seeds are dried under shade and not direct sunlight.

Utilization

In the last years, the distribution and uses of African nightshades and African eggplants as food and traditional medicines has been the subject of several publications (Table Ia, -Ib).

African Eggplant

The leaves are appreciated for their slightly bitter taste and are eaten separately or in sauces. The young leaves and fruits are cooked and consumed as a vegetable. In Uganda, the leaves of the Shum group are either steamed or fried in oil with onions or consumed with sauces. The pea-sized bitter fruited *S. anguivi* cultivars are collected from the wild or cultivated, and consumed in fresh or dried form.

In Ghana and Cote d'Ivoire, the big fruits of the Gilo group cultivars are boiled to thicken sauces and medium to large fruit of the Kumba group are used in Senegal and Mali as condiments just like any other vegetable in rice-based dishes. Small fruited cultivars of *S. macrocarpon* are consumed as leafy vegetables in Benin and the large fruits are eaten in other West African countries.

African eggplants are also used as ornamentals (Kumba and Aculeatum groups and *S. macrocarpon*) and as medicinal plants (chewed leaves are used to treat sore throat, heart disease, constipation and the roots are used to treat worms) (*13*).

African Nightshades

Vegetable nightshades (*Solanum* section *Solanum*) are grouped among the high priority African indigenous leafy vegetables with potential nutritional and economic benefits. While some reports indicate high nutritional value, others indicate toxicity in nightshades. However, nomenclatural confusion is these reports makes it difficult to clearly distinguish between evaluations of nutritious and toxic species. The true *S. nigrum* is reportedly poisonous but, generally, the vegetable nightshades are confused with the deadly nightshade (*Atropa belladonna*), which has highly toxic leaves and berries. This confusion has caused all nightshade reports to be treated with some caution (*16*, *20*). The status of nightshades, however, is completely different in many parts of Africa, where they have long been used as leafy vegetables and for various medicinal uses. Berries are used as a substitute for raisins in pies, jams and preserves and food-colorant (*13*, *16*).

Over the years, African nightshades production has shifted from leaf collection from plants in the wild, to kitchen garden cultivation for domestic use and local sale, and now to cultivation in large commercial plots for sale in supermarkets and hotels. The leaves provide a useful green vegetable when boiled like spinach, with a much-liked unique flavor that accompanies the predominantly starch/carbohydrate-based staples (*13*, *15*).

Nightshade is recommended for malaria patients, newly circumcised initiates, and pregnant or lactating mothers (*23*). Medicinally, Kenyans use the nightshade leaves to treat duodenal ulcers and stomach upsets, boils, swollen glands, and they are often rubbed on the gums of young children when teeth are coming in crooked (*16*, *24*). Young shoots, leaves and stem of *S. scabrum*, *S villosum* and *S. americanum* are used whereas only the ripe yellow fruit of *S. villosum* are eaten, especially by children (*4*).

Nutritional Values

Data on the nutrient content of African nightshade and African eggplant are scarce in literature. Only one citation by K'Opondo et al. (*25*) provided a nutrient composition table for African nightshade. According to K'Opondo's data, 100 g fresh portion of African nightshade contains 87.2g water, 1.0mg iron, 4.3g protein, 38g calories, 5.7g carbohydrates, 1.4g fibre, 20mg ascorbic acid, 442mg calcium, 75mg phosphorus, 3660 μg β-carotene, and 0.59mg riboflavin. For African eggplant, only the calcium and zinc data were measured for the leaves. Research by Oboh et al. (*26*) showed that the leaves contained 32.6mg/kg calcium and 8.2mg/kg zinc. Due to the lack of nutrient data for these two AIVs, nutritional analysis is vital to provide accurate information for the future promotion of these vegetables.

Through several research projects funded by the Council of Agriculture (COA) in Taiwan, the Ministry for Economic Cooperation and Development (BMZ) of Germany, and the United States Agency for International Development (USAID), we evaluated five species of African nightshades and 10 species of

African eggplants for their nutrient contents Figure 4. Seeds were collected in Taiwan or transferred from Tanzania, Kenya and Cameroon with proper MTA (material transfer agreements). Plants were grown at the AVRDC – The World Vegetable Center in Taiwan and measured for nutritional and functional properties. Nutritional values of fruits and leaves of selected Solanum crops are presented in Table IIa and IIb. Flavonoid contents are listed in Table III. The group means of nutritional values of African nightshades (AN) and African eggplants (AE) are presented in Figure 3 in the background comparison with 120 plant species studied previously in AVRDC for their nutritional values (*27*).

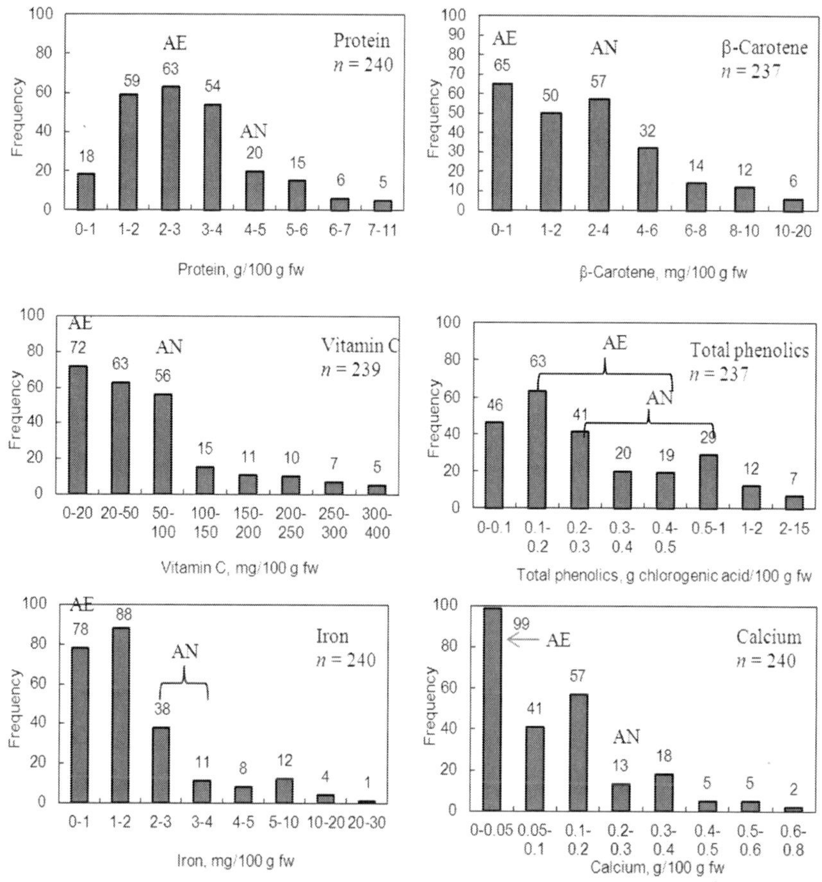

Figure 3. Distribution of 120 vegetable species for contents of protein, vitamins, minerals and phytochemicals on fresh weight basis (adapted from Yang et al (27). Nutrient values of African nightshades (AN) and African eggplants (AE) are classified according to the distribution.

Figure 4. African nightshades and African eggplants grown in fields at the AVRDC–The World Vegetable Center, Taiwan for analysis at the Nutrition Laboratory, AVRDC-The World Vegetable Center. Photos: RY Yang and Ruby Shiao, AVRDC, Taiwan.

Table Ia. Distribution and Utilization of African Solanum Vegetables. Data source: *(1)*. WA, CA, SA, EA: west, central, south and east Africa

Names	*Distribution*	*Part consumed and use*	*Traditional medicinal application*
African eggplant			
S aethiopicum L. African eggplant Garden egg Scarlet eggplant Bitter tomato	Domesticated from wild *S anguivi* Tropical Africa, Brazil	Immature fruit (garden egg) in WA, leaves and immature fruit in savanna area, leaves in EA (Uganda) Eaten raw or cooked as a vegetable Occasionally as a rootstock for tomato and eggplant	Roots, fruits and leaves: as carminative and sedative, treat colic and high blood pressure, uterine complaint Alcohol extract of leaves: as sedative, anti-emetic, treat tetanus after abortion Crushed fruit: enema
*S. anonalun*Thonn. ex Schumach Children's tomato	Sierra Leone to southern Nigeria, Cameroon and DR Congo Grown in wild	Ripe red fruit cooked in soup and sauce, or eaten fresh (mostly elderly people due to bitterness) Leaves are not eaten	Fruit and leaves juice: treat leprosy (WA) Leaves: as a laxative and digestive Fruit in soup: as an appetizer for sick persons Fruit juice: applied to relieve sores and pain
S. macrocarpon L. African eggplant Gboma	Spiny wild form: tropical non-arid Africa Cultivated form (fruit): gboma in WA Local cultivar (for leaves): WA, CA	Young leaves and young fruits cooked and consumed as a vegetable Less bitterness is preferred	Heated leaves chewed to treat throat troubles, stomach trouble Fruit as a laxative, treat cardiac diseases Flowers and fruits: chewed to clean teeth Boiled root juice: treat hookworm infection

Continued on next page.

Table Ia. (Continued). Distribution and Utilization of African Solanum Vegetables

Names	*Distribution*	*Part consumed and use*	*Traditional medicinal application*
African eggplant			
S. melongena L. eggplant brinjal	Cultivated worldwide	Immature fruits with immature seeds Grilled, fried, steamed, stewed	Various plant parts used in decoction, as powder for curing ailments such as diabetes, cholera, toothache, skin infections, asthenia and hemorrhoids. Eggplant is also ascribed to have narcotic, anti-asthmatic and anti-rheumatic properties.
S. torvum Sw. Pea eggplant, Cherry eggplant Turkey berry	Originated from Central and South America Pan tropical weed A kitchen garden crop in CA and WA[1] Cultivated in Asia, popular in Thailand	Bitter fruits used in soup and sauce and mixed with eggplant and tomato	Fruits and leaves to control a wide range of microbial activity The glycoalkaloid (solasodine) in the manufacture of steroidal sex hormones Leaves: applied to cuts and wounds, for anti-diabetic in India Fruits in decoction: as a cold medicine for children, treat sore throat, stomach ache

Table Ib. Distribution and Utilization of African Solanum Vegetables. Data source: *(1)*. WA, CA, SA, EA: west, central, south and east Africa

African nightshade			
S. americium Mill Glossy nightshade	Pan tropic and subtropics	Shoots and younger leaves, ripe fruits (some), boiled Cooking time in water depends on bitterness, especially for children	Leaves cooked with milk to prevent malnutrition Leaves extract: relieve chronic conjunctivitis and related inflammations, treat sores, heart pain, skin problems Treat worms in chicken
S anguivi Lam. African nightshade	Non-arid areas throughout Africa Grow in wild	Green fruit eaten as a vegetable and appetizer	Bitter small green fruit used for high blood pressure
S. scabrum Mill. African nightshade Black nightshade Garden huckleberry	A cultivated vegetable Lowland and highland in WA, CA, EA[1] Outside Africa: Europe, Asia, Australia New Zealand, North America	• shoots and leaves, cooked as a vegetable • fodder for cattle and goats • leaves and fruits: dye or ink • bitter taste removed by discarding cooking water	• leaf extract: treat diarrhea in children, eye infections, jaundice • raw fruit: treat stomach ulcer or stomach ache • leaves and seeds: rubbed on the gums of children for crooked tooth
S. villosum Mill. • red-fruited nightshade • hairy nightshade	• south European origin • used as a vegetable most popular in EA	• leaves and young shoots are used as leafy vegetable, boiled or fried • ripe fruits (orange, yellow and red)	• unripe fruits: soothe toothache • leaves: treat stomach ache • extract of leaves and fruits: treat tonsillitis • boiled roots in milk as a tonic for children • apply leaves to swellings, fruit juice to calm sore eye

Table IIa. Nutrient content (per 100 g Fresh Weight) of Fruits and Leaves of Selected African *Solanum* Crops. Data source: Yang et al, unpublished data, AVRDC Nutrition laboratory; values were data from 1 accession or means of 2- 8 accessions of a species; plant materials were obtained from AVRDC germplasm and grown at AVRDC, Taiwan

	Stg	*n*	*Clr*	*DM*[1]	*PR*[1]	*CF*[1]	*Oil*[1]	*VI*[2]	*NE*[2]	*LU*[2]	*αC*[2]
Fruits											
S aethiopicum	Y	5	G	8.4	1.2	1.6	1.4	nd	nd	nd	nd
S aethiopicum	M	8	R	9.5	1.2	1.8	5.0	nd	nd	nd	0.2
S anguivi	Y	3	G	11.5	1.7	2.8	3.0	nd	nd	0.23	0.03
S anguivi	M	3	R	12.8	1.8	3.6	6.3	nd	Nd	0.15	0.36
S ferox	Y	1	G	11.5	2.0	2.9	3.7				
S indicum		1	R	16.6	2.4	5.2	nt				
S macrocapon		2	W	8.3	1.0	1.0	nt	nd	Nd	0.01	nd
S macrocarpon		1	Y	11.2	1.4	1.7	1.5	nd	Nd	0.19	nd
S mammosum		1	Y	16.3	2.2	5.6	nt				
S melongena		1	P	7.5	1.5	0.9	0.2				
S torvum	Y	1	G	20.4	2.7	7.2	4.5				
S xanthocarpum	Y	3	G	15.9	2.3	4.6	10.0				
S xanthocarpum	Y	3	Y	17.9	2.6	6.0	9.6				

	Stg	*n*	*Clr*	*DM*[1]	*PR*[1]	*CF*[1]	*Oil*[1]	*VI*[2]	*NE*[2]	*LU*[2]	*αC*[2]
Fruits											
S zuccagnianum	Y	1	G	14.3	1.7	3.6	2.6	nd	Nd	0.07	nd
S zuccagnianum	M	1	R	13.1	1.6	3.9	5.2	nd	Nd	0.58	0.49
Leaves											
S florulentum	Y	1		8.0	3.4			2.4	1.3	3.19	0.08
S sarrachoides	Y	1		8.4	3.7			1.9	1.1	2.63	0.08
S scabrum	Y	2		10.5	4.4	1.0	3.1	5.0	2.8	7.38	0.24
S villosum	Y	1		11.1	4.2	1.3	1.9	3.7	1.9	5.58	0.10
S zuccagnianum	Y	1		14.2	4.8	1.8	2.5	1.5	1.7	5.85	0.08

Nutrient contents are means of accessions within the same types and species; n: number of accessions Stage: Y, young, M, mature. Color: G, gree, R, red, W, white, Y, yellow, P, purple. DM: dry matter; PR: protein; CF: crude fiber; VI: Violaxanthin; NE: Neoxanthin; LU: lutein;αC: α-carotene; βC: β-carotene; γE: γ-tocopherol; αE: α-tocopherol; AA: antioxidant activity; TE: trolox equivalent (μmol); TP: total phenol; CE: chlorogenic acid equivalent (mg); OX: oxalate. 1. Grams, 2. Milligrams nd: below detectable level; missing value: not tested.

Table IIb. Nutrient content (per 100 g fresh weight) of fruits and leaves of selected African *Solanum* crops. Data source: Yang et al, unpublished data, AVRDC Nutrition laboratory; values were data from 1 accession or means of 2- 8 accessions of a species; plant materials were obtained from AVRDC germplasm and grown at AVRDC, Taiwan

	βCa^2	VC^2	γE^2	αE^2	F^2	C^2	Fe^2	Zn^2	*AA TE*	*TP CE*	OX^2
Fruits											
S aethiopicum	nd	13	0.07	0.95	31	20.9	0.6	0.21	265	60	24
S aethiopicum	0.32	10	0.14	0.66	15	14.1	0.9	0.12	467	87	21
S anguivi	0.09	18	0.08	1.04	62	50.8	1.0	0.37	641	177	34
S anguivi	0.38	9	0.40	2.39	14	53.9	1.0	0.38	1489	214	16
S ferox	0.19	11	0.10	0.30		28.2	0.9		699	150	
S indicum	0.60	1				135.1	1.2		1604	260	
S macrocapon	0.02	13	0.00	0.40	6	7.6	0.6	0.23	389	77	74
S macrocarpon	0.07	27	0.29	0.33	23	14.9	1.0	0.25	604	72	22
S mammosum	0.07	20	0.34	0.92		10.5	0.7		1246	228	
S melongena	nd	13	0.00	0.03	5	9.5	0.9		276	44	22
S torvum	0.16	15	0.12	0.57		81.6	1.0		1373	280	
S xanthocarpum	0.07	10	0.50	0.16		21.9	0.7		791	162	
S xanthocarpum	nd	15	0.45	0.27		22.6	0.8		1171	235	

	βCa^2	VC^2	γE^2	αE^2	F^2	C^2	Fe^2	Zn^2	*AA* *TE*	*TP* *CE*	OX^2
Fruits											
S zuccagnianum	nd	14	0.06	2.29	117	79.9	0.8	0.58	4132	419	23
S zuccagnianum	1.57	11	0.08	3.31	67	48.6	0.4	0.42	1891	548	24
Leaves											
S florulentum	1.90	55				193	2.6	0.52	513	113	
S sarrachoides	1.82	72				232	2.4	0.52	660	133	
S scabrum	5.76	125	nd	2.35	70	194.4	3.0	0.47	1202	275	33
S villosum	2.95	79	nd	2.14	61	175.0	3.3	0.80	768	159	68
S zuccagnianum	5.82	44	0.50	6.81	9	241.3	2.5	0.81	2430	722	118

Nutrient contents are means of accessions within the same types and species; n: number of accessions Stage: Y, young, M, mature. Color: G, gree, R, red, W, white, Y, yellow, P, purple. DM: dry matter; PR: protein; CF: crude fiber; VI: Violaxanthin; NE: Neoxanthin; LU: lutein;αC: α-carotene; βC: β-carotene; γE: γ-tocopherol; αE: α-tocopherol; AA: antioxidant activity; TE: trolox equivalent (μmol); TP: total phenol; CE: chlorogenic acid equivalent (mg); OX: oxalate. 1. Grams, 2. Milligrams nd: below detectable level; missing value: not tested.

Table III. Flavonoid Content (per 100 g Fresh Weight) of Fruits and Leaves of Selected African *Solanum* Vegetables. Source of data: (*32*)

	Part	*DM* *g*	*Quercetin* *mg*	*Kaempferol* *mg*	*Isorhamnetin* *mg*	*Luteolin* *mg*	*Apigenin* *mg*	*Total* *mg*
S aethiopicum	furit	9.7	0.3	0.3	nd	nd	nd	0.6
S nighrum	young leaf	7.8	3.7	1.0	nd	nd	nd	4.7
S scabrum	young leaf	11.7	23.2	2.0	0.4	nd	nd	25.5
S villosum	young leaf	12.4	18.1	0.9	nd	nd	nd	19.0
S zuccagnianum	leaf	15.7	6.1	29.8	nd	nd	nd	35.8
S zuccagnianum	fruit	19.9	2.3	1.9	nd	nd	nd	4.1
S zuccagnianum	young fruit	14.6	nd	nd	nd	nd	nd	nd

nd: below detection limit (<0.05 mg/100gfw).

Table IV. Glycoalkaloid Content (mg/ 100g fw) of Worldwide and African *Solanum* Vegetables

Name	*Part*	*α-chaco nine*	*α-sola nine*	*Dehydro tomatine*	*α-tomatine*	*Solamar gine*		*α-solaso nine*	*Total*	*ref*
						α-	*β*			
Worldwide potato, tomato and eggplant										
S. tuberosum[1]	T	0.5-12	0.3-6.8						0.7-18.7	(*36*)
S. andigena[2]	T	2-154	1-56						3-210	(*43*)
S. tuberosum[3]	L	93-211	33-103						127-314	(*36*)
S. lycopersicum	F			48 (S1)[a], nd (S9-11)	361 (S1) nd (S9-11)				nd (S9-11)	(*52*)
S. melongena	RF					0.9-1.6		0.2-0.4	1.1-2	(*53*)
African eggplant										
S.aethiopicum	F					0.6-4.9		0.4-1.0	1-5	(*53*)
S.macrocarpon	F					124 – 198		16-23	140-221	(*53*)
S.torvum	F								4	(*54*)
African nightshade										
S.americanum	L		32.9			19.6	nd	35.0	87.5	(*38*)
S.anguivi	L,F					detected				(*55*)
S. spp.	L		47.0			50.3	16.9	58.0	172.2	(*38*)
S.villosum	L		15.0			11.7	98.0	20.1	144.8	(*38*)

1. commercial source). 2. cultivated potato. 3. tomato. 4. eggplant Part: F, Fruit, L, leaves, RF, ripen fruit, T, tuber nd: not detected; S1, S9-11: tomato fruit maturation stage 1 and 9 – 11 a S1 – S11: tomato fruit ripening stage.

The nutrient values of African nightshades (*S. scabrum* and *S. villosum*) in leaves are comparable to those among the top 20-30% of 240 plant samples we studied previously (*5*, *27*). Theses nutrients include protein, β-carotene, vitamin C, vitamin E, folates, iron, antioxidant activity and total phenols (Table IIa and IIb). For African eggplants, high antioxidant activity and total phenol contents were detected. The nutrient values of African eggplant fruit are generally low, except for oil content. High oil concentrations (9 - 10 g/100g) were measured in the young and mature fruits of *S. xanthocarpum*. High crude fiber likely contributed to the rough and fibrous taste of *S. aethiopicum* that was also noted in fruit of *S. xanthocarpum* and *S. torvum*.

Flavonoids are present in most plant tissues, and serve many functions: color definition to attract insect pollinators and seed dispersers; as antioxidants to protect plants against UV-radiation, as insect feeding attractants in host-species recognition, as signal molecules in inducible defense systems against bacterial and fungal attack and as providers of bitter or astringent tastes to repel birds and another animals (*28*). Flavonoids are recognized as beneficial in human diets as antioxidants, anticancer, antioxidant, anti-proliferative, antihypertensive and anti-inflammatory activities, free radical scavenging, coronary heart disease prevention, immunity enhancement and Type 2 diabetes prevention (*29–31*).

The major types of flavonoids in the leafy nightshades (*S. scabrum* and *S. villosum*) were quercetin, followed by kampferol (Table III), a pattern similar to those reported for most leafy vegetables (*32*). Higher flavonoid concentrations were found in leaves (19 – 36 mg/100g) than in fruits for the Solanum crops (Table III). Flavonoids in unripe fruits of *S. zuccagnianum* were below detectable limits (<0.05 mg/ 100g fw). In our previous study of 115 edible plants for nine flavonoids, 86 samples had total flavonoid contents ranging from 0.5 - 254 mg / 100g with a group mean of 33 mg/ 100g (*32*). The other 29 crops had very low total flavonoid concentrations of less than 0.5 mg. Overall, Solanum fruits were low in flavonoids and contents in the leaves were comparable to the group mean (33 mg) of all tested crops.

Glycoalkaloids in Solanum Vegetables

Alkaloids are a diverse group of low molecular weight, nitrogen-containing compounds that are mostly derived from amino acids and found in about 40% of plant families and 20% of plant species (*33*, *34*). They are secondary metabolites, not all necessarily required for plant growth and function, but play a defensive role against herbivores and pathogens. They can be present throughout the plant or restricted to certain tissues. The concentration can vary from less than 0.1% to 12% dry weight (*34*).

Steroidal glycoalkaloids are found in many agricultural products obtained from members of the Solanaceae, including potato, tomato, and eggplant (*35*). The steroidal alkaloids are triterpenoid in origin, being derived biosynthetically from six isoprene units. They might equally well be placed close to the phytosterols, except they do contain nitrogen and hence are bases, like the other alkaloids.

Solanum glycoalkaloids have two structural components: an aglycone unit and a glycosidic unit. The aglycone unit (or steroidal alkamine) consists of a hydrophobic 27-carbon skeleton of cholestane with nitrogen incorporated into the F ring. The glycosidic unit is a carbohydrate side chain composed of various combinations of D-glucose, D-galactose, and L-rhaminose, and attached at the 3-OH position of the aglycone. The glycoalkaloids are refereed to as α-compounds, and the cleavage of the individual sugars of the glycoside leads to β-,γ- and δ-compounds, depending on the number of sugars in the side chain.

At least 90 steroidal glycoalkaloids were identified in over 350 Solanum species with many wild species serving as genetic sources for varietal improvement (*34*, *35*). The primary glycoalkaloids in commercial potato cultivars are α-chaconine (solanidine - chacotriose) and α-solanine (solanidine - solatriose) which have the same aglycon but conjugated with different sugar side chains. The major glycoalkaloids in unripe tomato are α-tomatine (tomatidine–lycotetraose) and dehydrotomatine (tomatidennol–lycotetraose). α-Solamargine and α-solasonine occur as the two principal glycoalkaloids in eggplant (*S. melogena*) and African eggplants (*S. aethiopicum*, *S. macrocarpon* and *S. anguivi*). These two compounds have the same solasodine aglycone, but different glycosidic units of chacotriose for α-solamargine and solatriose for α-solasonine, respectively (*36*, *37*). For African nightshade (*S. villosum*), α-solamargine, β-solamargin, α-solasonine and α-solanine are major alkaloids found in leaves (*38*).

The biosynthetic pathway of Solanum glycoalkaloids has not yet been fully elucidated. It is generally accepted that the aglycone moiety of the glycoalkaloids is derived from simvastatin and cholesterol (*35*). Toxicity of glycoalkaloids in humans is documented (*35*, *36*, *39–42*) especially for wild potato, potato sprouts and the weedy nightshades (*S. nigrum*) in Western countries. The potential negative effects led to the establishment of guidelines limiting glycoalkaloid content (< 200 mg / kg or <20 g/100g fw in potato tubers) in new potato cultivars before they can be released for commercial use (*37*). Glycoalkaloid content of commercial potato varieties were about 1 - 20 mg/ 100g (*43*); however, in potato leaves, the concentration could be 10 times more (>200 mg/100g) than in the tubers (*35*, *36*, *44*). In many countries people eat potato almost daily and glycoalkaloid toxicology issues and studies were vigorous in the 1970's-1990's. More recent reports on alkaloid contents in Solanum crops indicate potential health benefits, such as anti-ulcerogenic activity (*45*); anti-hypertension (*46*), anti-cancer (*47–50*), anti-diabetes (*51*), and others (*37*) but depending on dose and conditions of use.

We have summarized glycoalkaloid contents of worldwide and African Solanum crops from recent reports (Table IV). Glycoalkaloid levels in commercial cultivars of potato are generally undetectable or very low (< 5 mg/ 100g fw). Glycoalkaloids in ripe eggplant and tomato fruits are below detectable levels (*52*). Glycoalkaloid data for African nightshade and African eggplant are scarce. High glycoalkaloid concentrations (140 -220 mg/ 100g fresh weight) were found in green and white fruits of *S. macrocarpon* originating from the Ivory Coast and Ghana, but low in other African eggplants (1-5 mg/ 100 g fw), including *S. aethiopicum,* from Ivory Coast and *S. melongena* from Spain (*53*). Mohy-Ud-Din et al (*38*) reported high concentration of glycoalkaloids in leaves of the *Solanum nigrum* complex, ranging from 80 – 150 mg/ 100g fw. This complex includes species of *S. americanum, chenopodioides, nigrum, retroflexum, and villosum*, and the plant materials for the study were obtained from the Pakistan botanical garden for traditional medicinal use rather than cultivars grown for human consumption (*38*).

As indicated in the taxonomy section of this report and by Mohy-Ud-Din et al (*38*), the taxonomy of *Solanum* are complex, and the "deadly nightshade" reputation of *S. nigrum* and other Solanum weeds or medicinal plants has extended to the edible nightshades (African nightshades: *S. scabrum* and *S. villosum*) which has confused researchers and people unfamiliar with these plants, taxon and their morphological differences.

A preliminary test of glycoalkaloids in leaves of African nightshades (*S. scabrum* and S. *villosum)* produced from seeds of Tanzanian varieties and planted at AVRDC, Taiwan was conducted recently in our laboratory using LC/MS method. We found low concentration of glycoalkaloids in African nightshade samples. The analyses of alkaloids in African nightshades and African eggplants are on-going.

Conclusions and Recommendations

Morphological and Molecular Characterization of Genebank Material

The AVRDC-RCA holds the largest collection of traditional African Solanaceae in the world. The collection includes traditional landraces that are important genetic resources for plant breeders because of their considerable genotypic variations, but are largely uncharacterized. A combined molecular and morphological characterization of the germplasm giving their cytology and population structure analysis would help accessing the diversity of the collections and eventually in parent selection for breeding.

Variety Development and Quality Seed Availability

In order to improve the productivity of vegetables, good quality seed be made available to farmers. AVRDC scientists and collaborators in Sub-Saharan African have made initial efforts in breeding IVs resulting in the official release of IV varieties including African nightshades and eggplants. There is a great need for more IV breeding and seed system development by public and private sectors and institutions throughout Africa.

Studies of the Positive and Potential Negative Health Outcomes

There are clearly great potential public health benefits from increased consumption of African eggplant and African nightshade. Additional assessment of glycoalkaloid content of a greater range of types and cultivar groups as well as the health outcomes from consumption of African nightshade and African eggplants should be conducted prior to large-scale promotion of these crops.

References

1. *PROTA-Plant Resources of Tropical Africa 2. Vegetables*; Grubben, G. J. H., Denton, O. A., Eds.; PROTA Foundation: Wageningen, Netherlands, 2004.
2. *Lost crops of Africa: Vegetables*; National Research Council of the National Academies Press: Washington, DC, 2006.
3. Weinberger, K.; Msuya, J. *Indigenous vegetables in Tanzania - Significance and prospects*; AVRDC-The World Vegetable Center: Shanhua, Taiwan, 2004.
4. Keding, G.; Weinberger, K.; Swai, L.; Mndiga, H. *Diversity, traits and use of traditional vegetables in Tanzania*; Technical Bulletin No. 40; AVRDC-The World Vegetable Center: Shanhua, Taiwan, 2007.
5. Yang, R. Y.; Keding, G. B. Nutritional contribution of important African vegetables. In *African Indigenous Vegetables in Urban Agriculture*; Shackleton, C. M., Pasquini, M. W., Drescher, A. W., Eds.; Earthscan: London, U.K., 2009.
6. Ojiewo, C. O.; Agong, S. G.; Murakami, K.; Masuda, M. *J. Hortic. Sci. Biotechnol.* **2006**, *81*, 183–188.
7. *PROSEA – Plant resources of South-East Asia 8 – Vegetables*, 2nd Ed.; Siemonsma, J. S., Kasem, P., Eds.; Prosea Foundation: Indonesia, 1996.
8. Lehmann, C.; Biela, C.; Töpfl, S.; Jansen, G.; Vögel, R. *Euphytica* **2007**, *158*, 189–199.
9. Mwai, G. N. *Morphological and molecular characterization and yield evaluation of vegetable African nightshades (SolanumL. section Solanum) under varying cultural practices*. Ph.D. Thesis, Maseno University, Maseno, Kenya, 2007.
10. Lebecka, R. *Eur. J. Plant Pathol.* **2008**, *120*, 233–240.
11. Campos, M. A; Ribeiroa, S. G.; Rigden, D. J.; Montea, D. C.; Grossi de saa, M. F. *Physiol. Mol. Plant Pathol.* **2002**, *61*, 205–216.
12. Porcher, M. H. *Know your eggplant. Part 1*; http.//www.plantnames.unimelb.edu.au./new/sorting/CATALOUE/ EGGPLANTS-intro.html (accessed 2009).
13. Schippers, R. R. *African Indigenous Vegetables: an overview of the cultivated species*, 2nd ed.; Natural Resources Institute International, University of Greenwich and Horticultural Development Services LLP: Chatham, U.K., 2002.
14. Lester, R. N. *Acta Hortic.* **1986**, *182*, 125–132.
15. Lester, R. N.; Daunay, M. C. *Rudolf Mansfeld and plant genetic resources* **2003**, *22*, 137–152.
16. Edmonds, J. M.; Chweya, J. A. Black nightshade: *Solanum nigrum* L. and related species. In *Promoting the conservation and use of underutilized and neglected crops*; International Plant Genetic Resources Institute: Rome, Italy, 1997.
17. Masinde, P. W.; Onyango, C. A.; Abukutsa-Onyango, M. O.; Wesonga, J. M.; Ojiewo, C. O.; Agong, S. G. *Afr. J. Hortic. Sci.* **2010**, *3*, 34–47.
18. Olet, E. A. *Taxonomy of Solanum L. section Solanum in Uganda.* Ph.D. Thesis, Agricultural University of Norway, Norway, 2004.

19. Mwai, G. N.; Schippers, R. R. Solanum tarderemotum Bitter. In *Plant resources of Tropical Africa 2. Vegetables*; Grubben, G. J. H., Denton, O. A., Eds.; PROTA foundation: Wageningen, Netherlands, 2004; pp 498−501.
20. Olet, E. A.; Heun, M.; Lye, K. A. *Afr. J. Ecol.* **2005**, *43*, 158–161.
21. Seck, A. *Acta Hortic.* **2000**, *522*, 195–208.
22. *AVRDC Annual Report 2003*; Asian Vegetable Research and Development Center: Tainan. Taiwan, 2004.
23. *Traditional Food plants of Kenya*; Maundu, P. M., Ngugi, G. W., Kabuye, C. H., Eds.; National Museums of Kenya: Nairobi, Kenya, 1999.
24. *The Biodiversity of Traditional Leafy Vegetables*; Chweya, J. A., Eyzaguirre, P. B., Eds.; International Plant Genetic Resources Institute: Rome, Italy, 1999.
25. K'opondo, F. B. O; Muasya, R. M.; OK, K. A. A review on the seed production and handling of indigenous vegetables (Spider plant, Jute mallow, and African nightshade complex). In *Proceedings of the Third Horticulture Workshop on Sustainable Horticultural Production in the Tropics*; Abukutsa-Onyango, M. O., Muriithi, A. N., Anjichi, V. E., Eds.; Maseno University: Maseno, Kenya, 2005; pp 42−48.
26. Oboh, G.; Ekperigin, M. M.; Kazeem, M. I. *J. Food Comp. Anal.* **2005**, *18*, 153–160.
27. Yang, R. Y.; Wu, J.; Chen, D.; Lin, M.; Shiao, R.; Kuo, G. *Acta Hortic.* **2007**, *752*, 115–122.
28. Croteau, R.; Kutchan, T. M.; Lewis, N. G. Natural products (secondary metabolites). In *Biochemistry & Molecular Biology of Plants*; Buchanan B. B., Gruissem, W., Jones R. L., Eds.; American Society of Plants Physiologists: Rockville, MD, 2000; pp 1250−1318.
29. Havsteen, B. H. *Pharmacol. Ther.* **2002**, *96*, 67–202.
30. Hodgson, J. M.; Croft, K. D. *Mol. Aspects Med.* **2010**, *31*, 495–502.
31. Xiao, Z. P.; Peng, Z. Y. *Mini-Rev. Med. Chem.* **2011**, *11*, 169–177.
32. Yang, R. Y.; Lin, S.; Kuo, G. *Asia Pac. J. Clin. Nutr.* **2008**, *17*, 275–279.
33. Zulak, K. G.; Liscombe, D. K.; Ashihara, H.; Facchini, P. J. Alkaloids. In *Plant secondary metabolites: Occurrence, structure and role in the human diet*; Crozier, A., Clifford, M. N., Ashihara, H., Eds.; Blackwell Publishing: Oxford, 2006; pp 102−136.
34. *Phytochemical dictionary: A handbook of bioactive compounds from plants*, 2nd ed.; Harborne, J. B., Baxter, H., Moss, G. P., Eds.; Taylor & Francis: London, U.K., 1999.
35. Milner, S. E.; Brunton, N. P.; Jones, P. W.; O'Brien, N. M.; Collins, S. G.; Maguire, A. R. *J. Agric. Food Chem.* **2011**dx.doi.org/10.1021/jf200439q.
36. Friedman, M. *J. Chromatogr., A* **2004**, *1054*, 143–155.
37. Friedman, M. *J. Agric. Food Chem.* **2006**, *54*, 8655–8681.
38. Mohy-ud-din, A.; Khan, Z.; Ahmad, M.; Kashmiri, M. A. *Pak. J. Bot.* **2010**, *41*, 653–660.
39. Chami, L. A.; Méndez, R.; Chataing, B.; O'Callaghan, J.; Usubillaga, A.; LaCruz, L. *Phytother. Res.* **2003**, *17*, 254–258.

40. Mensinga, T. T.; Sips, A. J. A. M.; Rompelberg, C. J. M.; van Twillert, K.; Meulenbelt, J.; van den Top, H. J.; van Egmond, H. P. *Regul. Toxicol. Pharmacol.* **2005**, *41*, 66–72.
41. Abdel-Aziz, H.; Fawzy, N.; Ismail, A. I.; El-Askary, H. *Food Chem. Toxicol.* **2011**, *49*, 903–909.
42. Nema, P. K.; Ramayya, N.; Duncan, E.; Niranjan, K. *J. Sci. Food Agric.* **2008**, *88*, 1869–1881.
43. Kozukue, N.; Yoon, K.-S.; Byun, G.-I.; Misoo, S.; Levin, C. E.; Friedman, M. *J. Agric. Food Chem.* **2008**, *56*, 11920–11928.
44. Dao, L.; Friedman, M. *J. Agri. Food Chem.* **1996**, *44*, 2287–2291.
45. Nguelefack, T. B.; Feumebo, C. B.; Ateufack, G.; Watcho, P.; Tatsimo, S.; Atsamo, A. D.; Tane, P.; Kamanyi, A. *J. Ethnopharmacol.* **2008**, *119*, 135–140.
46. Mohan, M.; Jaiswal, B. S.; Kasture, S. *J. Ethnopharmacol.* **2009**, *126*, 86–89.
47. Liu, L.-F.; Liang, C.-H.; Shiu, L.-Y.; Lin, W.-L.; Lin, C.-C.; Kuo, K.-W. *FEBS Lett.* **2004**, *577*, 67–74.
48. Shiu, L. Y.; Liang, C. H.; Chang, L. C.; Sheu, H. M.; Tsai, E. M.; Kuo, K. W. *Biosci. Rep.* **2009**, *29*, 35–45.
49. Shiu, L. Y.; Liang, C. H.; Huang, Y. S.; Sheu, H. M.; Kuo, K. W. *Cell Biol. Toxicol.* **2008**, *24*, 1–10.
50. Gao, J.; Li, X.; Gu, G.; Sun, B.; Cui, M.; Ji, M.; Lou, H.-X. *Bioorg. Med. Chem. Lett.* **2011**, *21*, 622–627.
51. Huang, W.-H.; Hsu, C.-W.; Fang, J.-T. *Clin. Toxicol.* **2008**, *46*, 293–296.
52. Choi, S.-H.; Lee, S.-H.; Kim, H.-J.; Lee, I.-S.; Kozukue, N.; Levin, C. E.; Friedman, M. *J. Agric. Food Chem.* **2010**, *58*, 7547–7556.
53. Sánchez-Mata, M.-C.; Yokoyama, W. E.; Hong, Y.-J.; Prohens, J. *J. Agric. Food Chem.* **2010**, *58*, 5502–5508.
54. Pérez-Amador, M. C.; Ocotero, V. M.; Castañeda, J. M. G.; Esquinca, A. R. G. *Int. J. Exp. Bot.* **2007**, *76*, 39–45.
55. Ripperger, H.; Himmelreich, U. *Phytochemistry* **1994**, *37*, 1725–1727.

Chapter 12

Quality Characteristics of Shea Butter, *Vitellaria paradoxa*

Hee Seung Nahm,[1] H. Rodolfo Juliani,[*,2] and James E. Simon[2]

[1]Department of Food Science, Rutgers, The State University of New Jersey, New Brunswick, New Jersey 08901
[2]New Use Agriculture and Natural Plant Products Program, Department of Plant Biology and Pathology, Rutgers, The State University of New Jersey, New Brunswick, New Jersey 08901
[*]E-mail: hjuliani@rci.rutgers.edu.

Shea butter is a plant fat extracted from kernels of seeds of shea trees (*Vitellaria paradoxa*) native to sub-saharan Africa. Shea butter has long been used in Africa as a versatile fat and used as cocoa butter equivalent. Shea butter has gained increasing popularity as an ingredient for cosmetics and personal care products in international markets. The objective of this work is to assess the chemistry and quality of West African shea butter samples to contribute to assessment of the quality of locally available shea butter and development of quality control standards.

Botany of Shea Trees

Shea butter is a plant fat extracted from kernels (embryo) contained in the seeds of *Vitellaria paradoxa,* a tree belonging to *Sapotaceae* family. Shea trees grow wild across a savanna belt extending from Senegal to Ethiopia encompassing Guinea, Mali, Côte d' Ivoire, Burkina Faso, Ghana, Togo, Benin, Niger, Nigeria, Cameroon, Chad, Sudan, and Uganda (*1–3*). This savanna belt is also known as a 'shea belt'among traders (*4*).

Shea trees are divided into two subspecies: West African subspecies '*paradoxa*' and East African '*nilotica*' (*4–6*). The *V. paradoxa* subsp. *nilotica* is grown in east African region mostly in southern Sudan and Northern Uganda, while *V. paradoxa* subsp. *paradoxa* grows in most west African countries. In international markets, it is hard to find shea butter made from east African subspecies since the major production countries, Sudan and Uganda have been regions where political instability or other issues have led to inconsistent supply (*4*).

In the wild, shea trees grow up to 9-12 m (30-40 ft) in height, and begin to bear commercial quantities of fruit after approximately 20-50 years (*7*, *8*). Shea trees do not reach maturity until 45 years and after getting mature, they can continue produce shea nuts for many decades with some reports suggesting up to 200 years (*7*, *9*). The entire shea industry has been relying on shea nuts or shea butter collected from naturally growing shea trees in many countries.

Shea trees blossom from February to March and the fruits become mature from June to July (*9*). A shea fruit is similar to a fig in size with a diameter of 5-8 cm (2-3 inches) and shows light green color (*1*). Shea fruit consist of a green epicarp (the outer part), a fleshy mesocarp (pulp), and a relatively hard endocarp (shell) surrounding the kernel or embryo (*10*). Most shea fruits contain 1-2 kernels but some can occasionally contain 3-4 kernels from which shea butter is extracted (*9*).

Extraction of Shea Butter

Traditionally, shea butter has been extracted at the village level, where shea butter is sold at local markets. Shea butter extraction has been done mostly by shea nut collectors who are African women following their own traditional methods and passing their methods to their daughters. In recent years, dried shea kernels have been exported to processing countries in Europe, Japan, and India where shea butter is extracted in large-scale plants (*11*). Commercial extraction is usually conducted by pressing or solvent extraction with further refining steps. Currently, however, with the increasing interests in naturally derived products, the efforts has been made to industrially produce shea butter by following the traditional extraction methods.

Either at the village or industrial level, shea butter is extracted from dried shea kernels. In Africa, once shea fruits dropped from trees to the ground, African rural women collect them and manually remove pulps from the fruit to get shea nuts (*1*, *12*). After the collection of shea nuts, shea nuts are treated by one of the following methods: West African boiling, West African oven, and East African raw methods (*11*). In West African boiling method, shea nuts are boiled to kill the embryo and thus prevent germination of the seeds as well as to make shells easier to break. This method can result in final shea butter products with lower hydrolytic degradation levels since the heating step can deactivate lipases, enzymes responsible for hydrolysis. However, boiling can also cause oxidation due to high temperature and water involved (*3*, *11*, *13*). The boiled nuts are then dried under the sun (*12*). In West African oven method, shea nuts are roasted or smoked on ovens, which can cause high amounts of polycyclic aromatic hydrocarbons (PAHs) known to be carcinogenic (*11*). While the West African methods involve heating procedure in treating nuts, East African method involves sundrying steps without heating and thus final shea butter extracted can have higher amounts of free fatty acids (FFAs) due to less chance of deactivating lipases (*12*).

The nuts are further cracked to separate kernels from shells and the kernels are further dried by roasting or sun-drying (*11*, *14*). The dried nuts are then subjected to pounding or wet-milling with water to make a paste which are further boiled to separate shea oil from water and shea cakes (*12*, *14*). The shea oil scooped up after boiling are then filtered through a filter cloth and placed in a cool place to solidify into a shea butter (*14*).

Uses of Shea Butter

In Africa, traditionally, shea butter has long been used, dating back to ancient Egypt based on the record showing caravans carried clay jars of shea butter for cosmetic uses during the Cleopatra's Egypt (*2*). Shea butter has been used for ethnopharmacological purposes. Local healers have used shea butter as a treatment for rheumatism, inflammation of nostrils, nasal congestion, leprosy, cough, and minor bone dislocation (*2*, *8*, *10*, *14*). In addition, shea butter has been used to prevent stretch marks among African pregnant women, and massage newly born babies, and shea butter has also been used to protect against *Simulium* infection as insect repellent (*2*). Besides the ethnopharmacological uses, shea butter has served as edible oil in African cuisine due to its highly nutritional value and affordable price (*1*, *10*, *16*, *17*). Shea butter, for example, is used as the base of soup and condiments (*1*). The uses of shea butter are very extensive and thus shea butter is also used as lamp and heating oils, lubricants and in weather-proofing roofs and soap manufacturing in African local communities (*2*, *10*, *11*). Shea butter also benefits domestic animals and is applied to dogs to protect their skin and paws against harsh sand and salt (*2*).

In the international market, shea butter has primarily served as cocoa butter equivalents (CBEs) since 1960s in the manufacture of chocolate especially in Europe due to similarity of chemical and physical properties with more expensive cocoa butter (*9*, *18*). CBEs are plant fats containing no lauric acid that have similar physical and chemical properties to cocoa butter and mixable with cocoa butter in every amount without altering the properties of cocoa butter (*19*). The European Union (EU) allows to use a non-cocoa fat in the manufacture of chocolate up to 5% under the current European Union Chocolate Directive, while the US does not allow to label as chocolate on the products containing any cocoa butter alternatives including CBEs (*9*, *11*). Besides price competitiveness, shea butter also has lower melting point compared to cocoa butter and thus can be easier to make hard candy coatings (*1*). More recently, shea butter has gained increasing popularity as a natural ingredient for cosmetics and personal care products mostly due to its highly moisturizing and emollient properties. In addition, high unsaponifiable levels in shea butter responsible for medicinal properties such as anti-oxidant, anti-inflammatory, and other purported activities can bring more potential for shea butter or shea butter's unsaponifiable fraction to be used in pharmaceutical industries (*9*, *20–23*). Recently, unsaponifiable fraction of shea butter has been used to develop treatments for arthritis, eczema, and herpes lesions and to lower cholesterol levels by a pharmaceutical company, BSP Pharma (*3*).

Benefits of Shea Butter

Shea butter has semi-solid characteristics and buttery consistency, and thus itself can be a good moisturizer or emollient without further processing but shea butter are also found in many moisturizing products as an active ingredient (*24*). In addition, fractionated shea butter, especially olein fraction that is rich in oleic acid and thus in liquid state at room temperature, is formulated in creams or surfactant-based products such as bath products and shampoo to provide the skin, scalp, and hair with well-maintained or increased moisture levels as well as to deliver antioxidants in unsaponifiable fraction to the skin and hair (*9*, *25*). While these emollient and moisturizing properties are due to shea butter's characteristic fatty acid composition, many other medicinal properties are related to shea butter's unsaponifiable constituents which take minor parts but considered valuable.

Shea butter products usually contain claims referring to the anti-aging properties in the skin, which maybe due to the presence of α-amyrin and lupeol. These two triterpenes contribute to the inactivation of proteases that are responsible for degradation of structural proteins such as collagen and elastin (*9*). Shea butter also has some sun-screen function, though providing a weak protection against UV radiation (*9*, *22*). The sunscreening property of shea butter is also related to shea butter's triterpenoids, especially cinnamate esters of triterpene alcohols that are known to have strong absorbance of ultraviolet (UV) radiation in the 250-300nm range (*9*, *22*). Considering that UV-B radiation which is known to cause photocarcinogenesis (*26*) ranges from 290 to 320 nm, these triterpenes can somewhat protect the skin from UV-B radiation.

Another important health benefit of shea butter is its anti-inflammatory effect. A study including eight triterpenes isolated from shea butter revealed significant antiinflamatory activities (acetate esters of α-amyrin, β-amyrin, lupeol, and butyrospermol and cinammate esters of amyrin, β-amyrin, lupeol, and butyrospermol). The activities of these terpenes on the model 12-*O*-tetradecanoylphorbol-13-acetate (TPA)-induced inflammation in mice ear edema, showed that they ranged between 0.15 to 0.75 μmol/ear when compare to indometachin (ID_{50} as 0.91), a commercially available anti-inflammatory drug (*27*).

Shea butter also showed hypersensitivity-alleviating effects. It was found that pharmaceutical composition containing at least 5 % of shea butter's triterpenes such as butyrospermol, lupeol, parkeol, germanicol, dammaradienol, 24-methylene-dammarenol, and α- and β-amyrins effectively suppressed hypersensitivity reaction such as the Immunoglobulin E (IgE)-mediated allergic reaction and autoimmune reaction in mammals (*28*).

Chemical Composition of Shea Butter

Triglyceride Fraction

As a plant fat, shea butter consists of approximately 90% or more of triglycerides and a minor unsaponifiable fraction. Shea butter's fatty acids are composed dominantly of palmitic (16:0), stearic (18:0), oleic (18:1) and linoleic (18:2) acids. Stearic and oleic acids are the major fatty acids ranging from 40-45% of total fatty acids, respectively, while palmitic acid and linoleic acid are present in lower amounts (5-10% and 4%, respectively) (*9*). Compared to plant oils which are in liquid state at room temperature such as grape seed oil (total saturated fatty acids: 10.4-14.3% of total fatty acids), olive oil (12.7-16.2%), and canola oil (5.5-7.7%) (*29–31*), shea butter has relatively high amounts of saturated fatty acids (stearic and palmitic), thus characterized by a solid to semi-solid state at room temperature.

Triglyceride fraction consists of fatty acids (acyl chains) attached to a glycerol backbone. In shea butter, the most common triglycerides are SOS (stearic-oleic-stearic acids, making up to 40 % of total triglycerides), SOO (stearic-oleic-oleic acids, making up to 27 %), POS (palmitic-oleic-stearic acids), POP (palmitic-oleic-palmitic), and OOO (oleic-oleic-oleic acids) with regional variation (*5*, *9*).

Unsaponifiable Fraction

The unsaponifiable fraction contains many bioactive substances including hydrocarbons, tocopherols, sterols, and alcohols and thus provides plant oils and fats with medicinal properties as well as better stability (*32*, *33*). Shea butter has been characterized to contain unusually high amounts of unsaponifiables ranging from 4 to 11% (*9*, *18–20*, *34*), compared to many other plant oils and fats that usually contain 0-2 % of unsaponifiable fraction (*18*, *20*, *34–36*) (Table I).

The main unsaponifiables of shea butter are triterpene alcohols which take approximately 68-75 % of total unsaponifiable fraction. They are commonly found esterified with cinnamic acid or fatty acids rather than free forms (*18*, *19*, *34*). Among the triterpenes, α-amyrin was reported to be a major triterpene ranging from 27.6 to 54.6 % of total triterpenes in shea butter (*18*, *34*, *36*). Other triterpenes found in shea butter include butyrospermol (12.3-26 %), lupeol (16-22.6 %), β-amyrin (7.1-10.6 %), and minor amounts of germanicol, ψ-taraxasterol, parkeol, cycloartenol, 24-methylene-lanost-9(11)-en-3-ol, dammaradienol, and 24-methylene-dammarenol (*9*, *18*, *34*, *36*, *40*).

The second dominant constituents are hydrocarbons, of which karitene was the major hydrocarbon, ranging from 17-27 % (*19*, *34*) to 2-5% of total unsaponifiable fraction (*9*).

Phytosterols were also reported from shea butter's unsaponifiable fraction ranging from 5 to 8 % (*19*, *34*) including α-spinasterol, Δ-7-stigmasterol, Δ-7-avenasterol, β-sitosterol, stigmasterol, campesterol, 24-methyl-cholest-7-enol (*9*, *15*, *18*, *34*, *40*).

Table I. Amounts of Unsaponifiables in Many Plant Fats and Oils Reviewed in Previous Studies

Plant Fat and Oil	*Amounts of Unsaponifiables(%)*	*Family*	*References*
Olive extracted	~2.5	Oleaceae	(*20*)
Coconut	0-0.5	Arecaceae	(*20, 35*)
Cocoa butter	0.1-1.2	Sterculiaceae	(*20*)
Linseed	0.1-1.7	Linaceae	(*20*)
Groundnut	0.2-0.8	Fabaceae	(*20*)
Palm kernel	0.2-0.8	Arecaceae	(*20*)
Babassu kernel	0.2-0.9	Arecaceae	(*20*)
Cotton seed	0.2-1.5	Malvaceae	(*20, 35*)
Rapeseed	0.2-2.0	Brassicaceae	(*20*)
Palm	0.3-1.2	Arecaceae	(*20*)
Safflower seed	0.3-1.3	Asteraceae	(*20*)
Sunflower seed	0.3-1.3	Asteraceae	(*20*)
Camellia seed	0.3-0.4	Theaceae	(*34, 36*)
Kapok seed	0.5-1.0	Malvaceae	(*20*)
Walnut	0.5-1.0	Juglandaceae	(*20*)
Soy bean	0.5-1.7	Fabaceae	(*20, 35*)
Corn (Maize)	0.5-2.8	Poaceae	(*20*)
Tea seed	0.6-0.8	Theaceae	(*34, 36*)
Olive pressed	0.7-1.1	Oleaceae	(*20*)
Mustard seed	0.7-1.5	Brassicaceae	(*20*)
Illepe	0.7-2.0	Dipterocarpaceae	(*20*)
Olive flesh	0.8-1.5	Oleaceae	(*35*)
Pentadesma	0.8-1.8	Clusiaceae	(*37*)
Sesame seed	0.9-2.0	Pedaliaceae	(*20*)
Olive pit's kernel	1.5	Oleaceae	(*35*)
Date palm seed	1.8	Arecaceae	(*38*)
Mange kernel	2.0	Anacardiaceae	(*39*)
Sal	2.0	Dipterocarpaceae	(*39*)
Wheat germ	2-5	Poaceae	(*20*)
Pokeweed seed	2.1	Phytolaccaceae	(*36*)

Continued on next page.

Table I. (Continued). Amounts of Unsaponifiables in Many Plant Fats and Oils Reviewed in Previous Studies

Plant Fat and Oil	*Amounts of Unsaponifiables(%)*	*Family*	*References*
Kokum	2.3	Clusioideae	(*20*)
Mahua	2.4	Sapotaceae	(*39*)
Spinach	2.9	Amaranthaceae	(*34*)
Rice bran	3-7	Poaceae	(*20*)
Alfafa	3.3	Fabaceae	(*34*)
Shea nut	**1.2**	**Sapotaceae**	(*17*)
	4-11	**Sapotaceae**	(*9, 12, 19, 20, 34, 36*)
Avocado flesh	4.8-12.2	Lauraceae	(*35*)
Olive pit's shell	4.9	Oleaceae	(*35*)
Garden balsam	5.6	Balsaminaceae	(*34*)
Avocado kernel	55.5	Lauraceae	(*35*)

Besides triterpenes, hydrocarbons, and phytosterols, unsaponifiable fraction of shea butter also contain minor amounts of tocopherols and phenols as bioactives. The four isomers α, β, γ, and Δ-tocopherols were found in trace amounts in shea, ranging from 0.0029-0.0805% on a weight basis (*23*), with α-tocopherol as the major tocopherol (64% of total tocopherols) (*23*). Phenols were also found in the unsaponifiables, though they are found in high amounts in the kernels (4000 ppm). However, the majority (97%) of phenols are lost during hexane extraction and thus shea butter was found to contain only trace amounts (97 ppm), which shows that phenolics are highly polar and thus less soluble in fats (*40*).

Fractionation and Chemical Composition

Shea butter is subjected to further fractionation after extraction to yield 'hard stearin', a fraction rich in stearic acid and thus solid at room temperature, and 'liquid olein', a fraction rich in oleic acid and thus liquid at room temperature, from shea butter. While stearin fraction is usually used in the manufacture of chocolate as a cocoa butter equivalent (CBE), liquid fraction is preferred in cosmetics and personal care product industries (*9*). As the level of fractionation increases from semi-solid unrefined shea butter to liquid shea butter concentrate, triglyceride fraction decreases due to loss of stearic acids in great amounts and as a result unsaponifiable fraction and amounts of tocopherol increase (*22*).

Factors Causing Variability of Shea Butter Quality

Regional Variability and Chemical Composition

Shea butters from different origins can have varied range of chemical composition due to genetic or environmental factors. Generally, west African subspecies '*paradoxa*' and east African subspecies '*nilotica*' have major difference in chemical compositions. The East African variety yields shea butter generally with higher amounts of oleic acids compared to the West African variety. Several studies found that shea butters from Uganda (subsp. *nilotica*) and Chad were found to contain lowe levels of stearic (22.5-28.9%) and higher levels of of oleic acid (57.8-68.0%). Shea butter samples from Mali, Côte d' Ivoire, Burkina Faso, Benin, and Nigeria (subsp. *paradoxa*) typically contained higher levels of stearic (31.1-46.8%) and similar percentages of oleic acid (39.3-47.5%) (*5*, *15*, *19*, *41*) (Table II). Shea butter samples from the same country can however also exhibit varied composition of fatty acids possibly due to genetic variations within populations, ages of the nuts or different processing practices. Thus, shea butter samples from Mali were characterized by low levels of palmitc (3.3%) and high levels of stearic acid (43.3%) while a sample another place showed higher amounts of palmitic acid (19%) and lower levels of stearic acid (31.1%) (Table II).

Regional variability of fatty acid composition also affect the triglyceride composition of shea butter. Ugandan shea butter samples with relatively larger amounts of oleic acid and lower amount of stearic acid exhibited a triaglycride fraction composed of 19.9% of SOS, 33.4% of SOO, and 19.0% of OOO while the triglyceride fraction of shea butter samples from Mali, Burkina Faso, and Nigeria with relatively larger amounts of stearic acid and lower amounts of oleic acid consisted of 40 % of SOO, 26% of SOO, and 10% of OOO (*5*).

While regional variability in the triglycerol fraction of shea butter is closely related to subspecific variation, unsaponifiable fraction seems to be largely influenced by the environment. Concentration of phenolic compounds especially catechins in shea kernels have been linked to environmental stress. Shea kernels from the hottest and driest areas (e.g., the lake Chad basin area) as well as from much cooler and wetter areas (e.g., Guinea and west Cameroon) all showed the highest amounts of total catechins, compared to shea kernels from regions characterized by moderate rainfall and temperature (e.g., Burkina Faso) (*40*). However, since phenolics were mostly lost during hexane extraction of kernels in the same study, the relation between concentration of gallic acid in shea butter and climate was much weaker than those found in shea kernels (*40*). Unlike phenolics in shea butter, tocopherol contents especially α-tocopherol, the most dominant tocopherol in shea butter, were proportionally correlated to temperature (*23*). Shea butter extracted from shea nuts grown in hot, dry areas (e.g., N'Djamena in Chad) contained much higher amounts of α-tocopherol (414 μg/g) than those from cooler areas of northern Ghana (29μg/g) (*23*).

Regional variation in shea butter was also found in the amount of triterpenes of unsaponifiable. West African shea butters from Nigeria (12.6%), Mali (9.6%), and Burkina Faso (7.1%) contained more triterpene alcohos than the east African Ugandan shea butter (3.7%) (*5*).

Table II. Regional Variability of Fatty Acid Composition (% total fatty acids) of Shea Butter

Origin	*Palmitic acid*	*Stearic acid*	*Oleic acid*	*Linoleic acid*	*References*
Benin	3.8	44.1	43.8	6.65	(*19*)[1]
Burkina Faso	12.1	42.5	39.3	4.5	(*41*)[2]
Burkina Faso	3.3	43.5	44.5	5.9	(*5*)
Côte d' Ivoire	6.6	46.8	51.4	8.4	(*19*)[1]
Mali	3.3	43.3	44.6	6.0	(*5*)
Mali	19	31.1	42.6	5.7	(*41*)[2]
Nigeria	4	46	41	7	(*15*)
Nigeria	3.4	43.8	44.3	5.8	(*5*)
Nigeria	3.2	38.9	47.5	6.5	(*41*)[2]
Chad	4.2	22.5	68.0	4.9	(*19*)[1]
Uganda	4.2	28.9	57.8	6.3	(*5*)
Uganda	6.5	26.4	59.3	6.2	(*41*)[2]

[1] NOTE: Reviewed the study conducted by Jacobsbers *et al.*, 1977 (Title: Causes de I' adification du beurre du karité au cours de la préparation et du stockage des amandes. *Oléagineux*. **1977,** 32, 529-533); [2] Reviewed the study conducted by the Ben Gurion University, Israel.

Post-Harvest Processing and Quality Characteristics of Shea Butter

Various extraction methods can result in shea butters with variable qualities. Initial gathering and processing procedures for shea nuts are presumed to be the major cause leading to inconsistent quality of shea butter extracted from different places. For example, if shea butter is extracted by hexane extraction or extracted from the nuts which were previously parboiled, the resulting shea butter can show almost half levels of free fatty acids compared to shea butter extracted without parboiling of shea nuts (*6*). The study conducted by Mbaiguinam *et al.* (2007), found shea butters from hexane extraction or extraction involving parboiling of shea nuts showed acidity of 5.1-5.5 while the butters extracted from sun-dried nuts which were not parboiled showed acidity of 10.3-10.6, much higher level (*6*).

Solvents used in extracting shea butter from kernels also affect the quality of shea butter. Kar *et al.* (1981) found extraction using water yielded almost half or less than half amounts of shea butter from kernels compared to extraction methods using petroleum ether, n-haxane, chloroform, or benzene (*16*). In addition, while shea butter extracted with these organic solvents contained nearly 0.01 % of vitamin E and showed no detectable levels of peroxides, those extracted with water had no detectable levels of Vitamin E and showed 5.0-8.3 mEq/kg of peroxides (*16*).

Temperature involved in extraction is also an important factor affecting shea butter's quality. Olaniyan *et al.* (2007) reported when shea kernels were previously heated to 50, 70, 90, and 110 °C, respectively, resulting extracted shea butters with dry extraction showed decreased moisture content as the kernel heating temperature increased (*10*). However, too high temperature above 90 °C was deleterious, resulting in lower specific gravity due to expansion of the oil and thus the volume, and also high levels of free fatty acids, higher peroxide value and rancidity value, and decreased ester value that can harm palatability (*10*). In effect, heating step can prevent hydrolytic degradation by deactivating lipase, but too much high temperature can also increase lipase activity by destructing oil cells (*10*). Nahm et al (2011) showed shea butter was greatly protected from oxidation with the addition of antioxidants. No significant increase in peroxide value, conjugated dienes, and TBARS, and no significant decrease in major fatty acids were found in most samples with antioxidants over the storage period while there was significant increase in oxidation parameters and decrease in the major unsaturated fatty acids, oleic and linoleic acids, were found. The effectiveness of natural antioxidants (rosmarinic and gallic acids) was almost the same as the synthetic BHT comparisonone.

Quality Analysis of Locally Processed West African Unrefined Shea Butter

Fair Trade and/or Organic certifications of shea butter products are supporting local processing of shea butter that is extracted with traditional extraction methods, natural extraction methods using no solvent. However, without quality control systems in place, shea butter processed by local African women can lead to high variations in quality. In the present study, seven west African shea butter samples were analyzed. Five samples were from Ghana (TPC01-03, GHA01-02), one sample was from Benin (BEN01), and another sample was from a local market in the US with no information on specific origin (West Africa) (WAF01).

All of these shea butter samples showed light yellow to yellow color (*43*) which was in agreement with previous reports (*2*, *12*). The yellow color might be due to the presence of β-carotene (*20*). Aroma of the samples revealed shea butter's characteristic aroma but some samples showed hint of chocolate-like aroma.

The samples were subjected to routine quality analysis. Specific gravity was 0.91 in all samples which was in accordance with specific gravity range of 0.87-0.97 from previous studies (*10*, *17*, *44*) (Table III). Refractive index ranged from 1.463 to 1.466 which was within the typical refractive index range of shea butter, 1.463-1.467 (*20*) (Table III). Different melting points were tested with the samples to determine their applicability in assessing the quality of shea butter. The dropping point seemed to be the best with the narrower range of variation (31-34°C) (Table III). This character makes shea butter melt once applied on the skin while shea is usually in a semi-solid to solid state at room temperature. Another melting point procedure tested was the clear point defined as the temperature at which shea becomes a clear liquid oil without cloudiness. The results showed that shea butter samples melted into clear liquid oil at the range of 46-59°C, showing a high variation in the temperature range (Table III). The results suggested that some shea butter samples contained fatty acids that melt at very high temperature (e.g., 59°C) while others samples become clear at relatively low temperatures (e.g., 46°C).

Moisture content ranged from 0.01 to 0.20 % (Table III). Lower moisture content is better regaring to quality of shea butter in that high moisture content can lead to hydrolytic and oxidative degradation. Insoluble impurities of shea butter samples were measured as 0.12-0.15%, and this parameter was also needed to be controlled since insoluble impurities lower shea butter quality in that they may include metals which can accelerate oxidation (Table III). Free fatty acids were found to be the most variable quality parameter of the shea butter samples, ranging from 1.07 to 8.56 % (Table III). This range was much higher than free fatty acids of commercially available vegetable oils in a range of 0.02-1.38% (*45, 46*). Peroxide value was measured as 2.15-15.32 mEq/kg (Table III). Peroxide values were quite low enough in most samples, which were less than 10 mEq/kg. One sample showed, however, very high levels of peroxides as 15.32 mEq/kg and considering that the sample was placed in very hot place in summer without air-conditioning system when purchased, the shea butter must have been oxidized in great extent before arrival to our laboratory. There was no clear correlation among moisture, insoluble impurities, free fatty acids, and peroxides due to small samlple size, but it is resulted that for analyzing quality, all factors should be considered collectively.

Shea butter is characterized to have exceptionally high amounts of unsaponifiables compared to other plant oils mostly containing ~2% of unsaponifiables (Table I). Our results showed shea butter samples contained 2.21-4.18% of unsaponifiable fraction, which was lower than previously reported value, 4-11% (*9*, *18–20*, *34*) (Table III). However, considering shea butter samples still contained higher amounts of unsaponifiables than most plant oils, the use of shea butter or shea butter's unsaponifiable fraction is promising as a medicinally active ingredient. Gas chromatographic (GC) analysis of unsaponifiable fraction , especially the composition of triterpenes and sterols found α-amyrin was the major triterpene that takes almost 60 % of total triterpenes and sterols. α-amyrin was reported to exhibit anti-inflammatory property (*47–49*).

GC analysis on fatty acid composition of west African shea butter showed most shea butter samples shared consistent composition of fatty acids. Major fatty acids were palmitic, stearic, oleic, and linoleic acids which composed 3.36-4.44, 39.74-44.62, 40.71-44.48, and 5.73-6.41% of total fatty acids, respectively (Table III). The consistent composition of fatty acids are related to consistent physical characterietics such as density, refractive index, and melting point among west African shea butters.

Table III. Quality Characteristics of Seven Locally Processed West African Shea Butter

Property	*TPC 01*	*TPC 02*	*GHA 01*	*GHA 02*	*WAF 01*	*TPC 03*	*BEN 01*
Physicochemical parameters							
Specific gravity							
	0.91	0.91	0.91	0.91	0.91	0.91	0.91
Refractive index							
	1.466	1.464	1.463	1.465	1.464	1.464	1.465
Melting Point (°C)							
Dropping	34	31	31	32	31	33	32
Clear	54	52	55	47	52	46	59
Moisture content (%)							
	0.06	0.09	0.20	0.06	0.12	0.01	0.06
Insoluble impurities (%)							
	0.14	0.12	0.15	0.12	0.12	0.12	0.12
Free fatty acids (%)							
	1.94	3.92	8.56	1.46	7.91	2.11	1.07
Peroxide value (mEq/kg)							
	2.15	6.01	8.73	15.32	8.94	4.46	6.79
Unsaponifiables (%)							
	3.99	4.18	2.69	2.21	2.77	4.17	2.56
Fatty acid composition (% total fatty acids)							
Palmitic	3.69	3.95	4.44	3.62	3.63	3.58	3.36
Stesric	42.67	43.82	39.74	43.49	43.62	43.97	44.62
Oleic	41.53	40.71	44.48	41.91	42.61	42.46	43.40
Linoleic	5.84	5.77	6.41	5.86	6.02	6.06	5.73

Currently Available Shea Butter Standards

Quality standard for unrefined shea butter has been developed by ProKarité, a project managed by the World Agroforestry Centre and funded by Common Fund for Commofities/ Food and Agriculture Organization (CFC/FAO) that has been approved by Union Economique Monétaire Ouest Africaine (UEMOA) (*50*). This regional board has proposed sensory, physical, and chemical parameters that define the shea butter's quality (e.g., color, odor, taste, free fatty acids, peroxide value, insoluble impurities, moisture, volatile matters, soap content, relative density, refractive index, saponification value, unsaponifiable matters, iodine value, and melting point) (*50*, *51*). Among these characters, grading system has developed that determine the quality of shea butter according to moisture, insoluble impurities, free fatty acids, and peroxide value (Table IV). In this grading system, Grade 1 is the highest quality shea butter that can be used by cosmetics and pharmaceuticals, and for direct consumption. Grade 2 is the second greart quality shea butter that can be used in food industry for manufacturing confectionary, chocolate, edible oil, and basis for margarines. Grade 3 represents the shea butter that is recommended to be used in soap-making or further refined for direct consumption.

Table IV. Quality Characteristics and Grading System for Unrefined Shea Butter

Parameters	*Grade 1*	*Grade 2*	*Grade 3*
	Min. Max.	*Min. Max.*	*Min. Max.*
Moisture (%)	~ 0.05	> 0.05 ~ 0.2	> 0.2 ~ 2.0
Insoluble impurities (%)	~ 0.09	> 0.09 ~ 0.2	> 0.2 ~ 2.0
Free fatty acids (%)	~ 1.0	> 1.0 ~ 3.0	> 3.0 ~ 8.0
Peroxide value (mEq/kg)	~ 10.0	> 10.0 ~ 15.0	> 15.0 ~ 50.0

The American Shea Butter Institute (ASBI) is a US-based organization that also serves grading of shea butter with its own standard, where shea butter is graded from Grade A to Grade F (*52*). ASBI regards unsaponifiables as another important parameter in grading along with many physicochemical properties (*52*).

Ways To Produce Shea Butter with Consistent Quality and Extend Shelf-Life

Efforts Made on Locally Processed Shea Butter To Improve the Quality

During the last decades, efforts have been made to enhance the quality of in-country processed African shea butter. The efforts were mostly made with partnership between 'university, non-government organizations (NGOs), and/or companies' and 'African private business sectors or unions of undividual shea butter procedures' by setting up quality control (QC) systems and providing tranings . These efforts ultimately aimed at increasing market value of locally processed African shea butter for the benefits that returns to local shea nut collectors and producers.

The Northern Uganda Shea Processors Association (NUSPA) that established with funding from the United States Agency for International Development (USAID) has been producing pharmaceutical-grade shea butter for exports to the

US, Europe, and Japan (*53*). In addition, NUSPA assisted Sudan by providing trainings and by helping to establish its processing centers involving rural women (*53*).

Agribusiness in Sustainable Natural African Plant Products (ASNAPP), a non-profit organization formed with funding from USAID has working with a Ghanaian private sector and woman shea nut collectors. ASNAPP especially has assisted The Pure Company which has just started to industrially produce shea butter with collaboration with Rutgers, the State University of New Jersey. The New Use Agriculture and Natural Plant Products (NUANPP) Program in Rutgers University has provided technical support to develop QC protocols, train lab technicians, and conduct cross validation of the results from quality tests on shea butter by doing the same tests in labs in Rutgers University and The Pure Company, Ghana at the same time.

L'Occitane, a France-based cosmetics company which is famous for various shea butter products, buys shea butter directly from Burkina Faso through "Union des groupements Kiswendsida (UGK), a network of shea butter producers (*53*). This has been able to provide better returns to the producers due to no middleman involved (*53*). The company provide tranings in quality control, and pay for the shea butter in advance (*53*), which can motivate the African woman to produce better quality shea butter and ultimately contribute to the livelihood and development of African women.

Currently, an international shea alliance was launched. In effect, this alliance will be officially launched at 'Shea 2011: Sustainable Solutions', the international shea industry conference, which will be held on April 4-7, 2011, in Accra, Ghana. This alliance consists of more than 50 major shea industry stakeholders uncluding traders, exporters, service providers, and NGOs (*54*, *55*) and is aimed at promoting shea butter worldwide as well as establishing industry standards for quality and sustainable sourcing, and exchanging information (*54*).

Addition of Antioxidants To Protect Shea Butter from Oxidation

As a plant fat containing large amounts of unsaturated fatty acids, shea butter is also at risk to be oxidized when processed or stored inappropriately just as many other edible oils even though shea butter has natural antioxidants in it such as tocopherols. Lipid oxidation is a major problem in edible lipids, resulting in chemical, sensory, and nutritional deterioration (*42*). It is needed to extract shea butter using appropriate temperature which can effectively prevent hydrolytic degradation while does not cause oxidative degradation.

As many previous studies on plant oils and fats proved, addition of natural antioxidants can be one way to maintain shea butter's quality and thus extend shelf-life. Addition of natural antioxidants or plant extracts (e.g., ascorbyl palmitate, caffeic acid, olive leave extracts and hydrolysate, protocatechuic acid, quercetin, rosemary extracts, sesame extracts, soy lecithin, tocopherols) were found to be effective in protecting various plant oils and fats (e.g., camellia, corn, olive, palm, rapeseed, soybean, sunflower, walnut oils and margarine made of sunflower oil and palm stearine) against oxidation (*56–61*).

Our study on oxidative stability of shea butter using butylatedhydroxy toluene (BHT), rosmarinic acid, and gallic acid also found the natural antioxidants used (e.g., rosmarinic and gallic acids) were almost equally or more effective in protecting shea butter against oxidation compared to synthetic antioxidant (e.g., BHT) (*62*). Compared to the control, shea butter samples without antioxidants, shea butters added with antoxidant showed significantly lower amounts of peroxide value, conjugated dienes, and thiobarbituric acid reactive substances. In addition, the fatty acids were well kept in oxidation-accelerating condition in shea butters with antioxidant while control lost significant amounts of major unsaturated fatty acids, oleic and linoleic acids.

Conclusions

Shea butter is an invaluable and versatile plant fat beloved by sub-saharan Africans which has gained increasing popularity as a natural ingredient of cosmetics and personal care products in the international market during last decades. The uses of shea butter has also stretched to food and confectionary industries as cocoa butter equivalents (CBEs) and bases for margarines and pharmaceutical industry as active ingredients in treatments especially targeting inflammation. However, locally produced shea butter have limitation that is inconsistent quality brought by variable collecting, treatment of shea nuts, extraction methods, and storage condition without strict quality control systems. The inconsistency can be bottleneck to access to the international market that requires consistent supply of quality shea butter and also can discourage shea butter's inherent great bioactivities such as anti-infammation or anti-oxidant. Therefore, the future study on shea butter needs to be focused on 'how to keep or improve the quality of shea butter'. As our study found natural antioxidants were significantly effective in protecting shea butter from oxidation, other natural plant extracts or mixture of them and/or natural antioxidants can be studied on shea butter's stability. In addition, better production procedures and storage condition should be studied to find optimum condition (e.g., temperature, packaging materials, etc.) where degradations, especially hydrolytic and oxidative degradations, are minimum and thus shea butter keeps its quality including bioactivities in maximum level.

Acknowledgments

We thank ASNAPP-Ghana, and in particular Julie Asante-Dartey, and Victor Attafua and John Addaquay, the Pure Company (Buipe, Northern Ghana) for providing many of the shea samples we have used in our research. We acknowledge the work, traditional knowledge and expertise of the African women who collect, process and trade in shea and thank those that spoke with us and shared how they process and work with shea. Funds from the New Use Agriculture and Natural Plant Products Program and the New Jersey Agricultural Experiment Station are acknowledged.

References

1. Chalfin, B. *Shea Butter Republic*; Routledge: New York, 2004; pp 1−40.
2. Goreja, W. G. *Shea Butter: The Nourishing Properties of Africa's Best-Kept Natural Beauty Secret*; Amazing Herbs Press: New York, 2004; pp 3−4.
3. Masters, E. T.; Yidana, J. A.; Lovett, P. N. *Unasylva* **2004**, *210*, 46–52.
4. Ferris, R. S. B.; Collinson, C.; Wanda, K.; Jagwe, J.; Wright P. *Submitted to USAID, The United States Agency for International Development*; 2001; www.foodnet.cgiar.org/Projects/Sheanut_Rep.pdf, accessed on 03/08/12.
5. Di Vincenzo, D.; Maranz, S.; Serraiocco, A.; Vito, R.; Wiesman, Z.; Bianchi, G. *J. Agric. Food Chem.* **2005**, *53*, 7473–7479.
6. Mbaiguinam, M.; Mbayhoudel, K.; Djekota, C. *Asian J. Biochem.* **2007**, *2*, 101–110.
7. Anonymous. 2002; http://unctad.org/infocomm/francais/karite/docs/world3.pdf, accessed on 03/08/11.
8. Tella, A. *J. Clin Pharmacol.* **1979**, *7*, 495–497.
9. Alander, J. *Lipid Technol.* **2004**, *16*, 202–205.
10. Olaniyan, A. M.; Oje, K. *J. Food Sci. Technol.* **2007**, *44*, 404–407.
11. Lovett, P. *The Shea Butter Value Chain. Production, Transformation and Marketing in West Africa*; WATH Technical Report No. 2 (publication produced for review by the United States Agency for International Development), 2004; felmart.com/valuechain.pdf, accessed on 03/09/11.
12. Moharram, H.; Ray, J.; Ozbas, S.; Juliani, H.; Simon, J. In *Herbs: Challenges in Chemistry and Biology*; Wang, M., Shengmin, S., Hwang, L. S., Ho, C. T., Eds.; ACS Symposium Series 925; ACS: Washington DC, 2006; pp 326−340
13. Bail, S.; Krist, S.; Masters, E.; Unterweger, H.; Buchbauer, G. *J. Food Comp. Anal.* **2009**, *22*, 738–744.
14. Addaquay, J. *The Shea Butter Value Chain. Refining in West Africa*; WATH Technical Report No. 3 (publication produced for review by the United States Agency for International Development), 2004.
15. Badifu, G. I. O. *J. Food Comp. Anal.* **1989**, *2*, 238–244.
16. Kar, A.; Mital, H. C. *Qual. Plant. - Plant Foods Hum. Nutr.* **1981**, *31*, 67–69.
17. Njoku, O. U.; Eneh, F. U.; Ononogbu, I. C.; Adikwu, M. U. *J. Nutraceut Funct. Med. Foods* **2000**, *2*, 33–39(currently, *J. Diet. Suppl.*).
18. Peers, K. E. *J. Sci. Food Agric.* **1977**, *28*, 1000–1009.
19. Lipp, M.; Anklam, E. *Food Chem.* **1998**, *62*, 73–97.
20. Rossell, J. B. *Analysis of Oils and Fats*; Hamilton, R. J., Rossell, J. B., Ed.; Elsevier Applied Science: New York, 1987; pp 1−90.
21. Rogers, S.; O'Lenick, A., Jr. U.S. Patent 7,544,824 B2, 2009.
22. Alander, J.; Andersson, A. C. *Cosmet. Toiletries Manuf. Worldwide* **2002**, *2002*, 202–205.
23. Maranz, S.; Wiesman, Z. *J. Agric. Food Chem.* **2004**, *52*, 2934–2937.
24. Kraft, J. N.; Lynde, C. W. *Skin Ther. Lett.* **2005**, *10*, 1–8.
25. Rogers, S.; O'Lenick, A., Jr. U.S. Patent 7,182,940 B1, 2007.
26. Velasco, M. V. R.; Sarruf, F. D.; Salgado-Santos, I. M. N.; Haroutiounian-Filho, C. A.; Kaneko, T. M.; Baby, A. R. *Int. J. Pharm.* **2008**, *363*, 50–57.

27. Akihisa, T.; Kojima, N.; Kikuchi, T.; Yasukawa, K.; Tokuda, H.; Masters, E. T.; Manosroi, A.; Manosroi, J. *J. Oleo Sci.* **2010**, *59*, 273–280.
28. Wiedner, M. S. U.S. 2008/0124413 A1, Issued May 29, 2008.
29. Baydar, N. G.; Özkan, G.; Cetin, E. S. *Grasas Aceites* **2007**, *58*, 29–33.
30. McClements, D. J.; Decker, E. A. In *Fennema's Food Chemistry*, 4th ed.; Damodaran, S., Parkin, K. L., Fennema, O. R., Eds.; CRC Press: Boca Raton, FL, 2008; pp 155−216.
31. Samman, S.; Chow, J. W. Y.; Foster, M. J.; Ahmad, Z. I.; Phuyal, J. L.; Petocz, P. *Food Chem.* **2008**, *109*, 670–674.
32. Esuoso, K. O.; Lutz, H.; Bayer, E.; Kutubuddin, M. *J. Agric. Food Chem.* **2000**, *48*, 231–234.
33. Shahidi, F.; Zhong, Y. *Chem. Soc. Rev.* **2010**, *39*, 4067–4079.
34. Itoh, T.; Tamura, T.; Matsumoto, T. *Lipids* **1974**, *9*, 173–184.
35. Gutfinger, T.; Letan, A. *Lipids* **1974**, *9*, 658–663.
36. Itoh, T.; Uetsuki, T.; Tamura, T.; Matsumoto, T. *Lipids* **1980**, *15*, 407–411.
37. Tchobo, F. P.; Natta, A. K.; Barea, B.; Barouh, N.; Piombo, G.; Pina, M.; Vileneuve, P.; Soumanou, M. M.; Sohounhloue, D. C. K. *J. Am. Oil Chem. Soc.* **2007**, 84, 755−760.
38. Nehdi, I.; Omri, S.; Khalil, M. I.; Al-Resayes, S. I. *Ind. Crops Prod.* **2010**, *32*, 360–365.
39. Dhara, R.; Bhattacharyya, D. K.; Ghosh, M. *J. Oleo Sci.* **2010**, *59*, 169–176.
40. Maranz, S.; Wiesman, Z.; Garti, N. *J. Agric. Food Chem.* **2003**, *51*, 6268–6273.
41. Ferris, R. S. B.; Collinsom, C.; Wanda, K.; Jagwe, J.; Wright, P. *Evaluating the Market Opportunities for Shea Nut and Shea Nut Processed Products in Uganda* (publication submitted to the United States Agency for International Development), 2001; www.foodnet.cgiar.org/projects/sheanut_rep.pdf, accessed on 03/14/11.
42. Velasco, J.; Dobarganes, C. *Eur. J. Lipid Sci. Technol.* **2002**, *104*, 661–676.
43. Nahm, H. S. M.S. Thesis, Rutgers, The State University of New Jersey, New Brunswick, NJ, May 2011.
44. Akingbala, J.; Falade, K.; Adebesi, E.; Baccus-Taylor, G.; Lambert, I. *West Indian J. Eng.* **2006**, *29*, 73–80.
45. Gan, H. L.; Che Man, Y. B.; Tan, C. P.; NorAini, I.; Nazimah, S. A. H. *Food Chem.* **2005**, *89*, 507–518.
46. Osawa, C. C.; Goncalves, L. A. G.; Ragazzi, S. *J. Food Compos. Anal.* **2007**, *20*, 523–528.
47. Otuki, M. F.; Ferreira, J.; Lima, F. V.; Meyre-Silva, C.; Malheiros, A.; Muller, L. A.; Cani, G. S.; Santos, A. R. S.; Yunes, R. A.; Calixto, J. B. *J. Pharmacol. Exp. Ther.* **2005**, *313*, 310–318.
48. Otuki, M. F.; Lima, F. V.; Malheiros, A.; Yunes, R. A.; Calixto, J. B. *Eur. J. Pharmacol.* **2005**, *507*, 253–259.
49. Holanda Pinto, S. A.; Pinto, L. M. S.; Cunha, G. M. A.; Chaves, M. H.; Santos, F. A.; Rao, V. S. *Immunopharmacology* **2008**, *16*, 48–52.
50. Lovett, P.; Miller, E.; Mensah, P., Adams, V.; Kannenberg, C. *Shea Butter Exporter Guide* (publication produced for review by the United States Agency for International Development, USAID), 2005.

51. Anonymous. *African Organisation for Standardisation*; Central Secretariat. Nairobi, Kenya. 2012. www.arso-oran.org/wp-content/uploads/2012/10/Africa-standards-for-shea-butter-unrefined.pdf, accessed on 05/03/2013.
52. *American Shea Butter Institute*; 2011; www.sheainstitute.com, accessed on 03/16/11.
53. Harsch, E. *Afr. Recovery* **2001**, *15*, 6.
54. Lamport, J. *International Shea Alliance Launched*; 2010; www.watradehub.com/activities/tradewinds/nov10/international-shea-alliance-launched, accessed on 04/17/12.
55. *Global Shea*; 2011; www.globalshea.com/about-us, accessed on 03/16/11.
56. Judde, A.; Villeneuve, P.; Rossignol-Castera, A.; Le Guillou, A. *J. Am. Oil Chem. Soc.* **2003**, *80*, 1209–1215.
57. Abramovic, H.; Abram, V. *Acta Agric. Slov.* **2006**, *87*, 255–261.
58. Kowalski, R. *Acta Chromatogr.* **2007**, *18*, 15–23.
59. Bouaziz, M.; Fki, I.; Jemai, H.; Ayadi, M.; Sayadi, S. *Food Chem.* **2008**, *108*, 253–262.
60. Azizkhani, M.; Zandi, P. *World Acad. Sci., Eng. Technol.* **2009**, *49*, 93–96.
61. Konsoula, Z.; Liakopoulou-Kyriakides, M. *LWT−Food Sci. Technol.* **2010**, *43*, 1379–1386.
62. Nahm, H. S.; Juliani, H. R.; Simon, J. E. *J. Medicinally Act. Plants* **2011**, *1* (2), 69–75**2011**.

Chapter 13

Guidelines for Quality Control during Preclinical Testing of African Traditional Medicines

Renee A. Street,*[,1] Gerhard Prinsloo,[2] and Nceba Gqaleni[1]

[1]University of KwaZulu-Natal, Traditional Medicine Laboratory, Durban, 4041, South Africa
[2]University of South Africa (UNISA), Department of Agriculture and Animal Health, Florida Campus, Florida, 1710, South Africa
***E-mail: street@ukzn.ac.za.**

Quality control of medicinal plants is becoming more critical in the anticipation of commercialization. Preclinical testing facilitates the compilation of safety, quality and efficacy data prior to commencement of clinical trials. Numerous pre- and postharvest factors influence such data. Coupled to the lack of regulatory guidelines for the testing of African medicinal plants, progress in the commercial development of many African medicinal plants has been impaired. Preclinical studies on traditional herbal based remedies available to the public are imperative to ensure safe products with proven efficacy. Challenges and scientific issues affecting safety, quality and efficacy are discussed.

The two biggest misconceptions with regards to African traditional medicines is that 'natural' equates to 'safe' and that 'traditional' equates to 'rigid/unchanging'. The public often deem traditional medicines harmless based on their extensive history of use. However, the lack of regulatory guidelines provides no such assurance and the difference between historical uses, including the specific preparations and plant tissues used, may be quite different than modern prepared extracts and products. Different methods of preparation and concentration of bioactive fractions may lead to different activities including different efficacy and safety.

African traditional medicines are constantly adapted to cater for the current health needs including epidemics and other public health concerns. A paradigm shift is evident from the use of single ingredients (herbs, minerals and animals) to combining them to generate new herbal remedies to enhance therapeutic properties (*1*). Even these single ingredients have complex chemistries. Under such circumstances, quality control is paramount since it can affect the safety and/or efficacy of the herbal products being used (*2*). The isolation and characterization of secondary metabolites (or natural plant products) in the extract and reisolation of known compounds for the standardization of plant extracts remains problematic (*3*). Natural products are by and large isolated in small quantities that are inadequate for lead optimization, lead development, and clinical trials (*4*). Despite the fact that the Sub-Saharan region as well as the Indian Ocean Islands, contain roughly a quarter of the world's total plant species, Africa has only contributed 83 of the world's 1,100 leading commercial medicinal plants (*5*). Quality assurance is the shared responsibility of manufacturers and regulatory bodies. Manufacturers must adhere to good agricultural and collection practice (GACP), good manufacturing practice (GMP) and good laboratory practice (GLP) standards, ascertain appropriate specifications for their products, intermediates and starting materials and compile well-structured all-inclusive documentation on pharmaceutical development and testing (*6*). Factors contributing to African natural plant products competing poorly in international markets include inappropriate methods of collection, processing and storage with undesirable contaminants in available products (*7*). Although numerous factors can be controlled by implementing standard operating procedures (SOP) leading to GACP, GLP and GMP for producing medicinal products (*8*), the current lack of regulatory guidelines on African traditional medicines causes a setback with regards to product development. For an extensive review of the regulatory situation in Africa and abroad, see 'WHO's Regulatory situation of herbal medicines: a worldwide review' (*9*). Recent work by the Association for African Medicinal Plant Standards (AAMPs) and their recently published African Herbal Pharmacopoeia contribute to clarification of those by providing monographs on selected African medicinals (*10*).

The sections below discuss the influence of several pre- and post-harvest factors which influence the safety, efficacy and quality of African traditional medicines.

Impact and Application of Good Agricultural Practice (GAP)

The growing demand for medicinal plants has not only resulted in an increased hazard for overexploitation of wild populations but also an increased interest in cultivation (*11*). Cultivation offers possibilities for sustainable development and protection of natural resources (*12*) as well as the opportunity to optimize yield and attain a more uniform, high quality product (*13*). Cultivation of medicinal plants in Europe, China, India and USA has been widely adopted but the most widespread practice in Africa is still to collect medicinal plants from the wild (*14*). Agricultural expansion could generate employment through small entrepreneurial endeavors (*12*) however until basic propagation and cultivation information on African medicinal plants is available, small-scale farmers cannot be expected to venture into growing medicinal plants as this could result in loss of income (*15*).

Successful propagation of medicinal plants is dependent on the understanding of basic principles of plant growth and manipulation under controlled conditions (*16*). Plant propagation from seed or by vegetative means both have pros and cons which directly affects the quality of the end product. When grown from seed, plants are generally stronger and express genetic diversity (*16*). However, low germination rate which is a common phenomenon with medicinal plant species is often a result of fungal infection or mechanical seed damage (*13*). This is further exacerbated by incorrect harvest times, coupled with poor seed storage conditions and often a lack of knowledge relative to after-ripening and other environmental treatments the seeds need to be exposed to facilitate germination. In addition, harvesting seeds from wild populations, which is the norm, in most developing countries, often leads to the depletion of already diminishing stocks (*15*). Vegetative propagation (multiplication through cuttings) ensures continuity of desirable traits (such as increased active constituent) as the plants are direct clones of their parent material and usually reach maturity faster than those grown from seed (*16*).

Propagation of medicinal plants through tissue culture can be divided into three broad categories. The most common approach is to isolate organized meristems such as the meristematic shoot tips or axillary buds and induce them to grow into complete plants. This system of propagation is commonly referred to as micropropagation. In the second approach, adventitious shoots are initiated on leaf, root and stem segments or on callus derived from those organs. The third system involves induction of somatic embryogenesis in cell and callus cultures. This system is theoretically most efficient as large numbers of somatic embryos can be obtained once the whole process is standardized. Biotechnology involving modern tissue culture, cell biology and molecular biology offers the opportunity to develop new germplasms that are well adapted to changing demands (*17*). Medicinal plant cultivation opens up the opportunity of using biotechnology to decipher common problems such as species misidentification, genetic and phenotypic variability, variability and instability of extracts, toxic elements and contaminants (*13*). In fact medicinal plants cannot be utilized on a large scale without modifying the characteristics of the available plant populations, be it wild or domesticated (*13*).

In vitro propagation of African medicinal plants has several advantages including production of a large number of microplants which can provide medicinal plant nurseries with seedlings that are in high demand (*18*). However, in several African countries, basic infrastructure and facilities for the simplest tissue culture techniques and even those just involving micropropagation are not available and is coupled with a shortage of skilled human resources in the plant sciences and biotechnology (*19*). There are reports on the micropropagation of African medicinal plants (*20*, *21*), however few result in field trials and information regarding the effect of propagation methods on the active constituents is lacking.

Cultivation techniques should take into consideration (1) site selection, (2) climate, (3) soil, (4) irrigation and drainage (5) plant maintenance and protection and (6) ecological environment and social impact (*22*). Secondary metabolism plays an important role in responding to environmental changes (*23*). Traditional Health Practitioners (THPs) are aware of the influence of environmental stresses on the potency of their medicinal plants, whilst scientists understand the significant effect on the chemistry of the plants (*24*). *Boophone disticha* (L.f.) Herb.is an example of a plant traditionally only harvested from certain locations. The plant is very poisonous, and by collecting it only from a specific location, the THP can ostensibly anticipate the desired therapeutic effect (*24*) however this has not been verified as of yet. Variation in harvest site and environmental conditions can have a pronounced effect on biological activity (*25*, *26*) as the active constituents of the plants adapt to fluctuating environmental conditions (*13*). Similarly soil nutrient content is a key factor which affects plant growth and development and in turn the production and accumulation of secondary metabolites (*27*). Soil and irrigation water should be within regional/national limits or free from potentially unsafe substances such as heavy metals, pesticides and herbicides (*22*). The geoclimatic/ environmental conditions of the region and the agricultural practices determine the level of contaminants that could accumulate in raw herbal materials. Natural resources may be contaminated due to excessive use and disposal of chemicals. These chemical residues have the capacity to cause abiotic and biotic changes at different levels of the ecosystem (*28*).

Few pragmatic data on agricultural practices of African traditional medicines are available, which slows down feasible improvements (*27*). Although there are numerous random studies on various species, plants need to be taken through the entire process from propagation and cultivation methods to plant collection as each step influences the quality of the end product. Medicinal plants grown in green houses under GAP would be an asset with regards to the reproducible quality needed for registration as medicines (*29*).

Social Impact of GAP

Biodiversity loss does not only entail a loss in biological resources, but may also involve a loss of cultural practices (*2*, *30*). Cultivation of medicinal plants can combine biodiversity conservation and poverty alleviation (*30*). There is a continuous concern about the acceptance of cultivated medicinal plants, as cultivated material is believed to lack the 'power' of healing (*31*). Although cultivated medicinal plants are accepted as a substitute in countries such as Swaziland and South Africa, more conservative THPs believe that plants grown under cultivation will have different healing properties from those harvested from wild populations (*11*). It was found however that beliefs concerning cultural uses are more enduring than the traditional use concerning cultivation. In trying to conserve the knowledge and use of medicinal plants other options are investigated to reduce pressures from wild harvesting. Cultivation of medicinal plants in nurseries, medicinal plant gardens and homestead gardens was explored to ensure that these plants would be grown and conserved in other localities as well (*15*, *30–32*).Working in close partnership with those THPs and directly with communities to assure them that the buyers will purchase and use- and in some cases may require that medicinal plants be cultivated and not gathered from the wild can go a long way toward the acceptability of introducing many medicinal plants into cultivation.

The social impact of cultivation on local communities should ensure that a negative effect on local livelihood is avoided (*22*). A number of factors need to be taken into consideration in linking of biodiversity conservation and poverty alleviation: (a) choice of specific target groups and the identification of the links between cultivation practices and livelihood conditions, (b) function of cultural factors in medicinal plant use and cultivation, and (c) cultivation by local people being not primarily based on local awareness of the loss of wild species, but on local perceptions about economically beneficial medicinal plants (*30*). Women hawkers spend most of their time at the market rather than at home, and thus cultivation competes with their trading activities (*30*). With regards to local income earning opportunities, small-scale cultivation is often preferable to large-scale production, especially if small-scale farmers are organized to market their products communally. If large-scale medicinal plant cultivation is or has been established, local communities should profit directly from, for example, fair wages, equal employment opportunities, and capital reinvestment (*22*). In South Africa, there is a lack of any notable investment into the development of certified traditional medicines and a rapid emergence of manufacturers of laissez faire traditional medicines that lay claim to cures for all types of health conditions, most of which are obtaining their raw materials from unsustainable plant supply sources (*33*).

Impact and Application of Good Collection Practice (GCP)

The development/implementation of a robust GCP should involve the following: (1) permission to collect, (2) technical planning, (3) selection of medicinal plants for collection, (4) collection, and (5) trained personnel (*22*). Ideally, collection permits should be required for any commercial collection of any plant (*14*). The advantage of the permit system (as opposed to banning), is that it provides flexibility to the issuing authority. For example permits can be suspended for one year if resources are limited. In doing this, the issuing authority can efficiently control the wild populations (*14*). Unfortunately most developing countries, especially those with rich tropical floras, do not have lists of their threatened species due to a lack of basic field information on the status of the individual plants (*14*). Awareness and understanding needs to be raised among THPs, collectors and traders to ensure that they have the correct collections permits and that the conditions are observed (*32*). A sustainable harvesting (pilot) project was successfully implemented in the Umzimkhulu Forests (KwaZulu-Natal, South Africa). The 'Sizamimpilo Harvesters Association' has been registered and legal harvesting permits have been issued to the members for the collection of indigenous medicinal plants (*34*).

Collection practices should guarantee the long-term survivability of wild populations and their habitats (*22*). In South Africa, unsustainable underground plant parts are commonly found at medicinal markets (*11*). Unfortunately little attention is paid to South African medicinal bulbous plants, and not much information is available on these slow growing plant species (*35*). The regional demand for *Merwilla plumbea* (Lindl.)Speta bulbs in KwaZulu-Natal (South Africa) is 300,000 bulbs per year, that are to be eight to ten years of age (*31*). Plant part substitution is a key conservation tactic, especially with regards to slow growing endangered species. However as phytochemicals amass in particular cells or organs it is important to identify and characterize these structures (*27*). Various studies have shown that biological activity can differ between plant parts of the same species (*36*, *37*). A recent study on the phytochemical composition and biological (antibacterial and anticandidal) activities of bulb and leaves of four widely used medicinal plant species (*Tulbaghia violacea*, *Hypoxis hemerocallidea*, *Drimia robusta* and *M. plumbea*) was conducted with the view of promoting the use of leaves as a conservation strategy (*38*). The study concluded that the leaves may be used as a substitute for bulbs in the treatment of bacterial and fungal ailments. Such studies are vital to optimize the utilization of available resources and to improve sustainability.

Medicinal plant species should be collected during a suitable season or time period to ensure the uppermost possible quality of both source materials and finished products (*22*, *39*). A study on the Eastern Nigeria mistletoe (*Loranthus micranthus*(Linn.)) showed an increase in anti-diabetic activity at the highest peak of the rainy season (July) compared to the onset of the rainy season (April). Lower levels of alkaloids, steroids, acidic compounds and carbohydrates were observed in July (*40*). Although the effect of seasonal variation on biological activity is an important aspect, only a few studies on African medicinal plants have been conducted (*39–42*). Information regarding seasonal variation is imperative in maximizing phytochemical yield.

The sustainable use of medicinal plants was facilitated in the past by several inadvertent or indirect controls and some intentional management practices. Plants were traditionally collected with a pointed wooden digging stick or small axe which tended to limit the quantity of bark or roots gathered. *Trichiliaemetica* (Vahl) is for instance conserved for their fruit although they are also used in traditional medicines. Harvesting of plants such as *Siphonochilus aethiopicus* (Schweif.)B.L. Burt and *Alepidea amatymbica* Eckl.&Zeyh., was restricted to winter collection to ensure seed set and multiplication during summer periods. Bark that is used for treating kidney diseases are sometimes only harvested from the eastern and western sides of the tree, traditionally resembling the kidneys thereby preventing ring-barking (*24*). These indirect control methods were purposefully incorporated into the use of medicinal plants to conserve the knowledge and plant material important in traditional healing.

As plants are collected from the wild or cultivated, the importance of botanical authentication is key. To accomplish this, plant voucher specimens need to be taken, prepared and submitted to a recognized herbarium. Voucher specimens are tangible, permanent, and verifiable scientific specimens preserved to support the result of a particular piece of research (*43*, *44*). Each specimen is a voucher, providing a record of the occurrence of a species at a particular geographical location and time; thus, specimens without associated data are of limited use (*44*). Even for the practicing botanical taxonomists, plants and their closely related species can be misidentified and verification of plant identification is a cornerstone in quality control programs. Should research ever be questioned or new information suggest the need for reappraisal, the voucher specimen can be critically examined, and its identity can be verified, contested, or disputed by other researchers. Even if shown to be misidentified, the voucher allows for correction by future researchers (*44*).

Impact and Application of Good Manufacturing Practice (GMP)

The production of traditional medicines that utilize materials of natural origin are prone to contamination, deterioration and deviation in quality (*45*) therefore the control of the starting materials, storage and processing of traditional medicines is vital (*46*). Generally, postharvest processing including primary cutting is covered by GACP however further comminuting carried out in the manufacturing processing should be covered by GMP (*46*).

According to the WHO GMP (*46*), incoming fresh herbal materials should be processed as soon as possible, unless specified otherwise. Drying is the most fundamental method for post-harvest preservation of medicinal plants because it increases the shelf life of the final product by slowing microbial growth and prevents certain biochemical reactions (*47*, *48*). The dryer design must correspond to the plant parts to be dried. For example, for seed, typical grain dryer types can be used however for the drying of flowers, fruits and roots, tray dryers or belt dryers are preferred (*48*).Inadequate drying may result in mold and decay whilst the use of high temperatures may volatize or degrade certain components such as essential oils (*49*). Unbalanced evaporation rates can result in deterioration of the final product (*50*) nonetheless response to temperature is a characteristic property of individual plant species (*48*). Drying may affect fragrance and/or appearance (*47*) and such changes are especially important for marketing of herbal tea as these criterion are directly apparent to the consumer (*48*). Ambient air-drying is the traditional technique used to preserve medicinal plant material as the low temperatures are thought to protect against degradation of the active components. However, this drying process is slow and metabolic processes may continue which may lead to quality loss of the plants and subsequently to the extracts, e.g. colour changes, loss in active ingredients (*51*). Other methods such as freeze-drying, oven drying and tray drying have been used to preserve medicinal plants (*52*). Although drying at higher temperatures results in shorter drying times it causes a reduction in the flavonoids and results in redder extracts.

The decrease in flavonoids and corresponding increase in condensed tannins may be due to polymerisation during high temperature drying. Freeze-drying, air-drying and oven or tray drying at 30°C yield extracts high in phenols, active ingredients and desirable colour for incorporation into beverages with potential anti-inflammatory properties. Therefore, tray drying medicinal herbs at low temperatures may decrease the drying time without having any major effects on the total phenols, bioactives and colour of the extracts (*52*). After reviewing the influence of drying process on the quality of medicinal plants (*53*) it was concluded that drying air temperatures between 50 and 60°C appear to be feasible for drying a large number of medicinal plant species. Drying signifies the greatest cost in the processing of medicinal plants, and, with the high moisture content in various plant parts, the choice of drying temperature is both an economic and ecological criterion (*48*). Although sun drying can be inexpensive, it takes a long time and it is environmentally dependent (*54*).

Despite the influence of drying methods on active constituents; this fundamental information on African traditional medicines is largely absent. Further research is necessary to investigate drying behavior of major medicinal plant species with the two main challenges being reducing both the drying time and expense.

Where materials are stored in bulk, to reduce the risk of mold formation or fermentation it is desirable to store them in aerated rooms or containers using natural or mechanical aeration and ventilation (*46*) as storage temperature, relative humidity and light can have an effect on the active constituents (*55*). Fresh medicinal plant material should be stored at suitably low temperatures, preferably at 2–8 °C; frozen products should be stored at less than −20°C (*22*). Stability data are always required to support the shelf-life proposed for the finished products (*46*). In a study by Stafford et al. (*56*) several South African medicinal plants were investigated to determine the effect of storage on the plants. The plants included in the study were *E. autumnalis, Bowiea volubilis, M. plumbea, Leonotis leonurus, S. aethiopicus, A. amatymbica, D. robusta, Vernonia colorata* and *Helichrysum cymosum*. From this study it was again shown that the biological activity is significantly changed during storage. Most of these plants showed an increase in activity after a couple of months in storage. Antimicrobial testing showed that after a storage period of 6 years, the biological activity of some medicinal plant species decreased (*57*). Interestingly, not all plant species lost their activity, with some species retaining activity against all of the tested pathogens. Traditional pharmacies in Nepal keep dried plants in bags (usually in cotton or paper) suspended in a second bag (thin plastic) from the ceiling. This storage method prevents plants from being exposed to sunlight, water, wind or insects, all of which are considered harmful to the medicinal value of the plant (*57*). According to medicinal plant traders in Nigeria, prolonged storage increases the potency of the drugs because the longer the storage the better the extraction of active ingredients into the liquid phase (*58*).

Enormous amount of plant material is either spoiled or contaminated by microorganisms especially fungi due to improper storage and packing (*58*, *59*). Most crude plant material is kept at ambient temperature which may satisfy the requirements for mycotoxin elaboration by most fungi, therefore some form of decontamination of stored plant parts before their use is necessary (*58*, *60*). Setting standards for toxigenic molds and mycotoxins in crude herbal drugs and medicinal plants is essential in order to reduce the risks for consumers health (*61*).

Contamination and Adulteration

From cultivation to the finished product, several possibilities exist to explain the presence of contaminants which may represent a very serious health and environmental problem (*62*). Contamination can be defined as 'the undesired introduction of impurities of a chemical or microbiological nature, or of foreign matter, into or on to a starting material or intermediate during production, sampling, packaging or repackaging, storage or transport' (*46*). The occurrence of microbes, heavy metals and pesticide residues in African traditional medicines has seriously affected the development and process of internationalization thereof (*63*). Materials of natural origin tend to show a much higher level of microbial contamination than synthetic material (*64*). Raw medicinal plant material normally carries a great number of molds, often from soil. Current practices of harvesting and preparation cause additional contamination and microbial growth (*62*). Microbiological parameters include total viable content, total mold content, total enterobacterial and their count (*65*). Contamination of raw plant materials with aflatoxins can cause potential carcinogenic effects if absorbed even in small amounts (*62*). Yeasts and molds can cause opportunistic infections in humans and are more significant in HIV+ patients (*66*). South African plants recommended for the treatments of HIV/AIDS showed high bacterial and fungal content indicating low environmental sanitation and low standard for processing during the preparation (*66*). Assessment of fungal and mycotoxin contamination of African herbal products sold in Cape Town and Tshwane (South Africa) revealed that of the 16 samples analyzed, 15 were contaminated with at least one of these three fungal genera: *Aspergillus, Fusarium,* and *Penicillium*. FumonisinB(1) was present in 13 of the samples in quantities ranging from 14 to 139 μg/kg. None of the samples were contaminated with aflatoxigenic fungi or aflatoxin (*67*).

According to WHO GMP, only permitted substances may be used for fumigation, and allowable limits for their residues together with specifications for the apparatus used should be set according to national regulations (*46*) in producing country as well as in the targeted country of sales and consumption. Nonetheless as pests are a recurrent problem for medicinal plant vendors, unregulated fumigation occurs in African medicinal plant shops. Shop owners, however, do not seem to be concerned about the consequences of potentially toxic residues on the plant material being sold to their patients (*51*). Due to the diversity in the types of pesticide residues in otherwise complex plant-based mixtures, it is difficult to find a method for the removal of pesticide residues whereby you have to take into consideration the loss of active ingredients and secondary pollution caused by the organic solvent (*63*). Of all the pesticides, only the chlorinated hydrocarbons and related pesticides (e.g. Aldrin, chlordane, DDT, dieldrin, HCH) and a few organophosphorous pesticides (e.g. carbophenothion) have a long residue action. Thus the WHO advocates that medicinal plants be tested for the presence of organically bound chlorine and phosphorous. Organochlorine pesticides are highly persistent insecticides. Although they were used in Egypt more than 20 years ago and most of them have been banned, their residues still appear as pollutants in food as well as in the environment (*62*). In South Africa, despite suspicions of organo-phosphate-like poisonings presumed to be caused by traditional medicine consumption (*68*) no studies have been done to identify and quantify pesticides found on commonly used medicinal plants. Pharmacopoeias of different nations have assay methods and residual limits for the organochlorine pesticides (*63*). However to date, guidelines regarding pesticides in African traditional medicines are lacking. This could be due to lack of tangible supporting evidence.

The uptake of heavy metals by medicinal plants and the ingestion thereof is a public health threat which is largely overlooked. Numerous African medicinal plants are heavy metal accumulators. Medicinally used *Datura metal* L. (Solanaceae) is an accumulator of cobalt and nickel and is recommended as a phytomonitor (*69*). Similarly, *Datura innoxia* Miller, Gard.is a metal tolerant species (*70*). Both *Datura* species are widely used in African traditional medicine even though their toxicity is well established (*71*). To date WHO (*22*) only has limits for Cd and Pb at 0.3 and 10 mg kg^{-1} respectively. Relatively high concentrations of Cd and Pb may be related to irrigation with contaminated water, as well as the addition of some fertilizers and herbicides (*62*). A study on Nigerian herbal remedies revealed that 100% of the samples contained elevated amounts of heavy metals, suggesting possible heavy metal toxicity from Nigerian herbal products (*72*). The screening of Egyptian medicinal plants revealed that Cd and Pb were detected in 43% and 80% of the plant samples respectively (*73*). Multiple metal contamination of African medicinal plant parts gives grounds for concern (*74*) particularly since it is not uncommon for inorganic metal powders to be deliberately added to plant based mixtures for alleged medicinal value. Such substances include copper sulphate and potassium dichromate.

One of the greatest risks to human health arises from economically motivated adulteration (*75*). This includes the intentional substitution with less expensive plant material (*76*). Examples of adulteration of medicinal plants with poisonous plants exhibit the need to establish GMP for starting materials (*22*). The only published report documenting the deliberate adulteration of traditional African herbal remedies showed two separate incidences in which South African traditional remedies were adulterated with western pharmaceuticals causing severe toxicity (*77*).

Good Practice in Quality and Safety Control of Traditional Medicine

Any assessment of traditional herbal medicines must be based on an unambiguous identification and characterization of the constituents (*6*). Consistent quality for plant-based medicines can only be assured if the starting materials are defined in a rigorous and detailed manner (*46*).

Preparation of Medicinal Plant Extracts

Plant preparation includes crushed or powdered plant materials, extracts, tinctures, fatty or essential oils, expressed juices and preparations whose production involves fractionation, purification or concentration (*78*). The general processing protocol describing the various operations performed on the plant material should be available on request (*46*). Most work on the chemistry of secondary plant components use dried plant material for the following reasons: (1) there are fewer problems associated with the large-scale extraction of dried plant material than with fresh material; (2) the time delay between collecting plant material and processing it makes it difficult to work with fresh material because differences in water content may affect solubility or subsequent separation by liquid–liquid extraction, (3) the secondary metabolic plant components should be relatively stable especially if it is to be used as an antimicrobial agent and (4) many, if not most plants are used in the dried form or as aqueous extract by THPs (*79*). Great emphasis should be placed on the ratio between the crude plat material and the extract obtained as the varying ratio might lead to differences in composition (*80*). As the target compounds may be non-polar, polar and/or thermally labile, the aptness of the extraction methods needs to be carefully considered. For hydrophilic compounds, polar solvents for instance ethanol, methanol or ethyl-acetate are often used. For extraction of more lipophilic compounds, dichloromethane or a combination of dichloromethane/methanol (1:1) may be selected (*3*). The rational for adopting sequential extraction is to ensure that the polarity of the solvents leach out compounds soluble in that particular solvent after which individual fractions obtained are commonly rotary evaporated to dryness to yield the solvent residue (*81*). The method of extraction used determines the type of compounds present in the final extract (*82*). Therefore it is imperative to document solvent(s) used, times and temperatures of extraction, concentration stages and method (*83*) in order for the study to be reproducible. Prospective active constituents should not be lost, altered or destroyed during such preparation (*3*). Fractionation of extracts habitually leads to a reduction or loss of biological activity by compound splits or loss of additive or synergistic effects flanked by analogue components (*3*). In addition, fractionation multiplies the number of samples to be tested (*84*). A concentration step is usually required for the identification of active compounds present in very small quantities in the extracts, and is based on evaporation of the solvent in vacuo. Extraction and evaporation should be performed at a low temperature so not to destroy any thermolabile constituent (*3*). New methods for the extraction of analytes in plant materials are constantly being developed which improve the extraction process and minimize the use of organic solvents (*2*).

Scientific Validation of Medicinal Plants

In vitro tests, generally known as biological assays, are used globally to scientifically validate traditionally used plants. Examples of commonly used assays include antibacterial, anti-inflammatory, antifungal, anthelmintic and antimalarial assays (*51*). The disease being targeted and the availability of practical and biologically validated laboratory models will influence the screening method (*3*). Primary bioassays are designed for rapid screening of large numbers of products or extracts. These assays are simple, easy to implement, generally qualitative and indicative of actions, rapid to produce results and are usually cost efficient. Compounds or extracts with a specific activity at a non-toxic dose (so-called "hits"), then need further evaluation in secondary or specialized *in vitro* bioassays and in animal models to define "lead" status (*3*). Most large pharmaceutical companies have reputable HTS (high-throughput screening) infrastructures which contain substantial combinatorial compound libraries, which cover a wide range of chemical diversity (*2*). Validation and selection of primary screening assays are critical for the selection of extracts or molecules with relevant pharmacological action and creditable following-up (*3*). Biological activity ascertained at a particular screening level should then be subjected to confirmation using a model in the next higher evaluation level. For example, subcellular (enzymatic) screening results should be confirmed against the whole organism, followed by confirmation in an animal model (*3*). Biochemistry, molecular biology, and cell biology are invaluable in establishing quantifiable and reproducible assays. *In vitro* models using the whole organism are the 'golden' standard and should be used if feasible (*3*). The more sophisticated the biological assay, the more likely that it is measuring a single parameter (*in vitro*) of what might be a complex set of reactions *in vivo* (*85*). Using *in vitro* tests can reduce the number of *in vivo* experiments (*78*).An effective way to reconcile *in vitro* and *in vivo* results is the use of animal models (*85*).

Research on non-standardized plant materials and/or their respective extracts using unsuitable criteria for activity, inappropriate or exclusion of controls and the use of unrealistically high assay dosages has received much criticism (*3*, *86–88*). Numerous papers are published which deduce *in vitro* results to argue *in vivo* activity without taking into account traditional methods of preparation, treatment of the extract prior to administration, the effect of other added substances or administered dosage (*86*). A further criticism is that although medicinal plants are often used in conjunction with one another, research on African traditional medicines is largely done on individual plant species. Studies verifying biological activity of single plants rarely continue towards the validation of polyherbal mixtures which are preparations largely consumed by the public. This may be due to limited expertise and/or financial resources or perhaps it is beyond the curiosity of the research team who are merely interested in which compounds contribute to the expected biological activity.

Unfortunately, potential interferences from complex medicinal plant mixtures have not been rigorously documented.

Use of Fingerprinting and Marker Compounds for Identification and Standardization

Chemical fingerprints can and need to be linked to biological assays to provide assurance of efficacy and consistency (*85*). Medicinal plants contain a large number of mostly unique, species-specific compounds (*89*, *90*). Such compounds may be found in very high levels, others at very low level, some very polar, or very non-polar, and some are simply very unstable (*91*). Markers are chemically defined constituents of plant material (*83*). The quantity of the chemical marker can be an indicator of the quality of a herbal product (*89*, *92*). In addition, they may also be useful indicators of adherence during clinical studies. Marker compounds like hypericins and hyperforins may be used as potency standards in the control of *H. perforatum*samples because they are characteristic markers for this genus and allow genuine products to be differentiated from adulterants. Hypericins absorb visible light with a maximum absorption at 588 nm and are highly fluorescent in methanol when exposed to UV light (*93*). An ideal chemical marker for a natural product should be not only a distinguishing component but also the therapeutic component (*94*). Frequently it is not well known which constituents in a herbal product are responsible for these activities (*94*). The chemical profiles of a plant extract should be compared to the results of biological assays in order to assure batch to batch consistency (*85*, *95*). Significant variation in antioxidant capacity and biomarker content of commercially available *Scutellaria*-containing herbal medicines has been reported. Such variation may explain the differing efficacies reported for particular plant based medicines and further highlights the need for regulation of such herbal medicines to address issues of quality, safety and efficacy (*96*).

The bioactive extract should be standardized on the basis of active principles or major compounds along with the chromatographic fingerprints (TLC, HPTLC, HPGC and GC) (*65*). The low acquisition, operational and maintenance costs needed to successfully perform thin TLC analytical technique are very important because it can provide product quality assessment capability in areas where laboratory facilities for pharmaceutical quality analysis are minimal or do not exist (*97*).

According to Srivastava and Mishra (*90*) DNA markers associated with a particular gene or trait may have several advantages over typical phenotype markers. DNA markers are reliable for informative polymorphisms as the genetic composition is unique for each species and is not affected by age, physiological conditions or environmental factors. In addition, DNA can be extracted from fresh or dried organic tissue of the botanical material; hence the physical form of the sample for assessment does not restrict detection trait (*90*).

Currently, chromatographic fingerprinting is recommended by the WHO as the most preferable technique to use in the quality control of herbal medicines (*78*) although it is recognized that this approach has limitations. Conversely, synergistic interactions between multiple component parts of the herbal medicine may be ignored by only applying fingerprinting for quality control (*96*). Nonetheless, the fingerprint methods used for the stability studies should be as similar as possible to those used for quality control purposes (*46*). Regardless of the chosen analytical method, the availability and robustness of the method must be considered (*6*). Although the chemical fingerprints of some commonly used African medicinal plants are available, the lack of a standardized analytical approach may give rise to ambiguity. Applicable monographs would ensure identification and authentication of appropriate plant species. A neglected area within chemical fingerprinting for standardization and quality control is in using the ratio of known compounds in plant species rather than only relying and/or looking at the concentration of one or two bioactives.

Safety Assessment of Traditional Medicines

Safety is an elementary principle in the prerequisite of herbal medicines and a critical component of quality control (*98*). In some cases, certain plants are added to reduce the toxicity of the more therapeutically effective plants (*84*). In Traditional Chinese medicine, a standard prescription contains one herb that is non-toxic and can be used long-term (this is the 'Master' or 'Superior' category), one or two that are non-toxic or slightly toxic and used in moderation ('Adviser' or 'Medium' category), and one or two that are toxic and should not be used for a long period ('Soldier' or 'Inferior' category) (*99*). Although multi-herbal products are common in African traditional medicine, no such epistemology has been documented to explain the alleviation of potentially toxic plant material. However this is not to say that such a theory does not exist.

Documented side-effects of a plant or herbal mixture, its closely related species, constituents of the plant and final product(s) should be taken into account when decisions are made about the need for new toxicological studies (*78*). The assessment should determine if there is sufficient information to guarantee safe use in vulnerable populations such as pregnant or lactating women and in small children (*6*). Suggested toxicological tests include immunotoxicity, carcinogenicity and reproductive toxicity (*78*) with particular attention given to effects that cannot be readily detected empirically e.g. genotoxicity (*6*). Based on their use in traditional medicines, 51 South African plant species were tested for genotoxic activity in human peripheral blood lymphocytes using the micronucleus test, with further testing of selected extracts using the alkaline comet assay. Screening results indicated the induction of noteworthy numbers of micronuclei by many of the plant extracts. Several samples also induced DNA damage in human white blood cells using the alkaline comet assay. Although a number of these plant species are recognized as toxic by THPs, several plants that are used in common remedies were found to be genotoxic and potentially dangerous (*100*). Well known plant toxins including cardiac glycosides (*101*) and pyrrolizidine alkaloids (*102*) have been reported in commonly used African medicinal plants both of which have caused severe toxicity (*103*, *104*). Despite these findings, the results of such studies are rarely relayed back to the THPs and consumers of traditional medicine. By and large once off toxicity studies are not done in parallel with studies on biological activity.

Pharmacovigilance of Traditional Medicines

Medicinal plants are often administered in combination with western medicine thus increasing the potential of herb-drug interactions. An extensive review of the literature identifies reported herb-drug interactions with clinical significance (*105*) however to date there is little information with regards to interaction between western and African traditional medicines. Two African medicinal plants, *H. hemerocallidea* (African potato) and *Sutherlandia frutescens* (L.) R.Br. often recommended for treatment of HIV/AIDS, showed a negative interaction with antiretroviral medication, thus risks to patients may include treatment failure, viral resistance or drug toxicity (*106*). THPs routinely enquire about allopathic medicine used by their patients. However the lack of scientific evidence with regards to herbal-drug interactions does not equip them to make informed decisions (Mirranda Javu, South African Traditional Health Practitioner, pers. comm.). According to WHO (*6*), pharmacovigilance units are necessary to collect and assess information on herbal and traditional medicines and each report on adverse drug reaction should be checked for a possible association with traditional or herbal medicines. This is of course only possible if and when the names of medicinally used plants and plant based remedies have been documented and characterized in a national pharmacopeia.

Formal nationally accepted documentation on African traditional medicine is currently non-existent.

Conclusion

'The chemical complexity of many natural products and the lack of assurance of a renewable supply have created a diminishing interest by the pharmaceutical industry, which in turn endorses the pivotal role of academia and public organizations in the protracted exploration and evaluation of natural products' (*3*). New strategies for developing medicinal plants as commercial crops is needed and other possibilities such as incorporating crops as an alternative for larger, commercial farms and for the small-scale farmer has been proposed (*15*). Plants that are in demand and that could be cultivated in short periods of time are especially attractive for cultivation practices (*32*). Cultivation has been suggested to be a solution to not only meet increased demand for medicinal plants, but also a tool for biodiversity conservation and poverty alleviation (*30*). Cultivation data however needs to be linked with pharmacological assessments of the crop during the development of protocols. It is important for the THPs to know and trust that the plants needed for traditional healing practice will be not only available but also affordable in the future. Medicinal plants are either grown/collected to use in their raw form or as a single constituent after extraction and isolation. With regards to the latter however the amount of plant material necessary to obtain an active extract or single biochemical constituent is often overlooked (*83*). Concentrations of active principles higher than 1% are not the rule, but the exception (*107*). Biotic and abiotic factors are effective ways to stimulate secondary metabolites due to both plant defense mechanism and metabolite production which are interrelated via secondary metabolism (*108*). Correct drying; storing and processing procedures of medicinal plants are necessary to ensure quality of the end product. Therefore all stages of postharvest preparation must be carefully documented and should include drying information (time, temperature, moisture content), nature of pulping methods (crushing, milling), size of post processed plant material and storage conditions (time, humidity).

For laboratory investigations, plant preparations may be produced by extraction, fractionation, purification, concentration, or other physical or biological procedure (*83*). Safety evaluation of medicinal plant products should include both *in vitro* and *in vivo* results from both scientific literature and medical reports involving adverse reaction. Such information needs be accessible to the public - consumers, manufacturers, scientists and medical practitioners. Insufficient information is often provided in reported cases of adverse reactions to herbal products, both those published in the scientific literature and those submitted to regulatory authorities (*75*). Quality and purity may be compromised by the presence of weeds, dirt, pesticides, toxic metals, radioactivity, bacteria, molds and mycotoxins, processing impurities and solvent residues (*75*). However, as few African countries have formal monitoring systems, quality control is non-existent and the testing of medicinal plants for contaminants such as heavy metals and pesticides is costly.

A product-tracking system for the cultivation, sustainable harvesting, shipping, quality control, and quality assurance of plant products, either nationally or internationally is imperative for the end product of African medicinal plants (*7*). WHO (*98*) has established a set of guidelines for safety monitoring of herbal medicines which urges the development of national and regional guidelines/policies. The largest glaring error in African traditional medicines is the lack of monographs and pharmacopeia. Monograph preparation involves identification of chemical as well as biological markers (*90*). Without sound methodic guidelines on traditional medicines research, the scientist, using common sense and research expertise, will need to define the set of criteria that will yield the greatest chances to attain a robust "proof-of-concept" (*3*). There is a need to approach scientific proof and clinical validation with chemical analysis, biological assays, animal models and clinical trials (*85*, *95*). However with the current haphazard approach of scientists carrying out such experiments, African traditional medicines remains unfocused and time, money and valuable plant resources are wasted. Ironically, this does not affect the thriving trade in African traditional medicines - this multi-million dollar informal industry continues to supply primary health care to millions of people with or without the contribution from researchers.

Acknowledgments

We thank the National Research Foundation (South Africa), Medical Research Council (South Africa) and the University of KwaZulu-Natal for their financial assistance.

References

1. Kong, D.; Li, X.; Zhang, H. Where is the hope for drug discovery? Let history tell the future. *Drug Discov. Today* **2009**, *14* (3-4), 115–119.

2. Mendonça-Filho, R. R. Bioactive Phytocompounds: New Approaches in the Phytosciences. In *Modern Phytomedicine: Turning Medicinal Plants into Drugs*; Aqil, A. F., Owais, M., Eds.; Wiley: New York, 2006; pp 1−24.
3. Cos, P.; et al. Anti-infective potential of natural products: How to develop a stronger in vitro 'proof-of-concept'. *J. Ethnopharmacol.* **2006**, *106* (3), 290–302.
4. Balunas, M. J.; Kinghorn, A. D. Drug discovery from medicinal plants. *Life Sci.* **2005**, *78* (5), 431–441.
5. Gurib-Fakim, A. An African herbal pharmacopeia could underpin the medicinal plant industry on the African continent. In World Congress on Medicinal and Aromatic Plants. *Afr. J. Tradit., Complementary Altern. Med.* **2008**.
6. WHO. *Guidelines on minimum requirements for the registration of herbal medicinal products in the Eastern Mediterranean Region*; Cairo, 2006.
7. Tadmor, Y. Quality assurance and quality control for African natural plant products from the ground up. In *Trends in New Crops and New Uses*; Janick, J., Whipkey, A., Eds.; ASHS Press: Alexandria, VA, 2002; pp 93–97.
8. Sharma, A. K.; et al. Problems associated with clinical trials of Ayurvedic medicines. *Braz. J. Pharmacogn.* **2010**, *20* (2), 276–281.
9. WHO. *Regulatory Situation of Herbal Medicines - A Worldwide Review*; 1998; p 49.
10. Brendler, T. African Herbal Pharmacopoeia. In *Association for African Medicinal Plant Standard*; Graphic Press: Los Angeles, 2010; p 289.
11. Mander, M. *The marketing of indigenous medicinal plants in South Africa: A case study in KwaZulu Natal*; FAO: Rome, 1998.
12. Sparg, S. G.; et al. Cultivating the medicinal plant Merwilla natalensis as a crop: a small-scale farming approach. *Outlook Agric.* **2005**, *34*, 116–120.
13. Canter, P. H.; Thomas, H.; Ernst, E. Bringing medicinal plants into cultivation: opportunities and challenges for biotechnology. *Trends Biotechnol.* **2005**, *23* (4), 180–185.
14. WHO/IUCN/WWF. *Guidelines on the Conservation of Medicinal Plants*; 1993.
15. Jäger, A. K.; Van Staden, J. The need for cultivation of medicinal plants in southern Africa. *Outlook Agric.* **2000**, *29*, 283–284.
16. Crouch, N. Symmonds, R. Propagation Technologies for Medicinal Plants. In *Commercialising Medicinal Plants: A Southern African Guide*; Diederichs, N., Ed.; African Sun Media: Stellenbosch, South Africa, 2006; p 53−66.
17. Rout, G. R.; Samantaray, S.; Das, P. In vitro manipulation and propagation of medicinal plants. *Biotechnol. Adv.* **2000**, *18* (2), 91−120.
18. Fennell, C.; Makunga, N.; van Staden, J. Medicinal plants and biotechnology: an integrated approach to healthcare. In *Floriculture, Ornamental and Plant Biotechnology – advances and topical issues*; da Silva, J. A. T., Ed.; Global Science Books: U.K., 2006; pp 41−53.
19. Brink, J. A.; Woodward, B. R.; DaSilva, E. J. Plant biotechnology: a tool for development in Africa. *Electron. J. Biotechnol.* **1998**, *1*, 14–15.
20. Prakash, S.; Van Staden, J. Micropropagation of Hoslundia opposita Vahl--a valuable medicinal plant. *S. Afr. J. Bot.* **2007**, *73* (1), 60–63.

21. Makunga, N. P.; Jager, A. K.; Van Staden, J. *Plant Cell Rep.* **2003**, *21*, 967–973.
22. WHO. *Guidelines on good agricultural and collection practices (GACP) for medicinal plants*; 2003.
23. Gang, W.; et al. The mutual responses of higher plants to environment: Physiological and microbiological aspects. *Colloids Surf., B* **2007**, *59*, 113–119.
24. Van Wyk, B.-E.; Van Oudtshoorn, B.; Gericke, N. *Medicinal Plants of South Africa*; Pretoria: Briza, 2000.
25. Shale, T. L.; Stirk, W. A.; van Staden, J. Variation in antibacterial and anti-inflammatory activity of different growth forms of Malva parviflora and evidence for synergism of the anti-inflammatory compounds. *J. Ethnopharmacol.* **2005**, *96* (1-2), 325–330.
26. Namdeo, A. G.; et al. Influence of Geographical and Climatic Conditions on Camptothecin Content of Nothapodytes nimmoniana. *Rec. Nat. Prod.* **2010**, *4*, 64–71.
27. Poutaraud, A.; Girardin, P. Improvement of medicinal plant quality: a Hypericum perforatum literature review as an example. *Plant Gen. Res.* **2005**, *3* (2), 178–189.
28. Barthwal, J.; Nair, S.; Kakkar, P. Heavy Metal Accumulation in Medicinal Plants Collected from Environmentally Different Sites. *Biomed. Environ. Sci.* **2008**, *21* (4), 319–324.
29. Verpoorte, R.; et al. Commentary: "A systems view on the future of medicine: Inspiration from Chinese medicine?". *J. Ethnopharmacol.* **2009**, *121* (3), 479–481.
30. Wiersum, K. F. Cultivation of medicinal plants as a tool for biodiversity conservation and poverty alleviation in the amatola region, South Africa. In *Medicinal and Aromatic Plants*; Bogers, R. J., Craker, L., Lange, D., Eds.; Springer: Netherlands, 2006; pp 43−57.
31. Cunningham, A. B., African medicinal plants: setting priorities at the interface between conservation and primary health care. In *People and Plants working paper 1*; UNESCO: Paris, 1993.
32. Keirungi, J.; Fabricius, C. Selecting medicinal plants for cultivation at Nqabara on the Eastern Cape Wild Coast, South Africa. *S. Afr. J. Sci.* **2005**, *101*, 497–501.
33. Mander, M. Economics of the traditional medicine trade in South Africa. In *S. Afr. Health Rev.* Health Systems Trust: Durban, 2007; pp 189−200.
34. Anonymous. *Medicinal Plant Trade Sector Analysis*; 2003.
35. Louw, C. A. M.; Regnier, T. J. C.; Korsten, L. Medicinal bulbous plants of South Africa and their traditional relevance in the control of infectious diseases. *J. Ethnopharmacol.* **2002**, *82* (2-3), 147–154.
36. Lewu, F. B.; Grierson, D. S.; Afolayan, A. J. The leaves of Pelargonium sidoides may substitute for its roots in the treatment of bacterial infections. *Biol. Conserv.* **2006**, *128* (4), 582–584.
37. Zschocke, S.; et al. Plant part substitution - a way to conserve endangered medicinal plants? *J. Ethnopharmacol.* **2000**, *71* (1-2), 281–292.

38. Ncube, B.; Finnie, J. F.; Van Staden, J. Seasonal variation in antimicrobial and phytochemical properties of frequently used medicinal bulbous plants from South Africa. *S. Afr. J. Bot.* **2011**, *77*, 387–396.
39. Buwa, L. V.; Van Staden, J. Effects of collection time on the antimicrobial activities of Harpephyllum caffrum bark. *S. Afr. J. Bot.* **2007**, *73* (2), 242–247.
40. Osadebe, P. O.; et al. Seasonal variation for the antidiabetic activity of Loranthus micranthus methanol extract. *Asian Pac. J. Trop. Med.* **2010**, *3* (3), 196–199.
41. Kamatou, G. P. P.; et al. Seasonal variation in essential oil composition, oil toxicity and the biological activity of solvent extracts of three South African Salvia species. *S. Afr. J. Bot.* **2008**, *74* (2), 230–237.
42. Van Vuuren, S. F.; et al. Seasonal and geographical variation of Heteropyxis natalensis essential oil and the effect thereof on the antimicrobial activity. *S. Afr. J. Bot.* **2007**, *73* (3), 441–448.
43. Huber, J. T. The importance of voucher specimens, with practical guidelines for preserving specimens of the major invertebrate phyla for identification. *J. Nat. Hist.* **1998**, *32* (3), 367–385.
44. Carter, R.; Bryson, C. T.; Darbyshire, S. J. Preparation and Use of Voucher Specimens for Documenting Research in Weed Science. *Weed Technol.* **2007**, *21* (4), 1101–1108.
45. Anonymous. *Guidelines on Good Manufacturing Practice for Traditional Medicines and Health Supplements*; Malaysia, 2007.
46. WHO. *Guidelines on good manufacturing practices (GMP) for herbal medicines*; 2007.
47. Diaz-Maroto, M. C.; et al. Influence of Drying on the Flavor Quality of Spearmint (Mentha spicata L.). *J. Agr. Food Chem.* **2003**, *51* (5), 1265–1269.
48. Muller, J.; Heindl, A. Drying of medicinal plants. In *Medicinal and Aromatic Plants*; Bogers, R. J., Craker, L. E. L. E., Lange, D., Eds.; Springer: New York, 2006; pp 237−252.
49. Whitten, G. *Herbal Harvest – commercial organic production of quality and dried herbs*; Bloomingham Books: Australia, 1997.
50. Tanko, H.; et al. Pre- and post-harvest processing of medicinal plants. *Plant Gen. Res.* **2005**, *3* (2), 304–313.
51. Fennell, C. W.; et al. Assessing African medicinal plants for efficacy and safety: agricultural and storage practices. *J. Ethnopharmacol.* **2004**, *95* (2-3), 113–121.
52. Harbourne, N.; et al. Effect of drying methods on the phenolic constituents of meadowsweet (Filipendula ulmaria) and willow (Salix alba). *LWT - Food Sci. Technol.* **2009**, *42* (9), 1468–1473.
53. Rocha, R. P.; E.C., M.; Radünzm, L. L. Influence of drying process on the quality of medicinal plants: A review. *J. Med. Plants Res.* **2011**, *5* (33).
54. Arslan, D.; Musa Özcan, M. Evaluation of drying methods with respect to drying kinetics, mineral content and colour characteristics of rosemary leaves. *Energ. Convers. Manag.* **2008**, *49* (5), 1258–1264.

55. Lin, S.-D.; Sung, J.-M.; Chen, C.-L. Effect of drying and storage conditions on caffeic acid derivatives and total phenolics of Echinacea Purpurea grown in Taiwan. *Food Chem.* **2011**, *125* (1), 226–231.
56. Stafford, G. I.; Jäger, A. K.; van Staden, J. Effect of storage on the chemical composition and biological activity of several popular South African medicinal plants. *J. Ethnopharmacol.* **2005**, *97* (1), 107–115.
57. Griggs, J. K.; et al. The effects of storage on the biological activity of medicinal plants from Nepal. *J. Ethnopharmacol.* **2001**, *77* (2-3), 247–252.
58. Efuntoye, M. Mycotoxins of fungal strains from stored herbal plants and mycotoxin contents of Nigerian crude herbal drugs. *Mycopathologia* **1999**, *147* (1), 43–48.
59. Khan, S.; et al. Fungal contamination of medicinal herbs during commercial storage in Punjab. *Mycopath* **2006**, *4*, 21–25.
60. Choudhary, A. K.; Kumari, P. Management of mycotoxin contamination in preharvest and postharvest crops: present status and future prospects. *J. Phytol.* **2010**, *2* (7), 37–52.
61. Bugno, A.; et al. Occurrence of toxigenic fungi in herbal drugs. *Braz. J. Microbiol.* **2006**, *37*, 47–51.
62. Abou-Arab, A. A. K.; et al. Quantity estimation of some contaminants in commonly used medicinal plants in the Egyptian market. *Food Chem.* **1999**, *67* (4), 357–363.
63. Sahoo, N.; Manchikanti, P.; Dey, S. Herbal drugs: Standards and regulation. *Fitoterapia* **2010**, *81* (6), 462–471.
64. Yee, S.-K.; et al. Regulatory control of Chinese proprietary medicines in Singapore. *Health Policy* **2005**, *71*, 133–149.
65. Kumar, K. J. Bottle necks in standardisation of traditional systems of medicine. *Res. J. Med. Plant* **2011**, *5*, 443–447.
66. Govender, S.; et al. *S. Afr. J. Sci.* **2006**, *102*, 253–255.
67. Katerere, D. R.; et al. A preliminary survey of mycological and fumonisin and aflatoxin contamination of African traditional herbal medicines sold in South Africa. *Hum Exp. Toxicol.* **2008**, *27*, 793–798.
68. Van Wyk, L.; Els, I. Neonatal cholinergic syndrome – organophosphate poisoning or herbal medicine intoxication? South African journal of Child Health 2, 26-27. *S. Afr. J. Child Health* **2008**, *2*, 26–27.
69. Bhattachrjee, S.; Kar, S.; Chakravarty, S. Mineral compositions of Datura: a traditional tropical medicinal plant. *Commun. Soil Sci. Plant Anal.* **2004**, *35*, 937–946.
70. Kelly, R. A.; Andrew, J. C.; Dewitt, J. G. An X-ray absorption spectroscopic investigation of the nature of the zinc complex accumulated in Datura innoxia plant tissue culture. *Microchem. J.* **2002**, *71*, 231–245.
71. Hutchings, A. *Zulu Medicinal Plants: an inventory*; University of Natal Press: Pietermaritzburg. 1996.
72. Obi, E.; et al. Heavy metal hazards of Nigerian herbal remedies. *Sci. Total Environ.* **2006**, *369*, 35–41.
73. Dogheim, S. M.; et al. Pesticides and heavy metal levels in Egyptian leafy vegetables and some aromatic medicinal plants. *Food Addit. Contam.* **2004**, *21*, 323–330.

74. Street, R. A.; et al. Variation in heavy metals and essential elements in South African medicinal plants obtained from informal markets. *Food Addit. Contam.* **2008**, *25*, 953–960.
75. Jordan, S. A.; Cunningham, D. G.; Marles, R. J. Assessment of herbal medicinal products: Challenges, and opportunities to increase the knowledge base for safety assessment. *Toxicol. Appl. Pharm.* **2010**, *243* (2), 198–216.
76. Yu, H.; et al. Pharmacognostical investigations on traditional chinese medicinal herbs: identification of four herbs from the UK market. *J. Pharm. Pharmacol.* **1995**, *47*, 1129.
77. Snyman, T.; et al. *Ther. Drug Monit.* **2007**, *27*, 86–89.
78. WHO. *General Guidelines for Methodologies on Research and Evaluation of Traditional Medicines*; Geneva, 2000.
79. Eloff, J. N. Which extractant should be used for the screening and isolation of antimicrobial components from plants? *J. Ethnopharmacol.* **1998**, *60* (1), 1–8.
80. Srinivasan, V. S. Challenges and scientific issues in the standardization of botanicals and their preparations. United States Pharmacopeia's dietary supplement verification program-A public health program. *Life Sci.* **2006**, *78* (18), 2039–2043.
81. Perumal Samy, R.; Ignacimuthu, S.; Sen, A. Screening of 34 Indian medicinal plants for antibacterial properties. *J. Ethnopharmacol.* **1998**, *62* (2), 173–181.
82. Verpoorte, R. Some phytochemical aspects of medicinal plant research. *J. Ethnopharmacol.* **1989**, *25* (1), 43–59.
83. Matteucci, E.; Giampietro, O. Proposal open for discussion: defining agreed diagnostic procedures in experimental diabetes research. *J. Ethnopharmacol.* **2008**, *115* (2), 163–172.
84. Taylor, J. L. S.; et al. Towards the scientific validation of traditional medicinal plants. *Plant Growth Regul.* **2001**, *34*, 23–37.
85. Yuan, R.; Lin, Y. Traditional Chinese medicine: an approach to scientific proof and clinical validation. *Pharmacol. Therapeut.* **2000**, *86* (2), 191–198.
86. Houghton, P. J.; et al. Uses and abuses of in vitro tests in ethnopharmacology: Visualizing an elephant. *J. Ethnopharmacol.* **2007**, *110* (3), 391–400.
87. Verpoorte, R. Setting standards! *J. Ethnopharmacol.* **2006**, *106* (3), 289–289.
88. Verpoorte, R.; et al. Editorial. *J. Ethnopharmacol.* **2006**, *103* (3), 309–310.
89. Xie, Y.; et al. Combinative method using HPLC quantitative and qualitative analyses for quality consistency assessment of a herbal medicinal preparation. *J. Pharmaceut. Biomed.* **2007**, *43* (1), 204–212.
90. Srivastava, S.; Mishra, N. Genetic Markers - A Cutting-Edge Technology in Herbal Drug Research. *J. Chem. Pharm. Res.* **2009**, *1* (1), 1–18.
91. Verpoorte, R.; Choi, Y. H.; Kim, H. K. Ethnopharmacology and systems biology: A perfect holistic match. *J. Ethnopharmacol.* **2005**, *100* (1-2), 53–56.
92. Li, S.; et al. Chemical markers for the quality control of herbal medicines: an overview. *Chin. Med.* **2008**, *3* (1), 7.

93. Gomez, M. R.; et al. Metal content monitoring in Hypericum perforatum pharmaceutical derivatives by atomic absorption and emission spectrometry. *J. Pharmaceut. Biomed.* **2004**, *34* (3), 569–576.
94. Song, J.-Z.; et al. A novel approach to rapidly explore analytical markers for quality control of Radix Salviae Miltiorrhizae extract granules by robust principal component analysis with ultra-high performance liquid chromatography-ultraviolet-quadrupole time-of-flight mass spectrometry. *J. Pharmaceut. Biomed.* **2010**, *53* (3), 279–286.
95. Ong, E. S. Extraction methods and chemical standardization of botanicals and herbal preparations. *J. Chromatogr., B* **2004**, *812* (1-2), 23–33.
96. Boyle, S. P.; et al. Evaluation of quality control strategies in Scutellaria herbal medicines. *J. Pharmaceut. Biomed.* **2011**, *54* (5), 951–957.
97. Kaale, E.; Risha, P.; Layloff, T. TLC for pharmaceutical analysis in resource limited countries. *J. Chromatogr., A* **2011**, *1218*, 2732–6.
98. WHO. *Guidelines on safety monitoring of herbal medicines in pharacovigilance systems*. Geneva, 2004.
99. Chan, K. Progress in traditional Chinese medicine. *Trends Pharmacol. Sci.* **1995**, *16*, 182–187.
100. Taylor, J. L. S.; et al. Investigating the safety of plants used in South African traditional medicine: Testing for genotoxicity in the micronucleus and alkaline comet assays. *Environ. Mol. Mutagen.* **2003**, *42* (3), 144–154.
101. Marx, J.; et al. *Environ. Toxicol. Phar.* **2005**, *20*, 26–34.
102. Steenkamp, V.; et al. *J. Ethnopharmacol.* **2001**, *78*, 751–58.
103. Steenkamp, V.; Stewart, M. J.; Zuckerman, M. *Ther. Drug Monit.* **2000**, *22*, 302–306.
104. McVann, A.; et al. *S. Afr. Med. J.* **1992**, *81*, 139–141.
105. Hu, Z. Herb-Drug Interactions: A Literature Review. *Drugs* **2005**, *65* (9), 1239–1282.
106. Mills, E.; et al. *AIDS* **2005**, *19*, 95–96.
107. Souza Brito, A. R. M. How to study the pharmacology of medicinal plants in underdeveloped countries. *J. Ethnopharmacol.* **1996**, *54* (2-3), 131–138.
108. Sun, M.; et al. Change of secondary metabolites in leaves of Ginkgo biloba L. in response to UV-B induction. *Innovative Food Sci. Emerging. Technol.* **2010**, *11* (4), 672–676.

Applications and Commercialization of African Natural Plants Products as Foods and for Health and Nutrition

Chapter 14

Hibiscus sabdariffa: Phytochemistry, Quality Control, and Health Properties

Tom Villani,[1,2] H. Rodolfo Juliani,[1] James E. Simon,[1,2] and Qing-Li Wu*,[1,2]

[1]New Use Agriculture and Natural Plant Products Program, School of Environmental and Biological Sciences, Rutgers, The State University of New Jersey, 59 Dudley Road, New Brunswick, New Jersey 08901
[2]Department of Medicinal Chemistry, Ernest Mario School of Pharmacy, 160 Frelinghuysen Road, Piscataway, New Jersey 08854
***E-mail: qlwu@aesop.rutgers.edu.**

Hibiscus sabdariffa is a plant of increasing interest for its applications in health and medicine, beverages and cosmetic products. The flowers or calyces are most popular for uses in beverages, natural pigments and bioactive phytochemicals. The leaves are also consumed as a fresh market indigenous 'vegetables' in many sub-Saharan African countries. The natural products responsible for the plants bioactivities are reviewed.

The genus known as *Hibiscus* is large and composed of several hundred species of flowering plants in the family Malvaceae. These species are known for their large, colorful flowers. One of the most widely known species of this genus is *Hibiscus sabdariffa* L., known commonly as red sorrel, roselle or simply hibiscus. *H. sabdariffa* is grown throughout parts of Central and West Africa, as well as South East Asia (*1*). From an African perspective, some of the leading hibiscus producers include Egypt, Sudan, Gambia, Mali, Nigeria, Senegal, Tanzania and Uganda (*6*). The plant is a branched annual shrub growing up to more than 2 meters tall. The stems are reddish in color, whereas leaves are dark green to red, divided into 4-7 lobes (*2*). The flowers are usually red, pink yet sometimes white, and the colors of which are used commonly as food coloring agents (*3*).

The calyces of the flower are consumed as a hot or cold tea in many parts of the world due to popular sourness and flavor. Hibiscus tea is consumed around the world, being popular in Latino cultures in Central and South America, in Arabic cultures in Egypt, and sub-Sahara Africa (Senegal, Sudan) and Asia (China, Thailand). In Mexico, a red beverage known as Jamaica is derived from *H. sabdariffa*, and also known as roselle. The calyces are known as karkade in Arabic, and are used in sauces, jams, jellies, and wines (*4*). In the West Indies, calyces are used to color and flavor rum, and the stalks and leaves are eaten as salad and to season curries. The calyces are also used in making syrup, gelatin, beverages, puddings, tea, marmalade, ice-cream/sherbets, butter, pies, tarts, and other deserts (*5*). Newer products from hibiscus including flavored carbonated water, cold teas and extracts that are going into an array of personal hygiene and cosmetic products, such as shampoos and skin lotions, are on the market. Instant freeze dried and sprayed dried formulations are used as final products for consumers and as concentrated flavorings and coloring agents (*6*). The red pigment from hibiscus is used in meat and poultry as a natural coloring agent. Another type of *H. sabdariffa* is cultivated for its fiber used for cords and ropes, as well as in the manufacture of burlap (*5*).

Hibiscus also has a long history of use in traditional medicine, being prescribed because of its purported wide variety of reported pharmacological activities. The flower has been reported to be "antiseptic, aphrodisiac, astringent, cholagogue, demulcent, digestive, diuretic, emollient, purgative, refrigerant, resolvent, sedative, stomachic, and tonic (*4*)." Hibiscus, also known as Roselle, has been offered as treatment for abscesses, bilious conditions, cancer, cough, debility, dyspepsia, dysuria, fever, hangover, heart ailments, hypertension, neurosis, scurvy, and strangury (*4*, *7*). Due to the wide variety of its uses, and its extensive cultivation throughout the world, hibiscus has been studied extensively as a nutraceutical product.

Phytochemicals

As with most medicinally active plants, there is a diverse matrix of phytochemicals contained within *H. sabdariffa*. The nature of these chemicals determines the activity and qualities for which the plant is used. Many studies have been performed to elaborate the chemical constituents of *H. sabdariffa*, and most of these studies have examined the calyx rather than the leaves and focus on only a small subset of classes of phytochemicals. In this review, we will discuss polyphenols, organic acids, fatty-acids, and oils/volatiles. A summary of the phytochemical constituents, including their molecular weights as described in the scientific literature is shown in Table 1, following this section.

2.1. Polyphenols

Much of the research that has been done in the area of *H. sabdariffa* is related to polyphenols including flavonoids, more specifically, to anthocyanins. Anthocyanins are the sugar-substituted derivatives of anthocyanidins, a derivative of the positively charged flavylium species, which is an oxygen containing aromatic heterocycle (*8*). Anthocyanins and anthocyanidins absorb strongly in visible wavelengths, typically around 520nm, and comprise many pigments that make up the myriad colors of fruits and vegetables, and are well studied for their medicinal activities (*8*, *9*).

The content of anthocyanins in hibiscus extracts and teas have been documented for more than half a century. The highest concentration of these compounds occurs in the calyces. Delphinidin-3-sambubioside, also known as hibiscin, has been found as a major constituent (*10*). Cyanidin-3-sambubioside was identified as the second most abundant component in the extract, as well as delphinidin-3-glucoside and cyanidin-3-glucoside (*11*) (see Figure 1). Other anthocyanins that have been reported include cyanidin-3,5-diglucoside, cyanidin-3-(2-glucosyl-rutinoside) (*12*). Experimenters have studied the biosynthesis of anthocyanins by *H. sabdariffa* after treatment with various auxins. When used in combination with kinetin, 2,4-dichlorophenoxyacetic acid, will affect a marked increase in anthocyanin biosynthesis (*13*).

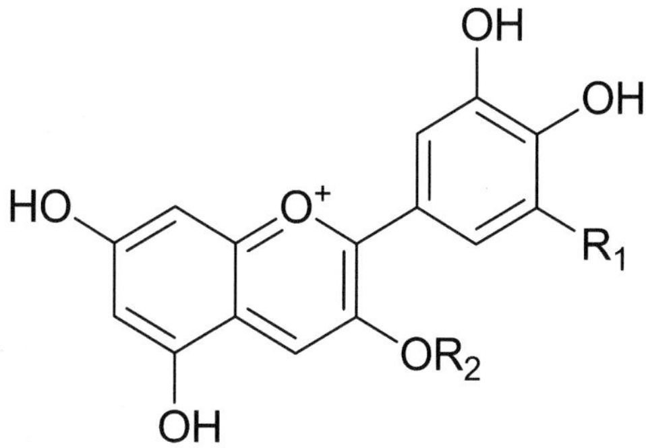

Cyanidin-3-sambubioside (R1=H, R2 = sambubiose)
Cyanidin-3-glucoside (R1=H, R2 = glucose)
Delphinidin-3-sambubioside (R1=OH, R2 = sambubiose)
Delphinidin-3-glucoside (R1=OH, R2 = glucose)
sambubiose = β-D-xylosyl-(1->2)-β-D-glucose

Figure 1. Structures of major anthocyanidin species found H. sabdariffa.

Flavonols are similar in structure to anthocyanins, but they are uncharged and they absorb mostly in the ultraviolet region. In hibiscus, the highest concentration of flavonols is in the flower petals, whereas a low concentration occurs in the calyces. The calyces of *H. sabdariffa* have been shown to contain the flavonol hibiscetrin, shown to be hibiscetin-3-glucoside (*14*). Also found to occur was gossyptrin, the 7-glucoside of gossypetin, as well as sabdaretrin, a glycoside of sabdaretin (*15*). Additional glucosides of gossypetin have been reported, including gossypetin-3-glucoside, gossypetin-7-glucoside, and gossypetin-8-glucoside (*12*). Using HPLC/MS techniques, a large number of flavonols were simultaneously identified in the calyces, including myricetin-3-arabinogalactoside, quercetin-3-sambubioside, quercetin-3-rutinoside, quercetin-3-glucoside, as well as kaempferol-3-O-rutinoside, and kaempferol-3-(p-coumarylglucoside) (*16*).

Other polyphenolic compounds have also been identified in the petals of the flower of *H. sabdariffa*. Protocatechuic acid, also known as hibiscus acid, a compound related to gallic acid, has been shown to occur in roselle flowers (*17*, *18*). Chlorogenic acid, a coumaric acid derivative, has also been found in the flowers (*18*).

One study was conducted to show the effect of ambient temperature, drying temperature and storage time on polyphenol content of hibiscus extracts. The researchers observed that long-time storage even at temperatures of 40°C for 15 weeks results in a drop of only a few percent in polyphenol content (*19*). Interestingly, anthocyanin content decreases from about 80% of total phenolics to about 50%, however there is a corresponding increase in the other polyphenolics. The authors hypothesized that this is due to the polymerization of monomeric anthocyanins during storage.

2.2. Organic Acids

The compounds which are responsible for giving sour taste in foods and beverages such as tea are organic acids. The major organic acids in hibiscus include citric acid and malic acid, with detectable levels of ascorbic acid (*20*, *21*). Tartaric acid was also found in Taiwanese-grown hibiscus calyces by paper chromatography (*22*). Other studies have shown these compounds, and in addition, shown the presence of hibiscus acid; all of these compounds are present in highest concentrations just before ripening of the calyx (*23*).

2.3. Fatty Acids, Oils, and Aromatic Volatiles

Oils and volatile components of *H. sabdariffa* have been well elaborated in the literature. More than 25 volatile hydrocarbons, alcohols, and aldehydes were detected in the seed oil of *H. sabdariffa* (*24*). A variety of sterols have also been detected in seed oil, such as cholesterol, campesterol, stigmasterol, β-sitosterol, α-spinasterol, and ergoterol (*25*). Another study has found *H. sabdariffa* seed oil contained the sterols: beta-sitosterol, campesterol, Delta-5-avenasterol, cholesterol, and clerosterol. Also detected were tocopherols at an average concentration of 2000 mg/kg dry weight; these including alpha-tocopherol (25%), gamma-tocopherol (74.5%), and delta-tocopherol (0.5%) (*26*).

In another study, more than 37 components were identified in the aromatic volatile constituents of *H. sabdariffa* tea. The major components in fresh samples were (Z)-3-hexenol, 2-hexenol and 1-hexenol, as well as α-terpineol, and eugenol (*27*). Another set of researchers evaluated the change in volatile components according to drying temperature and duration (*28*).

A number of particularly interesting fatty acid esters have been found in the pressed seed oil of *H. sabdariffa*. These derivatives contain cyclopropene moieties or epoxide functionality: malvalic acid, sterculic acid, and epoxy oleic acid (*29*, *30*). These oils are not removable by hydrogenation, and therefore present a problem in the processing of roselle oil. These exotic fatty acids impart a dark color and high viscosity, lowering the apparent quality of the seed oil. Removal of the sterculic and malvalic acids is accomplished on heating for 60 mins, although the epoxy-acids remained unchanged (*31*). Removal of these cyclopropene fatty acids is desired due to the inhibitory effect that these acids have on fatty acid desaturase enzymes in animals. Inhibition of these enzymes can lead to artherosclerosis (*32*).

Table 1. Phytochemicals found in *Hibiscus sabdariffa*

	Compound	*MW*	*Reference*
Phenolic acids	5-O-Caffeoylshikimic acid	336	(*16*)
	7-hydroxycoumarin	162	(*16*)
	Chlorogenic acid	354	(*16*)
Flavonols	Gossypetin-3-glucoside	480	(*12*)
	Gossypetin-7-glucoside	480	(*12*)
	Gossypetin-8-glucoside	480	(*12*)
	Gossyptrin	480	(*15*)
	Hibiscetin-3-glucoside	496	(*14*)
	Kaempferol 3-(p-coumarylglucoside)	594	(*16*)
	Kaempferol 3-O-rutinoside	594	(*16*)
	Myricetin-3-arabinogalactoside	612	(*16*)
	N-Feruloyltyramine	313	(*16*)
	Quercetin	302	(*16*)
	Quercetin 3-glucoside	464	(*16*)
	Quercetin 3-rutinoside	610	(*16*)
	Quercetin 3-sambubioside	596	(*16*)
	Sabdaretrin	499	(*15*)
Anthocyanins	Cyanidin 3-sambubioside	581	(*12*)
	Cyanidin-3-(2-glucosyl rutinoside)	773	(*12*)
	Cyanidin-3,5-diglucoside	611	(*12*)
	Cyanidin-3-glucoside	449	(*12*)
	Delphinidin 3-sambubioside	597	(*11*)
	Delphinidin-3-glucoside	465	(*11*)
Organic acids	Ascorbic acid	176	(*21*)

Continued on next page.

Table 1. (Continued). Phytochemicals found in *Hibiscus sabdariffa*

	Compound	*MW*	*Reference*
	Citric acid	192	(*20*)
	Hibiscus acid	190	(*16*)
	Hydroxycitric acid	208	(*14*)
	Malic acid	134	(*16*)
	Tartaric acid	150	(*22*)
Sterols	Campesterol	400	(*25*)
	Cholesterol	386	(*25*)
	Clerosterol	412	(*26*)
	delta-5-Avenasterol	412	(*26*)
	Ergosterol	396	(*25*)
	Stigmasterol	412	(*25*)
	α-spinasterol	412	(*25*)
	β-sitosterol	414	(*25*)
Tocopherols	α-tocopherol	430	(*26*)
	γ-tocopherol	416	(*26*)
	δ-tocopherol	402	(*26*)
Volatiles	α-terpinyl acetate	196	(*27, 28*)
	α-terpineol	154	(*27, 28*)
	α-farnesene	204	(*27, 28*)
	α,4-dimethyl-3-cyclohexyl-1-acetaldehyde	152	(*27*)
	tetrahydro-2,2-dimethyl-5-(1-methylpropyl)furan	156	(*27, 28*)
	p-Cymene	134	(*28*)
Volatiles	Octanal	128	(*27*)
	Nonanal	142	(*27*)
	Decanal	156	(*28*)
	Hexanal	100	(*27*)
	Heptanal	114	(*27*)
	Furfural	96	(*27*)
	Benzaldehyde	106	(*28*)
	Methyl Salicylate	152	(*27*)
	Linalool oxide	170	(*27, 28*)

Continued on next page.

Table 1. (Continued). Phytochemicals found in *Hibiscus sabdariffa*

	Compound	*MW*	*Reference*
	Linalool	154	(*27, 28*)
	Limonene	136	(*27*)
	exo-2-hydroxycineole	170	(*27*)
	Eugenol	164	(*27*)
	Caryophyllene	204	(*27*)
	Acetic acid	60	(*27*)
	6,10,14-trimethyl-2-pentadecanone	268	(*27*)
	5-methyl-2-furaldehyde	110	(*27*)
	4-methyl-1-(1-methylethyl)-3-cyclohexen-1-ol acetate	196	(*27*)
	2-pentylfuran	138	(*27*)
	2-methyl-6-methylene-7-octen-2-ol	154	(*27*)
	2-hexenol	100	(*27*)
	2-ethylfuran	96	(*27*)
	2-ethenyltetrahydro-2,6,6-trimethyl-2H-pyran	154	(*27*)
	2,6-dimethyl-5,7-octadien-2-ol	154	(*27*)
	2,3-dimethylbutane	86	(*27*)
	2,2-dimethylhexanal	128	(*27*)
	1-methyl-4-(1-methylethyl)-3-cyclohexenol	154	(*27*)
	1-hexanol	100	(*27*)
	1,8-cineole	154	(*27, 28*)
	1,4-cineole	154	(*27*)
	(Z)-3-hexenol	100	(*27*)
	(E)-2-hexenal	98	(*27*)
	(E)-2-heptenal	112	(*27*)
Fatty acids	epoxy oleic acid	298	(*30*)
	malvalic acid	280	(*29*)
	sterculic acid	294	(*29*)

2.4. Nutrition

Hibiscus can be viewed as a very nutritious and healthy product; the plant is consumed by many cultures worldwide as a nutritious source of many vitamins, organic acids, and minerals. The extract of *H. sabdariffa* was found to be an excellent source of Vitamin C (ascorbic acid), Calcium, and Phosphorus, containing 60%, 88%, and 391% of each respective nutrient compared to the content found in oranges (*33*). Another study evaluated the vitamin and mineral composition of the red and yellow calyces, identifying them as good sources for Calcium, Iron, Phosphorus, Zinc, β-Carotene, Thiamine, Riboflavin and Vitamin C (*34*). The seeds of *H. sabdariffa* have also been evaluated for their nutritional composition. The seeds were found to contain Phosphorus, Calcium, Zinc, Manganese, Magnesium, Copper, Riboflavin, as well as 18 amino acids (*35*).

Methods of Analysis

Many methods have been employed to analyze the chemical constituents of *H. sabdariffa*. High performance liquid chromatography (HPLC) techniques have been used extensively for the analysis of many water-soluble compounds due to the speed and versatility of these methods. Gas chromatography (GC) and capillary electrophoresis (CE) have also been used to analyze the volatile components of this species. Several assays have also been employed to analyze the chemical constituents and activity of *H. sabdariffa*.

3.1. Chromatographic Methods

Due to the water-solubility of the many interesting compounds in hibiscus, HPLC techniques have been routinely used in their analysis. Typically, reverse phase HPLC is undertaken utilizing a reversed phase column. A method for the separation of flavonoids and anthocyanins in *H. sabdariffa* extract utilizing a gradient of water and acetonitrile (ACN), using a modifier of 0.05% trifluoroacetic acid to compensate for the acid-base equilibrium of anthocyanins (*36*). This method utilizes UV detection at 520 nm for detection of anthocyanins, 370 nm for flavonoids, and 210 nm for detection of other compounds. In another study, anthocyanin and other flavonoid content in the calyces of *H. sabdariffa* was analyzed simultaneously using a gradient of water and ACN, with 5% acetic acid modifier. Using UV for detection, anthocyanins were observed at 520 nm, coumarates at 316 nm, flavan-3-ols at 280 nm, flavonols at 365 nm and hydroxybenzoates at 280 nm (*37*, *38*).

A preparative HPLC method for the isolation of anthocyanins derived from *H. sabdariffa* extract has also been developed (*39*). An HPLC-DAD-ESI-TOF-MS method was reported for the direct characterization of 17 major compounds in *H. sabdariffa* extract, including major anthocyanins, flavonoids, and phenolic acids (*40*). In a different study, taking advantage of the positive charge on anthocyanins, researchers utilized capillary electrophoresis for analysis. Using a CE-TOF-MS technique, the researchers were able to achieve rapid separation and identification (*41*).

In order to separate water-insoluble compounds, different techniques are required. Gas chromatography has been utilized to study the lipid-soluble contents in seed oil in *H. sabdariffa.* Researchers were able to simultaneously identify several tocopherols, sterols, and volatile oils in roselle seed oil (*42*). GC-MS was used to profile the compounds present in the aroma of hibiscus tea, identifying 37 compounds simultaneously (*43*).

3.2. Nonchromatographic Assays

A simple and rapid assay for the determination of total phenol content is known as the Folin–Ciocalteu reagent assay. This assay involves mixing a solution of this reagent with the analyte and then measuring absorbance in the visible spectrum at 765 nm. Total phenol content of *H. sabdariffa* has been measured using this method (*44*). Results are expressed as total phenol content in terms of gallic acid equivalents. In order to measure anti-oxidant capacity, several assays have been used. Ferric reducing ability of plasma (FRAP), oxygen radical absorbance capacity (ORAC), and total antioxidant status assays have been used to measure antioxidant power of hibiscus extracts (*45*). In another study, antioxidant power was measured using the β-carotene bleaching method, as well as evaluating the ability of *H. sabdariffa* extract to scavenge the 1,1-diphenyl-2-picrylhydrazyl (DPPH) free radical (*46*).

A study which focused on analytical protocols for the quality control of *H. sabdariffa* demonstrated the effectiveness of using pH differential UV-Vis spectrophotometric methodology to evaluate content of major anthocyanins found in *H. sabdariffa.* A close correlation (R^2=0.82) was found between the anthocyanin content of water extracts of *H. sabdariffa* quantified using pH differential UV-Vis spectrophotometry and HPLC/UV (*47*), demonstrating the effectiveness of this cheaper, quicker alternative to chromatographic analysis.

Medicinal Uses and Biological Studies

Due to its history in traditional medicine, *H. sabdariffa* has been extensively studied to elucidate and verify the medicinal activities. Hibiscus extracts show a wide variety of pharmacological properties, and there is great interest in the tea as a therapeutic agent. Hibiscus tea has been shown to have no significant effect on the kinetics of metabolic pathways of acetaminophen (*48*) or chloroquine (*49*). These drugs are metabolized by CYP450 enzymes in the liver, and these results suggest that hibiscus tea has no significant effect on these enzymes, and therefore should have no significant interactions with other drugs metabolized by these enzymes. Hibiscus extract has also been shown to be non-toxic. Onyenekwe et al. have demonstrated that the LD_{50} of the extract of the calyces in rats to be above 5000mg/kg, equivalent to a human consuming about 500g extract (*50*). This extremely low toxicity allows for the use of hibiscus as a daily treatment for chronic diseases.

4.1. Antihypertensive

Aqueous extracts of *H. sabdariffa* calyces were administered parentally to rats in a study by Adengunloye et al. which showed a dose dependent response to lowering blood pressure. This effect was diminished by atropine and H_1 receptor blockers, but was resistant to other standard receptor blocking agents, suggesting that the mechanism of reducing blood pressure is mediated through cholinergic and/or histaminergic pathways (*51*). Another study reported the effectiveness of *H. sabdariffa* extract in mediating sodium induced hypertension as well as nitric-oxide synthetase inhibition induced hypertension in rats (*52*). The anthocyanins in hibiscus extract have been proposed to be responsible for hypotensive activity by inhibition of angiotensin-converting enzyme (ACE) (*53*), although more research is needed to elucidate the biochemical mechanisms. Extracts of hibiscus were found to induce a vasodilator effect on the contracted aortic rings of rats. This effect was diminished by anticholinergics and nitric oxide synthetase inhibitors; the authors propose that effects are mediated through the nitric oxide-cGMP-relaxant pathway and through inhibition of Ca^{2+}-influx into vascular smooth muscle cells (*54*). The wide range of evidence suggests that it is likely that hibiscus extract acts in several synergistic ways to produce hypotensive and antihypertensive effects.

Several clinical studies have been published and demonstrated the utility of HSE in treatment of essential hypertension in people. Faraji and Tarkhani (*55*) demonstrated the effectiveness of hibiscus tea at lowering the blood pressure of a group of 31 late-middle-aged men and women (average age: 52.6±7.9 years). Excluded from the trial were patients with secondary hypertension and patients taking more than two antihypertensive drugs. Patients in the study did not use antihypertensive drugs throughout the observation. The study found an 11.2% lowering of the systolic blood pressure and a 10.7% decrease of diastolic pressure in the experimental group just 12 days after beginning the treatment. Two days after stopping treatment, blood pressures had increased back to the levels of the control group.

A controlled and randomized clinical trial has shown the effectiveness of hibiscus tea (anthocyanin content: 9.6 mg/day) as compared to a common antihypertensive agent, captopril (25 mg, twice per day) over four weeks (*56*). The trial consisted of 75 patients ranged from 30-80 years old, and were diagnosed with hypertension, or were without hypertension for at least one month. The study showed an 11.0% reduction in systolic blood pressure, and a 12.4% reduction of diastolic blood pressure of the hibiscus tea group, which shows no significant difference from the captopril group.

Herrera-Arellano and Miranda-Sánchez performed a randomized, double-blind clinical trial showing the comparable effectiveness of *H. sabdariffa* extract (250 mg anthocyanin content/day) to a common antihypertensive, lisinopril (10 mg/day) (*57*). This study confirmed the antihypertensive activity of hibiscus extract, showing a reduction in systolic BP of 11.58%, and a reduction in diastolic BP of 12.21%.

McKay et al. confirmed the antihypertensive activity of the hibiscus extract in pre-hypertensive and mildly-hypertensive adults. These results suggest the possibility of hibiscus tea as a dietary supplement to prevent and control hypertension in adults (*58*).

A wealth of research has been published into the antihypertensive effects of *H. sabdariffa*. There is clear evidence to support its use as a treatment/prevention for hypertension. For further information, Wahabi et al. have recently published a review which summarizes in greater depth the use of *H. sabdariffa* in the treatment of hypertension (*59*). Due to the interest in anthocyanins derived from *H. sabdariffa* extract, the pharmacokinetic profiles of hibiscus derived anthocyanins following consumption in humans were evaluated. In this paper, the half-life of total anthocyanins was 2.6 hours with a 150 mg dose (*60*).

4.2. Anti-Cholesterol/Anti-Obesity

Hibiscus also has a history of use in traditional-medicine relative to weight-loss and reducing cholesterol. Hibiscus tea has been recommended as a safe and natural alternative to many weight-loss supplements (*61*). Several studies have confirmed this activity.

Ethanolic extracts of hibiscus have been shown to reduce the serum-lipid profile of rats fed a high-lipid content diet. Rats fed a 5% (of total diet weight) supplement of hibiscus extract showed a significant decrease in LDL cholesterol and triacylglycerol levels, as well as total lipid levels, compared to control, while HDL and phospholipid levels did not change to a significant degree (*62*). Aqueous hibiscus extract has also been shown to reduce the levels of LDL and the ratio of LDL to HDL in rats (*63*). Also demonstrated was the reduction of weight gain by rats fed a high fat diet concurrent with hibiscus extract. This study also demonstrates the ability of hibiscus extract to inhibit LDL oxidation, which is linked to the development of artherosclerosis (*64*). Studies have shown potential for *H. sabdariffa* extract to be used as a treatment for artherosclerosis. Rabbits were fed a high-cholesterol diet to affect experimental artherosclerosis in the aortas of rabbits; this was significantly attenuated by treatment with *H. sabdariffa* extract (*65*). The authors posit that the preventative effect is related to reduction of serum-levels of lipids in the animals. The effect on lipid profiles has prompted hibiscus extract to be examined for its potential as a weight-loss aid. *Hibiscus sabdariffa* extract has been shown to significantly reduce weight gain in obese mice (obesity was induced by monosodium glutamate) (*66*).

Studies have also shown the ability for *H. sabdariffa* extract to effect human lipid serum levels. A clinical trial of 42 patients aged 18-75 with a serum cholesterol level of 175-327 mg/dL were examined. After 4 weeks, serum cholesterol levels had been reduced by 8.3-14.4%, showing the potential for *H. sabdariffa* extract to be used as a treatment for patients with hypercholesterolemia (*67*). Metabolic syndrome is considered the reversible stage preceding diabetes and coronary heart disease. Treatment of patients diagnosed with metabolic syndrome with powdered hibiscus extract has been shown to lower serum lipid profiles and decrease insulin resistance (*68*), this effect is augmented by exercise. *Hibiscus sabdariffa* extract has been indicated for use in individuals with metabolic syndrome to control lipid levels.

4.3. Antioxidant Activity

The antioxidant capacity of polyphenols has been extensively studied throughout the literature, so it follows that extracts of hibiscus exhibit these activities as well. Many biological studies have shown the effectiveness of *H. sabdariffa* extract (HSE) in protecting cells from oxidative damage and scavenging free radicals, especially in liver cells. A number of assays can be utilized to determine the total antioxidant capacity of HSE, including ferric reducing ability of plasma (FRAP) assay, oxygen radical absorbance capacity (ORAC) assay and total antioxidant status (TAS) assay (*69*). Farombi and Fakoya have shown that organic-soluble fractions of ethanolic *H. sabdariffa* extract act as better free-radical scavengers of hydrogen peroxide, superoxide anion radical, and hydroxyl radical than α-tocopherol, BHA, and quercetin (*70*). Prenesti et al. found the total antioxidant power of hibiscus decoctions were high and proposed that hibiscus beverages act as a protection against free-radical damage (*71*). Due to the high content of tocopherols, roselle seed oil also possesses strong antioxidant capacity with highly lipid-soluble extracts (*72*).

Many biological studies have shown *H. sabdariffa* extract to protect liver cells against toxic damage from a variety of causes. Several *in vivo* studies have shown a protective effect of hibiscus derived protocatechuic acid (*73*), hibiscus derived anthocyanins (*74*), and raw *H. sabdariffa* extract (*75*) against oxidative stress on the hepatocytes of rats. Hibiscus protocatechuic acid has also been shown to protect hepatocytes against lipopolysaccharide induced nitric oxide synthetase in rats. Treatment with protocatechuic acid significantly reduced serum concentrations of hepatic enzyme markers associated with hepatotoxicity (*76*). Hibiscus extract was shown to attenuate acetaminophen-induced toxicity to hepatocytes. High doses were able to restore levels of serum markers indicative of liver damage (*77*). Another study has shown that HSE administered by intraperitoneal injection demonstrates the ability to protect rat liver cells against CCl_4 induced fibrosis (*78*), significantly reducing serum concentration of marker enzymes associated with hepatocyte toxicity. Other researchers have demonstrated the ability of HSE to prevent lipid-peroxidation in the brain induced by $FeSO_4$, sodium nitroprusside, and quinolinic acid. This has been demonstrated in vivo to exhibit neuroprotective properties in rats (*79*).

4.4. Antitumor

Given the increased interest in identifying new methods of treating tumors and finding new therapeutic techniques, it is not surprising that interest in hibiscus and other anthocyanin bearing plants relative to this application have been studied. Modern antitumor therapies are expensive and exhibit a wide range of toxicities. Chemotherapeutic techniques are invasive and have a huge array of side effects, including nausea and hair loss. Anthocyanins and their corresponding aglycones have been screened against a number of common carcinomas, including stomach, colon, breast, and CNS cancer cell lines. The sugar substituted anthocyanins exhibited no significant inhibitory effect. However, the anthocyanin aglycones of cyanidin, delphinidin, malvidin, and pelargonidin exhibited marked inhibition of tumor growth (*80*).

The anthocyanins derived from *H. sabdariffa* have been screened against certain human cancer cell lines. The study showed that hibiscus anthocyanin extract induces apoptosis in human promyelocytic leukemia cells, thought to be mediated by the p38-FasL and Bid pathway (*81*). Chewonarin et al. have demonstrated the ability of ethanol hibiscus extract to prevent mutagenicity of various heterocyclic amines, known to be colon carcinogens, in rats (*82*). The extract reduced the mutagenicity caused by 2-amino-1-methyl-6-phenylimidazo[4,5-*b*]pyridine, 2-amino-3-methylimidazo[4,5-*f*]quinoline, and other such derivatives by about 60-90%. This chemopreventative effect has been confirmed in a study in gastric carcinoma cells. It was found that HSE induced apoptosis in AGS cancer cell line in a concentration dependent manner, proposed to be mediated by the JNK/p38 signaling cascade (*83*). Its interest regarding antitumor activity has prompted the filing of a patent issued in 2011 for the use of anthocyanins derived from *H. sabdariffa* for the inhibition of tumor growth (*84*).

4.5. Antibacterial/Antifungal/Antiparasitic

Hibiscus seed oil has been studied in vitro to show an inhibitory effect on several bacteria, such as *Bacillus anthracis*, as well as *Staphylococcus albus* (*85*). Antibacterial activity was confirmed in another study which examined the inhibitory effect of the calyx extract and protocatechuic acid on growth of methicillin-resistant *Staphylococcus aureus*, *Klebsiella pneumoniae, Pseudomonas aeruginosa* and *Acinetobacter baumannii* (*86*). The experimenters also showed that protocatechuic acid exhibits more activity than the raw calyx extract alone. Antibacterial activity of a methanolic extract of *H. sabdariffa* was shown against *Staphylococcus aureus, Bacillus stearothermophilus, Micrococcus luteus, Serratia mascences, Clostridium sporogenes, Escherichia coli, Klebsiella pneumoniae, Bacillus cereus,*and *Pseudomonas fluorescence* (*87*). Ethanolic extracts of the dried leaves of hibiscus sabdariffa show an inhibitory effect on the growth of certain fungi. *Aspergillus fumigatus, Rhizopus nigricans,* and *Tricophylon mentagrophytes* (*88*) are each inhibited by the extract. Aqueous extract of the sepals of *H. sabdariffa* has been shown to be an effective antiparasitic treatment against *Schistoma mansoni* (*89*). Interestingly, aqueous extract of the dried seeds showed no activity against the parasite, even at 100x greater dose.

4.6. Anti-Inflammatory/Diuretic

Extracts of *H. sabdariffa* have been long ascribed to diuretic properties in traditional medicine, as well as for treating inflammation and thinning the blood. Although the diuretic properties of *H. sabdariffa* extract have been stated in many publications, few have supported such claims with any direct evidence. One study by Mojiminiyi et al. confirmed the anti-inflammatory effect (*90*).

Anti-inflammatory activity has been shown in HSE, mediated by inhibition of cyclooxygenase enzymes 1 and 2. The extract showed higher inhibition of COX-1 than COX-2, indicating its potential for use as a blood thinner as well (*91*). A patent issued in 2004 listed extracts of *H. sabdariffa* as an ingredient in a dietary food supplement meant to treat inflammation by inhibition of COX-2 (*92*).

Uses in Food

Due to the high concentrations of colored anthocyanins in *H. sabdariffa* extracts, these vibrantly colored compounds are used as natural coloring agents for foods including meats, poultry, cheeses and other dairy-based products, deserts, baked goods and beverages (*93*). There has been an increase in demand for natural alternatives to synthetic food colorants, and research has been undertaken to examine the feasibility of these compounds in industrial food use. A 1% solution of anthocyanin extract has been used to impart a deep red color to beverages (*94*).

The major problem presented in using *H. sabdariffa* derived anthocyanins as pigments is the instability of these compounds over relatively short times (*95*). In one study, drinks containing anthocyanins from hibiscus were deemed unacceptable after just 56 days in storage (*96*). The authors also present a method to stabilize the pigments by addition of 3% maltodextrin (w/v) increased the half-life to over 90 days. An additional study sought to examine use of hibiscus anthocyanins as colorant for 2 dry-packed foods, a beverage mix, and a gelatin dessert. Results showed good stability for up to 4 weeks storage time (*97*). A patent was issued for a method to produce a beverage intermediate with improved acid stability and color shelf-life, using Hibiscus as a principle ingredient (*98*). The Frito-Lay Co. was awarded a patent in 1982 for a process to enhance stability of anthocyanin pigment extracts for use in food-coloring (*99*). Another patent, originating from India, described a process for manufacturing food-grade colors from flowers, typically Hibiscus (*100*).

Hibiscus has been used in the preparation of a functional food extract for use in beverages and/or foods. The extract is high in anthocyanins, as well as organic acids, and most commonly prepared as a tea—a patent for the procedure to produce this extract was filed in 2006 (*101*). New cosmetic and personal hygiene products are on the market taking advantage of the anthocyanin rich pigments of hibiscus. *H. sabdariffa* has become an important crop with an economic impact on developing countries due to the spread of its use as a food-colorant and ingredient for food manufacturing (*102*).

Concluding Remarks

Ample research exists which supports the use of *H. sabdariffa* for many medicinal activities. Several clinical studies have shown the efficacy of *H. sabdariffa* extract in treatment of hypertension, comparable to common pharmaceutical antihypertensive agents. Use of HSE has been shown to improve lipid-profiles and aid in reduction of cholesterol and artherosclerosis. HSE has been shown as an effective chemoprotective agent for the liver and brain, as well as an effective agent to induce apoptosis in a number of tumor cell lines. A major benefit in *H. sabdariffa* extract for use in medicine is the extremely low toxicity. The low toxicity associated with HSE makes it suitable for use in treatment of chronic diseases, as well as use in preventative medicine. While so much evidence exists demonstrating the effectiveness of *H. sabdariffa* extract, more research should be done into the mechanisms of action, pharmacodynamics, and pharmacokinetics of these extracts. While polyphenolic compounds are hypothesized to promote much of this activity, there is relatively little information which explores the connection between specific phytochemical constituents and pharmacological activity. Clearly a wide body of evidence exists which demonstrates the value of *H. sabdariffa* extract as a nutraceutical product.

Acknowledgments

We thank Babou Diouf and Diatta Malainy, ASNAPP-Senegal; Dan Acquaye and Julie Asante-Dartey, ASNAPP-Ghana, with whom we have closely worked with hibiscus for many years (www.asnapp.org). We thank the many Senegalese communities, and in particular the women's growers associations with whom our core hibiscus research began, and Association Education Santé (AES)-Senegal. We recognize and thank the USAID for their support of our projects in sub-Saharan Africa to strengthen in-country African natural products industry through their support of the Partnership for Food and Industry in Natural Products (PFID/NP), a university led program supported by the Office of Economic Growth, Agriculture and Trade (EGAT/AG) of the USAID (Contract Award No. AEG-A-00-04-00012-00) which under different funding continues today (www.pfidnp.org); to Jerry Brown, Carol Wilson for their support and encouragement; and to Kit Chin, Southern University with whom are collaborative closely on hibiscus via the USDA/CSREES grant awarded to the Southern University Agricultural Center for the research *Adaptability, Flavonoid and Value Addition of Hibiscus sabrariffa* (USDA/CSREES-2008-38814-04772, 2008-2011); and more recently the National Institute of Food and Agriculture and whose support of the Southern University Agricultural Center-led research initiative, *Alternative Utilization of Roselle Hibiscus as Small Farm and Niche Market Crop* (USDA/NIFA-2012-38821-20092). Lastly, we thank the New Jersey Agricultural Experiment Station and the School of Environmental and Biological Sciences at Rutgers.

References

1. Ali, B. H.; Al Wabel, N.; Blunden, G. Phytochemical, Pharmacological and Toxicological Aspects of *Hibiscus sabdariffa* L.: A Review. *Phytother. Res.* **2005**, *19*, 369–375.
2. Qi, Y.; Chin, K. L.; Malekian, F.; Berhane, M. Biological Characteristics, Nutritional and Medicinal Value of Roselle, *Hibiscus Sabdariffa. SUAREC CIRCULAR – Urban Forestry Natural Resources and Environment No. 604*; 2005; http://www.suagcenter.com/fact_sheets.html.

3. Esselen, W.; Sammy, G. Roselle: A natural red colorant for foods? *Food Prod. Dev.* **1975**, *7*, 80–82.
4. Duke, J. Hibiscus sabdariffa. *Handbook of Energy Crops*; Purdue University: West Lafayette, IN, 1983
5. Watt, J. M.; Breyer-Brandwijk, M. G. *The medicinal and poisonous plants of southern and eastern Africa*, 2nd ed.; E&S Livingstone: Edinburgh, 1962.
6. Simon, J. E.; Govindasamy, R. *Market Intelligence Study for Bissap (Hibiscus sabdariffa) Products from Senegal in the United States of America. Final Report*; report submitted by J. Austin Associates and the Interational Resources Group in Support for Accelerated Growth and Increased Competitiveness to USAID Economic Growth Program (IQC No. 685-1-00-06-00005-00 Task Order No. 685-1-01-000005-00), December 04, 2004; 38 pp.
7. Ross, I. Hibiscus Sabdariffa. *Medicinal Plants of the World*, 2nd ed.; Humana Press: Totowa, NJ, 2003; Vol. 1, pp 267−275.
8. Welch, C; Wu, Q. L.; Simon, J. E. Recent advances in anthocyanin analysis and characterization. *Curr. Anal. Chem.* **2008**, *4*, 75–101.
9. Zhang, Y.; Vareed, S. K.; Nair, M. G. Human tumor cell growth inhibition by nontoxic anthocyanidins, the pigments in fruits and vegetables. *Life Sci.* **2005**, *76* (13), 1465–72.
10. Seshadri, T; Thakur, R. Constitution of gossytrin, a new glycoside from the flower petals of *Hibiscus sabdariffa. Indian Chem. Soc.* **1961**, *38*, 649–651.
11. Du, C.; Francis, F. Anthocyanins of Rosella (*Hibiscus sabdariffa* L.). *J. Food Sci.* **1973**, *38*, 810–812.
12. Subramanian, S.; Nair, A. Flavonoids of four Malvaceous plants. *Phytochemistry* **1972**, *11*, 1518–1519.
13. Mizukami, H.; Tomita, K.; Ohashi, H.; Hiraoka, N. Anthocyanin production in callus cultures of roselle (*Hibiscus sabdariffa* L.). *Plant Cell Rep.* **1988**, *7* (7), 553–556.
14. Rao, S.; Seshadri, T. Isolation of Hibiscitrin from the flowers of *Hibiscus sabdariffa*; contitution of *hibiscetin. Proc. Indian Acad. Sci.* **1942**, *15A*, 148–153.
15. Rao, S.; Seshadri, T. Pigments of cotton flowers; position of the glucose resid*ue in gossyptrin. Proc. Indian Acad. Sci.* **1939**, *9A*, 177–180.
16. Rodríguez-Medinal, I. C.; Beltrán-Debón, R.; Molina, V. M.; Alonso-Villaverde, C.; Joven, K.; Menéndez, J. A.; Segura-Carretero, A.; Fernández-Gutiérrez, A. Direct characterization of aqueous extract of *Hibiscus sabdariffa* using HPLC with diode array detection coupled to ESI and ion trap MS. *J. Sep. Sci.* **2009**, *32* (20), 3441–8.
17. Tseng, T.-H.; Hsu, J.-D.; Lo, M.-H.; Chu, C.-Y.; Chou, F.-P.; Huang, C.-L.; Wang, C.-J. Inhibitory effect of Hibiscus protocatechuic acid on tumor promotion in mouse skin. *Cancer Lett.* **1998**, *126* (2), 199–207.
18. Salah, A. M.; Gathumbi, J.; Vier*ling*, W. *Inhibition of* int*e*stinal motility by methanol extracts of *Hibiscus sabdariffa* L. (Malvaceae) in rats. *Phytother. Res.* **2002**, *16* (3), 283–5.

19. Tsai, P.-J.; McIntosh, J.; Pearce, P.; Camden, B.; Jordan, B. R. Anthocyanin and antioxidant capacity in Roselle (*Hibiscus sabdariffa* L.) extract. *Food Res. Int.* **2002**, *35* (4), 351–356.
20. Buogo, G.; Picchinenna, D. *Chemical Characteristics* of *Ro*selle hemp. *Ann. Chim. Appl.* **1937**, *27*, 577.
21. Reaubourg, G.; Monceaux, R. The chem*ical, botanical, and pharmac*ological characteristics of the karkade (rosella) *Hibiscus sabdariffa*. *J. Pharm. Chim* **1940**, *1*, 292–305.
22. Lin, Y. The study of red pigments in Taiwan plants. *Proc. Natl. Sci. Counc.* **1975**, 133–137.
23. Khafaga, E.; Koch, H. Stage of maturity and quality of roselle. *Angew. Botanik* **1980**, *54*, 295.
24. Jirovetz, L.; Jager, W.; Remberg, G. Analysis of the volatiles in the seed oil of *Hibiscus sabdariffa* (Malvaceae) by means of GC-MS and GC-FTIR. *J. Agric. Food Chem.* **1992**, *40*, 1186.
25. Salama, R.; Ibrahim, S. Ergosterol in *Hibiscus sabdariffa* seed oil. *Planta Med.* **1979**, *36*, 221.
26. Mohamed, R.; Fernández, J.; Pineda, M.; Aguilar, M. Roselle (*Hibiscus sabdariffa*) seed oil is a rich source of gamma-tocopherol. *J. Food Sci.* **2007**, *72* (3), S207–11 doi:10.1111/j.1750-3841.2007.00285.x.
27. Chen, S.; Huang, T.; Ho, C.-T.; Tsai, P.-J. Extraction, analysis, and study on the volatiles in Roselle tea. *J. Agric. Food Chem.* **1998**, *46*, 1101.
28. Gonzalez- Palomares, S; Estarron-Espinosa, M; Gomez-Leyva, JF; Andrade-Gonazalez, I Effect of the Temperature on the Spray Drying of Roselle Extracts (*Hibiscus sabdariffa*). *Plant Foods Hum. Nutr.* **2009** March, *64* (1), 62–67.
29. Ahmad, M. U.; Husain, S. K.; Ahmad, I.; Osman, S. M. *Hibiscus sabdariffa* Seed Oil: A Re-investigation. *J. Sci. Food Agric.* **1979**, *30* (4), 424–428.
30. Earle, F. R.; Glass, C. A.; Geisinger, G. C.; Wolff, I. A.; Jones, Q. Search for new industrial oils. IV. *J. Am. Oil Chem. Soc.* **1960**, *37* (9), 440–447.
31. Sarojini, G.; Rao, K. C.; Tulpule, P. G.; Lakshminarayana, G. Effects of processing on physicochemical properties and fatty acid composition of *Hibiscus sabdariffa* seed oil. *J. Am. Oil Chem. Soc.* **1985**, *62* (4), 728–730.
32. Bao, X.; Katz, S.; Pollard, M.; Ohlrogge, J. Carbocyclic fatty acids in plants: Biochemical and molecular genetic characterization of cyclopropane fatty acid synthesis of Sterculia foetida. *Proc. Natl. Acad. Sci. U.S.A.* **2002**, *99* (1), 7172–7177.
33. Fasoyiro, SB; Babalola, SO; Owosibo, T Chemical Composition and Sensory Quality of Fruit-Flavoured Roselle (*Hibiscus sabdariffa*) Drinks. *World J. Agric. Sci.* **2005**, *1* (2), 161–1642005.
34. Nnam, N. M.; Onyeke, N. G. Chemical composition of two varieties of sorrel (*Hibiscus sabdariffa L.*), calyces and the drinks made from them. *Plant Foods Hum. Nutr.* **2003**, *58*, 1–72003.
35. Rao, P. U. Nutrient composition and biological evaluation of mesta (*Hibiscus sabdariffa*) seeds. *Plant Foods Hum. Nutr.* **1996**January, *49*, 27–341996.

36. Kolawole, J.; Maduenyi, A. Effect of zobo drink (*Hibiscus sabdariffa* water extract) on pharmacokinetics of acetaminophen in human volunteers. *Eur. J. Drug Pharmacokinet.* **2004**, *29*, 25.
37. Salah, A. M.; Gathumbi, J.; Vierling, W. *Inhibition of* in*te*stinal motility by methanol extracts of *Hibiscus sabdariffa* L. (Malvaceae) in rats. *Phytothe. Res.* **2002**, *16* (3), 283–5.
38. Tsai, P.-J.; McIntosh, J.; Pearce, P.; Camden, B.; Jordan, B. R. Anthocyanin and antioxidant capacity in Roselle (*Hibiscus sabdariffa* L.) extract. *Food Res. Int.* **2002**, *35* (4), 351–356.
39. Herrera-Arellano, A.; Miranda-Sánchez, J. Clinical Effects Produced by a Standardized Herbal Medicinal Product of *Hibiscus sabdariffa* on Patients with Hypertension. A Randomized, Double-blind, Lisinopril-Controlled Clinical Trial. *Planta Med.* **2007**, *73*, 6–12.
40. Rodríguez-Medinal, IC; Beltrán-Debón, R; Molina, VM; Alonso-Villaverde, C; Joven, K; Menéndez, JA; Segura-Carretero, A; Fernández-Gutiérrez, A Direct characterization of aqueous extract of *Hibiscus sabdariffa* using HPLC with diode array detection coupled to ESI and ion trap MS. *J. Sep. Sci.* **2009**, *32* (20), 3441–8.
41. Segura-Carretero, A; Puertas-Mejía, MA; Cortacero-Ramírez, S; Beltrán, R; Alonso-Villaverde, C; Joven, J; Dinelli, G; Fernández-Gutiérrez, A Selective extraction, separation, and identification of anthocyanins from *Hibiscus sabdariffa* L. using solid phase extraction-capillary electrophoresis-mass spectrometry (time-of-flight /ion trap). *Electrophoresis* **2008**, *29* (13), 2852–61.
42. Mohamed, R.; Fernández, J.; Pineda, M.; Aguilar, M. Roselle (*Hibiscus sabdariffa*) seed oil is a rich source of gamma-tocopherol. *J. Food Sc.* **2007**, *72* (3), S207–11 doi:10.1111/j.1750-3841.2007.00285.x.
43. Chen, S.; Huang, T.; Ho, C.-T.; Tsai, P.-J. Extraction, analysis, and study on the volatiles in Roselle tea. *J. Agric. Food Chem.* **1998**, *46*, 1101.
44. Kruawan, K.; Kangsadalampai, K. Antioxidant activity, phenolic compound contents and antimutagenic activity of some water extract of herbs. *Thai J. Pharm. Sci.* **2006**, *30*, 28–35.
45. Tsai, P.-J.; McIntosh, J.; Pearce, P.; Camden, B.; Jordan, B. R. Anthocyanin and antioxidant capacity in Roselle (*Hibiscus sabdariffa* L.) extract. *Food Res. Int.* **2002**, *35* (4), 351–356.
46. Kruawan, K.; Kangsadalampai, K. Antioxidant activity, phenolic compound contents and antimutagenic activity of some water extract of herbs. *Thai J. Pharm. Sci.* **2006**, *30*, 28–35.
47. Juliani, HR; Welch, CR; Wu, QL; Diouf, B; Malainy, D; Simon, JE Chemistry and Quality of Hibiscus (*Hibiscus sabdariffa*) for Developing the Natural-Product Industry in Senegal. *J. Food Sci.* **2009**, *74* (2), S113–121.
48. Puckhaber, ; Stipanovic, R. D.; Bost, G. A. Analyses for Flavonoid Aglycones in Fresh and Preserved Hibiscus Flowers. In *Trends in new crops and new uses*; Janick, J., Whipkey, A., Eds.; ASHS Press: Alexandria, VA, 2002.

49. Mahmoud, M.; Ali, H. Significant reduction in chloroquine bioavailability following coadministration with the Sudanese beverages Aradaib, Karkadi, and Lemon. *J. Antimicrob. Chemother.* **1994**, *33*, 1005.
50. Onyenekwe, P.; Ajani, E.; Ameh, D.; Gamammiel, K. Anti-hypertensive effect of roselle (*Hibiscus sabdariffa*) calyx infusion in spontaneously hypertensive rats and a comparison of its toxicity with that in Wistar rats. *Cell Biochem. Funct.* **1999**, *17*, 1999.
51. Adegunloye, B. J.; Omoniyi, J. O.; Owolabi, O. A.; Ajagbonna, O. P.; Sofola, O. A.; Coker, H. A. Mechanisms of the blood pressure lowering effect of the calyx extract of *Hibiscus sabdariffa* in rats. *Afr. J. Med. Med. Sci.* **1996**, *25* (3), 235.
52. Mojiminiyi, F. B.; Dikko, M.; Muhammad, B. Y.; Ojobor, P. D.; Ajagbonna, O. P.; Okolo, R. U.; Igbokwe, U. V.; Mojiminiyi, U. E.; Fagbemi, M. A.; Bello, S. O.; Anga, T. J. Antihypertensive effect of an aqueous extract of the calyx of *Hibiscus sabdariffa. Fitoterapia* **2007**, *78* (4), 292–7.
53. Ojeda, D.; Jiménez-Ferrer, E.; Zamilpa, A.; Herrera-Arellano, A.; Tortoriello, J.; Alvarez, L. Inhibition of angiotensin convertin enzyme (ACE) activity by the anthocyanins delphin*idin- and cyanidin*-3-*O-sambubios*ides from Hibiscus sabdariffa. *J. ethnopharmacology* **2010**, *127* (1), 7–10.
54. Ajay, M.; Chai, H. J.; Mustafa, A. M.; Gilani, A. H.; Mustafa, M. R. Mechanisms of the anti-hypertensive effect of *Hibiscus sabdariffa* L. calyces. *Journal of Ethnopharmacol.* **2007**, *109* (3), 388–93.
55. Haji Faraji, M.; Haji Tarkhani, A. H. The effect of sour tea (*Hibiscus sabdariffa*) on essential hypertension. *J. Ethnopharmacol.* **1999**, *65* (3), 231–236 doi:10.1016/S0378-8741(98)00157-3.
56. Herrera-Arellano, A.; Flores-Romero, S.; Chávez-Soto, M. A.; Tortoriello, J. Effectiveness and tolerability of a standardized extract from *Hibiscus sabdariffa* in patients with mild to moderate hypertension: a controlled and randomized clinical trial. *Phytomedicine* **2004**, *11* (5), 375–382.
57. Herrera-Arellano, A.; Miranda-Sánchez, J. Clinical Effects Produced by a Standardized Herbal Medicinal Product of *Hibiscus sabdariffa* on Patients with Hypertension. A Randomized, Double-blind, Lisinopril-Controlled Clinical Trial. *Planta Med.* **2007**, *73*, 6–12.
58. McKay, D. L.; Chen, C.-Y. O.; Saltzman, E.; Blumberg, J. B. *Hibiscus sabdariffa* L. tea (tisane) lowers blood pressure in prehypertensive and mildly hypertensive adults. *J. Nutr.* **2010**, *140* (2), 298–303.
59. Wahabi, H. A.; Alansary, L. A.; Al-Sabban, A. H.; Glasziuo, P. The effectiveness of *Hibiscus sabdariffa* in the treatment of hypertension: a systematic review. *Phytomedicine* **2010**, *17* (2), 83–6.
60. Frank, T.; Janssen, M.; Netzel, M.; Strass, G.; Kler, A.; Kriesl, E.; Bitsch, I. Pharmacokinetics of anthocyanidin-3-glycosides following consumption of *Hibiscus sabdariffa* L. extract. *J. Clin. Pharmacol.* **2005**, *45* (2), 203–10.
61. Brudnak, M. A. Weight-loss drugs and supplements: are there safer alternatives? *Med. Hypotheses* **2002**, *58* (1), 28–33.
62. Carvajal-Zarrabal, O.; Waliszewski, S. M.; Barradas-Dermitz, D. M.; Orta-Flores, Z.; Hayward-Jones, P. M.; Nolasco-Hipólito, C.;

Angulo-Guerrero, O.; Sánchez-Ricaño, R.; Infanzón, R. M.; Trujillo, P. R. The consumption of *Hibiscus sabdariffa* dried calyx ethanolic extract reduced lipid profile in rats. *Plant Foods Hum. Nutr.* **2005**, *60* (4), 153–9.
63. Chen, C.-C.; Chou, F.-P.; Ho, Y.-C.; Lin, W.-L.; Wang, C.-P.; Kao, E.-S.; Huang, A.-C.; Wang, C.-J. Inhibitory effects of *Hibiscus sabdariffa* L extract on low-density lipoprotein oxidation and anti-hyperlipidemia in fructose-fed and cholesterol-fed rats. *J. Sci. Food. Agric.* **2004**, *84* (15), 8.
64. Lin, H.-H.; Chen, J.-H.; Wand, C.-J. Chemopreventive Properties and Molecular Mechanisms of the Bioactive Compounds in *Hibiscus sabdariffa* Linne. *Curr. Med. Chem.* **2011**, *18* (8), 1245–1254(n.d.).
65. Chen, C.-C.; Hsu, J.-D.; Wang, S.-F.; Chiang, H.-C.; Yang, M.-Y.; Kao, E.-S.; Ho, Y.-C.; Wang, C.-J. *Hibiscus sabdariffa* extract inhibits the development of atherosclerosis in cholesterol-fed rabbits. *J. Agric. Food Chem.* **2003**, *51* (18), 5472–7.
66. Alarcon-Aguilar, F. J.; Zamilpa, A.; Perez-Garcia, M. D.; Almanza-Perez, J. C.; Romero-Nuñez, E.; Campos-Sepulveda, E. A.; Vazquez-Carrillo, L. I.; Roman-Ramos, R. Effect of *Hibiscus sabdariffa* on obesity *in MSG mice. J. Ethnopharmacol.* **2007**, *114* (1), 66–71.
67. Lin, T.-L.; Lin, H.-H.; Chen, C.-C.; Lin, M.-C.; Chou, M.-C.; Wang, C.-J. *Hibiscus sabdariffa* extract reduces serum cholesterol in men and women. *Nutr. Res.* **2007**, *27* (3), 140–145.
68. Gurrola-Díaz, C. M.; García-López, P. M.; Sánchez-Enríquez, S.; Troyo-Sanromán, R.; Andrade-González, I.; Gómez-Leyva, J. F. Effects of *Hibiscus sabdariffa* extract powder and preventive treatment (diet) on the lipid profiles of patients with metabolic syndrome (MeSy). *Phytomedicine* **2010**, *17* (7), 500–5.
69. Tsai, P.-J.; McIntosh, J.; Pearce, P.; Camden, B.; Jordan, B. R. Anthocyanin and antioxidant capacity in Roselle (*Hibiscus sabdariffa* L.) extract. *Food Res. Int.* **2002**, *35* (4), 351–356.
70. Farombi, E. O.; Fakoya, A. Free radical scavenging and antigenotoxic activities of natural phenolic compounds in dried flowers of *Hibiscus sabdariffa* L. *Mol. Nutr. Food Res.* **2005**, *49* (12).
71. Prenesti, E.; Berto, S.; Daniele, P. G.; Toso, S. Antioxidant power quantification of decoction and cold infusions of *Hibiscus sabdariffa flowers. Food Chem.* **2007**, *100* (2), 433–438.
72. Mohamed, R.; Fernández, J.; Pineda, M.; Aguilar, M. Roselle (*Hibiscus sabdariffa*) seed oil is a rich source of gamma-tocopherol. *J. Food Sci.* **2007**, *72* (3), S207–11 doi:10.1111/j.1750-3841.2007.00285.x.
73. Liu, C.-L.; Wang, J.-M.; Chu, C.-Y.; Cheng, M.-T.; Tseng, T.-H. In vivo protective effect of protocatechuic acid on tert-butyl hydroperoxide-induced rat hepatotoxicity. *Food Chem. Toxicol.* **2002**, *40* (5), 635–641.
74. Wang, C. Protective effect of Hibiscus anthocyanins against tert-butyl hydroperoxide-induced hepatic toxicity in rats. *Food Chem. Toxicol.* **2000**, *38* (5), 411–416.
75. Tseng, T.-H.; Kao, E.-S.; Chu, C.-Y.; Chou, F.-P.; Lin Wu, H.-W.; Wang, C.-J. Protective effects of dried flower extr*acts of Hibiscus sabdariffa* L. against

oxidative stress in rat primary hepatocytes. *Food Chem. Toxicol.* **1997**, *35* (12), 1159–1164.

76. Lin, W.-L.; Hsieh, Y.-J.; Chou, F.-P.; Wang, C.-J.; Cheng, M.-T.; Tseng, T.-H. Hibiscus protocatechuic acid inhibits lipopolysaccharide-induced rat hepatic damage. *Arch. Toxicol.* **2003**, *77* (1), 42–7.
77. Ali, B. H.; Mousa, H. M.; El-Mougy, S. The effect of a water extract and anthocyanins of hibiscus sabdariffa L on paracetamol-induced hepatoxicity in rats. *Phytother. Res.* **2003**, *17* (1), 56–9.
78. Liu, J.-Y.; Chen, C.-C.; Wang, W.-H.; Hsu, J.-D.; Yang, M.-Y.; Wang, C.-J. The protective effects of *Hibiscus sabdariffa* extract on CCl_4-induced liver fibrosis in rats. *Food Chem. Toxicol.* **2006**, *44* (3), 336–43doi:10.1016/j.fct.2005.08.003.
79. Oboh, G.; Rocha, J. B. T. Antioxidant and Neuroprotective Properties of Sour Tea (*Hibiscus sabdariffa*, calyx) and Green Tea (*Camellia sinensis*) on some Pro-oxidant-induced Lipid Peroxidation in Brain in vitro. *Food Biophys.* **2008**, *3* (4), 382–389.
80. Zhang, Y.; Vareed, S. K.; Nair, M. G. Human tumor cell growth inhibition by nontoxic anthocyanidins, the pigments in fruits and vegetables. *Life sciences* **2005**, *76* (13), 1465–72.
81. Chang, Y.-C.; Huang, H.-P.; Hsu, J.-D.; Yang, S.-F.; Wang, C.-J. Hibiscus anthocyanins rich extract-induced apoptotic cell death in human promyelocytic leukemia cells. *Toxicol. Appl. Pharmacol.* **2005**, *205* (3), 201–12.
82. Chewonarin, T. Effects of Roselle (*Hibiscus sabdariffa* Linn.), a Thai Medicinal Plant, on the Mutagenicity of Various Known Mutagens in Salmonella typhimurium and on Formation of Aberrant Crypt Foci Induced by the Colon Carcinogens Azoxymethane and 2-Amino-1-methyl-6-ph. *Food Chem. Toxicol.* **1999**, *37* (6), 591–601.
83. Lin, H.-H.; Chen, J.-H.; Kuo, W.-H.; Wang, C.-J. Chemopreventive properties of *Hibiscus sabdariffa* L. on human gastric carcinoma cells through apoptosis induction and JNK/p38 MAPK signaling activation. *Chem.-Biol. Interact.* **2007**, *165* (1), 59–75.
84. Wang, C.-J. Taichung. *Hibiscus Anthocyanins for Inhibiting Cancers*; U.S. Patent 8,007,835, filed Dec 18, 2007, and issued Aug 30, 2011.
85. Gangrade, H.; Mishra, S.; Kaushal, R. Antimicrobial activity of the oil and unsaponifiable matter of red roselle. *Indian Drugs* **1979**, *16*, 147–148.
86. Liu, K.-s.; Tsao, S.-m.; Yin, M.-c. In vitro antibacterial activity of roselle calyx and protocatechuic acid. *Phytother. Res.* **2005**, *19* (11), 942–5.
87. Olaleye, M. T. Cytotoxicity and antibacterial activity of Methanolic extract of *Hibiscus sabdariffa*. *J. Med. Plants Res.* **2007**, *1* (1), 9–13.
88. Guerin, J.; Reveillere, H. Antifungal activity of plant extracts used in therapy. *Ann. Pharm. Fr* **1984**, *42*, 553–559.
89. Elsheikh, S.; Bashir, A.; Sullman, S. Toxicity of certain plant Sudanese plant extracts on Cercariae and Miracidia of *Schistosoma mansoni*. *Int. J. Crude Drug Res.* **1990**, *28*, 241–245.

90. Mojiminiyi, F. B. O.; Adegunloye, B. J.; Egbeniyi, Y. A.; Okolo, R. U. An investigation of the diuretic effect of an aqueous extract of the petals of *Hibiscus Sabdariffa. Afr. J. Med. Med. Sci.* **2000**, *2* (1), 77–80.
91. Christian, K.; Nair, M.; Jackson, J. Antioxidant and cyclooxygenase inhibitory activity of sorrel (*Hibiscus sabdariffa*). *J. Food Compos. Anal.* **2006**, *19* (8), 778–783.
92. Nair, M. *Dietary food supplement containing natural cyclooxygenase inhibitors*. U.S. Patent 6,818,234, filed Feb 27, 2002, and issued Nov 16, 2004
93. Simon, J. E.; Govindasamy, R. *Market Intelligence Study for Bissap (Hibiscus sabdariffa) Products from Senegal in the United States of America. Final Report.* Report submitted by J. Austin Associates and the Interational Resources Group in Support for Accelerated Growth and Increased Competitiveness to USAID Economic Growth Program (IQC No. 685-1-00-06-00005-00 Task Order No. 685-1-01-000005-00), December 04, 2004; 38 pp.
94. Bridle, P.; Timberlake, C. F. Anthocyanins as natural food colours—selected aspects. *Food Chem.* **1997**, *58* (1-2), 103–109.
95. Gradinaru, G.; Biliaderis, C. G.; Kallithraka, S.; Kefalas, P.; Garcia-Viguera, C. Thermal stability of Hibiscus sabdariffa L. anthocyanins in solution and in solid state: effects of copigmentation and glass transition. *Food Chem.* **2003**, *83* (3), 423–436 doi:10.1016/S0308-8146(03)00125-0.
96. Kiattisak, D; Busararat, S; Suchitra, S Roselle anthocyanins as a natural food colorant and improvement of its colour stability. *AIC 2004 Color and Paints, Interim Meeting of the International Color Association.*
97. Clydesdale, F.; Main, J.; Francis, F. Roselle (Hibiscus sabdariffa L.) anthocyanins as colorants for beverages and gelatin desserts. *J. Food Protect.* **1979** March1.
98. Spisak, M. *Beverage Intermediate and method for coextracting tea leaves.* U.S. Patent 6,096,365, filed Nov 22, 1995, issued Aug 1, 2000.
99. Hilton, B. *Processed anthocyanin pigment extracts.* U.S. Patent 4,320,009, filed Mar 5, 1979, issued Mar 16, 1982.
100. Koyazounda, A. *Functional Food and method for preparation thereof.* U.S. Patent Appl. 11,569,647, filed Nov 27, 2006, patent pending.
101. Welch, C; Wu, Q. L.; Simon, J. E. Recent advances in anthocyanin analysis and characterization. *Curr. Anal. Chem.* **2008**, *4*, 75–101.
102. Simon, J. E.; Govindasamy, R. *Market Intelligence Study for Bissap (Hibiscus sabdariffa) Products from Senegal in the United States of America. Final Report.* Report submitted by J. Austin Associates and the Interational Resources Group in Support for Accelerated Growth and Increased Competitiveness to USAID Economic Growth Program (IQC No. 685-1-00-06-00005-00 Task Order No. 685-1-01-000005-00), December 04, 2004; 38 pp.

Chapter 15

Increasing Micronutrient Availability from Food in Sub-Saharan Africa with Indigenous Vegetables

Ray-Yu Yang,*,[1] Sahrah Fischer,[2] Peter M. Hanson,[1] and J. D. H. Keatinge[1]

[1]AVRDC – The World Vegetable Center, Taiwan, R.O.C.
[2]Department of Biology, University of Bayreuth, Germany
***E-mail: ray-yu.yang@worldveg.org.**

Inappropriate and imbalanced diets result in two contrasting human dimensions of malnutrition: undernutrition and obesity. While micronutrient deficiencies are rampant in Sub-Saharan African regions; overweight and non-communicable diseases are rapidly on the rise. Diverse foods provide different types of essential nutrients and phytochemicals not present in monotonous diets. Food diversity is strongly associated with better nutrition of population in developing countries. Indigenous vegetables are native to particular regions, and are grown locally on a small scale. Many are highly nutritious and contain a large range of bioactive phytochemicals. For various reasons, most African Indigenous Vegetables are neglected or under-utilized, though they have the the potential to diversify both production systems and diets and in this manner improve a families health and nutrition. This paper provides important quantitative evidence of the potential of African Indigenous Vegetables for enhanced human nutrition in Africa.

Micronutrient malnutrition is presently a major challenge in achieving certain targets of the Millennium Development Goals. Furthermore, it is closely associated with high rates of childhood morbidity, mortality and death from major Non-Communicable Diseases (NCDs) such as heart attacks and strokes (*1*). Increased consumption of nutrient-dense fruit and vegetables within energy-sufficient and balanced diets has been recommended (*2–5*). Indigenous vegetables (IVs) can make a significant contribution toward increased micronutrient supplies and diet diversification. This paper provides important quantitative evidence of the potential of African Indigenous Vegetables (AIV) for enhanced human nutrition in sub-Sahara Africa.

Malnutrition

Essential nutrients are chemical substances necessary for life, growth and tissue repair that the human body either cannot synthesize or synthesizes insufficient amounts and must be obtained from food. Bioactive substances, derived mainly from plants, are not essential for life, but are associated with reduced risks of degenerative and age related diseases and have the potential to improve human health and well-being. Inappropriate and imbalanced food intake can result in the two contrasting human dimensions of malnutrition: undernutrition and obesity.

Undernutrition

Undernutrition can be divided into protein-energy deficiency and micronutrient deficiencies. Risks of undernutrition increase with unbalanced diets and when dietary energy falls below the minimum requirement. In particular, undernutrition occurs when households replace animal products (meat, fish, dairy and egg) and fruits and vegetables with high carbohydrate staple foods in order to maintain dietary energy intake.

In 2010, one person out of six was not getting enough food to eat on a daily basis (*6*). The highest prevalence was in Sub-Saharan Africa where one person out of three suffered chronic hunger. Asia and the Pacific harbors the largest number of hungry people (578 million), followed by Sub-Saharan Africa (239 million), Latin America and the Caribbean (53 million), and the Near East and North Africa (37 million) (*6*).

The most prevalent micronutrient deficiencies in the world include vitamin A, iron, and zinc (*7*, *8*). These micronutrients are vital for the human body, especially for the development of the immune system and for normal growth of young children. Pregnant women and preschool children including babies who are younger than 1000 days old are particularly vulnerable to nutritional disorders due to poor quality diets insufficient in micronutrients.

Vitamin A

There are two levels of vitamin A deficiency: the subclinical level and the clinical level. The main markers of subclinical vitamin A deficiency are reduced vitamin A body stores, lower serum retinol level (≤ 0.70 μmol/l), and metaplasia with transformation of cells from a normal to an abnormal state (*9*). The clinical stage of vitamin A deficiency shows more serious physiological damage, including xerophthalmia (night-blindness), defects in the formation of ground tissues (for bones, cartilage and teeth) and growth, and increased susceptibility to infection (*10*).

Sources of dietary vitamin A include preformed vitamin A (all-trans-retinol) from animal products and provitamin A from plant foods, including α-carotene, β-carotene, β-cryptoxanthin and other types of carotenoids with provitamin A activity. Carotenoid provitamin A can be converted into the Retinol Equivalent (RE) by the conversion rate of 6 mg β-carotene (or 12 α-carotene) to 1 mg retinol, or the Retinol Activity Equivalent (RAE) with 12 β-carotene (or 24 α-carotene) to 1 mg retinol (*11*). However, the question of retinol conversion rates is still under investigation and debate because carotenoid conversions are influenced by food source, type of food preparation, dietary composition and the nutritional status of the individual (*12–14*).

Iron

The development of iron deficiency is classified by sequential changes in iron content in various iron compartments of the body, including tissue iron, serum ferritin, serum iron, transferrin saturation, and hemoglobin. In the final stage of exhausted iron stores, red blood cell production is drastically reduced and microcytic hypochromic anemia develops (*10*). Consequences of iron deficiency include impaired cognitive functioning and memory, compromised growth and development, and increased risk of pregnancy complications, including prematurity and fetal growth retardation.

Two broad categories of iron are present in food: (1) Heme iron from hemoglobin and myoglobin, which are derived from animal products and which have an iron bioavailability of about 25%, (2) non-heme iron derived from both animal and plant foods and with a relatively low iron bioavailability of about 5-10 % (*15*). Absorption of non-heme iron is greatly influenced by the food matrix and presence of iron inhibitors such as phytate, oxalate and crude fibers, or iron enhancers e.g. vitamin C, other organic acids and heme-iron (*16–18*).

Zinc

Zinc is ubiquitous in the human body, can be measured in the serum, and is required for the normal function of more than three hundred enzymes. However, zinc deficiency is difficult to identify due to its efficient homeostasis mechanism and the lack of a main zinc store in the body (*15*, *19*). Zinc deficiency results in depressed immunity, impaired taste and smell, impaired memory, and decreased spermatogenesis; zinc deficiencies in young children often cause frequent infections, diarrhea, alopecia, and mental disturbances, contributing substantially to increased morbidity and mortality throughout the world (*19*, *20*).

The main sources for zinc are lean red meat, whole-grain cereals, pulses, and legumes. Fish, roots, tubers, and green leafy vegetables have moderate amounts of zinc (*15*). Like iron, zinc bioavailability from plant foods depends on the presence and amounts of inhibiting and enhancing factors, which can either impair or enhance iron absorption (*21*).

Regional Trends of Micronutrient Deficiencies

Global: In all developing countries, about 30% of children (163 million) are vitamin A deficient with low serum retinol (*22*). South and Central Asia (including India) has the highest prevalence with about two thirds of children affected. Central and West Africa has a prevalence of more than 40%. South and Central America and the Caribbean have the lowest prevalence of approximately 10% (*22*).

Iron deficiency anemia in women is a particularly persistent problem. An estimated 500 million or more women are anemic. About 40% of those women live in Asia and Africa. Even in South America and the Caribbean, 25% of women are anemic.

The global prevalence of zinc deficiency was estimated at 31%, ranging from 4–73% across subregions (*19*). The associated loss of disability-adjusted life years (DALYs) attributable to zinc deficiency amounts to more than 28 million. The burden of disease due to zinc deficiency is borne most heavily in Africa, the Eastern Mediterranean and South-East Asia (*20*).

Africa: Tables I and II present the nutritional status of women and pre-school children in sub-Saharan Africa by regions. Iron deficiency is the most widespread nutrition problem with an average of 45% of women iron deficient in tropical Africa. The worst situations are in West and East Africa where iron deficiency affects more than half of the female population, and the highest frequency of iron deficiency is found in West Africa with an average of 61%. In southern Africa, zinc deficiency affects an average of 32% of women. Zinc deficiency shows high variability among different African countries: frequency can be very high, for example in the Democratic Republic of Congo (45.7%), or it is very low, for example in Mali. Occasionally the estimated prevalence of zinc deficiency is almost as high as iron deficiency. For example in countries such as Sierra Leone, Liberia, and Rwanda: Here the severity of zinc deficiency is evident. Vitamin A deficiency in women is less prevalent than iron and zinc deficiencies yet still with an average of 17% vitamin A deficient adult females.

Table I. Nutritional Status[a] of Women in Sub-Saharan Africa

	Over-weight (%)	*Under nutrition (%)*	*Vitamin A deficiency (%)*	*Fe deficiency (%)*	*Zn deficiency (%)*
West Africa (16)[b]					
mean	19	22	18	62	25
range	6-43	6-50	5-34	41-75	9-59
n[c]	13	16	16	16	16
Central Africa (10)					
mean	17	27	16	47	35
range	7-30	8-86	6-20	11-67	19-58
n	7	10	9	10	9
East Africa (8)					
mean	14	15	17	55	29
range	4-23	9-23	13-23	41-63	17-38
n	5	5	8	7	7
Southern Africa (7)					
mean	20	22	17	39	36
range	14-28	7-51	14-20	19-57	14-61
n	5	7	7	7	7
Indian Ocean Islands (6)					
mean	34	10	19	42	33
range	7-68	4-15	14-24	25-55	19-50
n	4	4	4	4	4

[a] Data source: (*19*, *54*). [b] Numbers of countries in the region according to WHO. [c] Number of countries with data.

Overweight and Non-Communicable Diseases

The global epidemic of overweight and obesity has become a serious public health concern (*23*). The shift toward obesity and overweight has been linked to less physical activity than to the overconsumption of poor quality food high in energy and low in nutrient contents (*24*). Obesity and overweight pose a major risk for serious diet-related NCD, principally cardiovascular diseases, diabetes, cancers and chronic respiratory diseases (*3*).

Overweight affects an average of 20% of African women, but incidence is particularly high in the Indian Ocean Islands, with an average of 34%. Countries with relatively high proportions of overweight females include the Seychelles

(68%), and Mauritania (42.7%). Overweight is an important indicator as it presents another important public health issue contributing to chronic diseases (*3*).

Table II. Nutritional Status[a] of Children Under 5 in Sub-Saharan Africa

	Underweight (%)	*Vitamin A deficiency (%)*	*Iron deficiency (%)*	*Zinc deficiency (%)*
West Africa (16)[b]				
mean	24	52	74	35
range	9-43	2-76	50-92	12-47
n[c]	16	16	16	16
Central Africa (10)				
mean	22	40	58	39
range	9-39	6-96	37-84	25-53
n	9	10	10	9
East Africa (8)				
mean	26	41	71	42
range	4-40	21-84	64-85	35-51
n	6	8	7	6
Southern Africa (7)				
mean	18	47	47	37
range	13-21	18-69	19-75	29-53
n	7	7	7	7
Indian Ocean Islands (6)				
mean	34	20	44	28
range	25-42	8-42	17-68	5-53
n	2	4	4	4

[a] Data source: (*19*, *54*) [b] Numbers of countries in the region according to WHO [c] Number of countries with data.

Of the 57 million global deaths in 2008, 36 million, or 63%, were due to NCDs (*3*). Nearly 80% of NCD deaths occur in low-and middle-income countries. In African nations, NCDs are rising rapidly and are projected to exceed communicable, maternal, perinatal, and nutritional diseases as the most common causes of death by 2030.

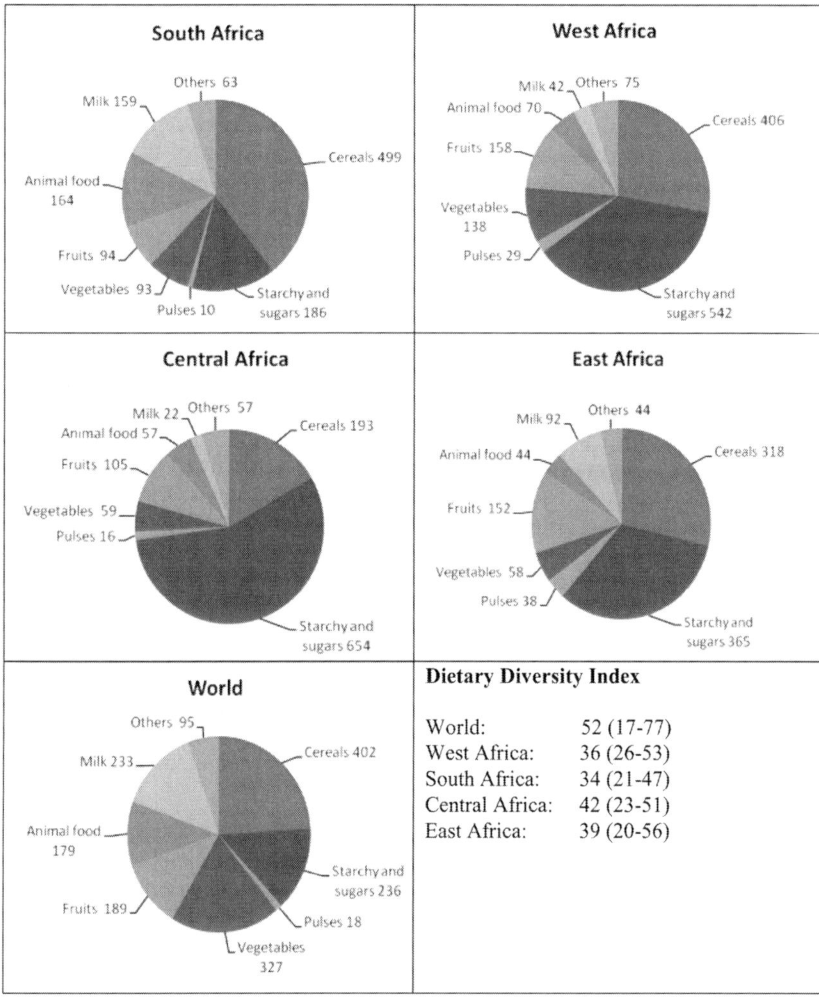

Figure 1. Dietary pattern in the world and regions of Sub-Saharan Africa. Data source: (15). (see color insert)

Dietary Pattern and Diversity

The Chicago Council on Global Affairs have reported in 2011 that agriculture must play an important role in the prevention and mitigation of the growing "epidemic" of NCDs associated with poor and imbalanced diets (*25*). Food diversity is clearly associated with the better nutritional status of populations in developing countries (*26*). Dietary diversity increases the amount of absorbed micronutrients and diverse foods provide different types of essential nutrients and phytochemicals not present in monotonous diets. A dietary diversity index has been developed to assess dietary diversity and represents the ratio of different foods consumed by the population of a particular country: a high index score (e.g. 76 for the United States, 71 for Germany and 69 for United Kingdom) indicates greater food diversity in diets while a lower index score in the lower the diversity (e.g. 27 for Burkina Faso, 24 for Zambia, 25 for Cambodia, and 27 for Nepal) indicates a more monotonous or less diverse diet.

Dietary patterns and diversity indexes among sub-Saharan African regions are compared in Figure 1. Average dietary diversity index in tropical Africa is 38% which is lower than the average of the global food diversity index (52%). Among sub-Saharan African countries, the highest diversity index is found in the Seychelles with 57%, and Uganda with 56% (country data not shown). The countries with the lowest diversity index values are Ethiopia (20%) and Madagascar and Mozambique (both 21%). The dietary diversity index is extremely low in most African countries.

Staple foods such as cereals, starchy foods, and pulses are the most commonly consumed in tropical Africa with an average of 718 g/person/day. Foods from animal products (meat, fish, dairy and egg) are the least consumed foods with an average of 80 g/person/day. Vegetables are the second-least consumed food in tropical Africa with an average of 80 g/person/day which is far below the recommend minimum level of 200 g/person/ day (*2*). Southern Africa is the region with the lowest consumption of vegetables with an average of only 47 g/person/day. Tomatoes and onions are the most consumed vegetables globally and also in Africa. The percentage share of total vegetable consumption in African regions of these crops ranges from 15 – 24% for tomato and 7-16% for onion (Figure 2).

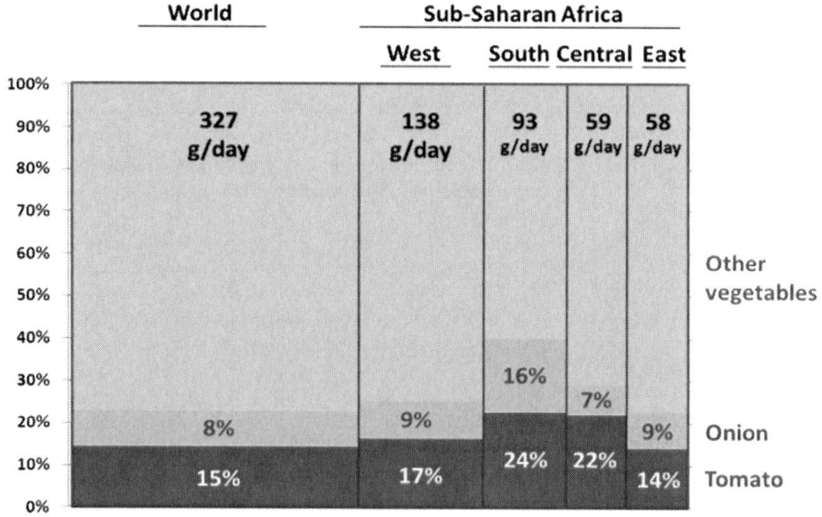

Figure 2. Share of vegetable availability (g/ person/day) in Sub-Saharan Africa. Data source: (1). (see color insert)

Vegetable consumption varies greatly among countries, but is generally low. The countries with low vegetable consumption such as Mozambique are also low in dietary diversity. Fruit consumption varies among African countries but in most regions is relatively higher than both vegetables and animal products with a total average of 138 g/person/day.

Nutritional Values of Indigenous Vegetables

Vegetables are a major source of essential nutrients, including vitamins and minerals, which help prevent nutritional disorders, and are an important source of diverse dietary phytochemicals with beneficial health effects. It is known that higher consumption of vegetables are associated with reduced risks of micronutrient deficiencies and chronic diseases (*27–31*).

Globally, there are hundreds of plant species consumed as vegetables, but only about 20 crops are produced in intensive cropping systems (*32*). Indigenous vegetables (IVs) are native to a particular region or were introduced to the region from another geographical area historically (*33*). They are grown locally on a small scale, are often less affected by disease and pest problems, and at this point in time after being grown for so many generations are often well adapted to local environments and may still be relatively nondomesticated suggesting a wide genetic background that could be used for breeding and improvement as needed. Importantly, many of these indigenous vegetables are highly nutritious and contain a large range of phytochemicals. For various reasons, most IVs are neglected or under-utilized and in some areas in sub-Sahara Africa are considered famine foods to eat when exotic or introduced European vegetables are not available. African IVs have potential for introduction and for greater use as cash crops in peri-urban systems, vegetables for daily sustenance in home gardens, and as a means to diversify production systems and diets (*34*).

Vegetable species are diverse in Africa. The Plant Resource of Tropical Africa (PROTA) database has documented that 528 plant species are used for food, medicinal or ornamental purposes in Africa; among them, at least 275 plant species are primarily used as vegetables (*35*, *36*). The primary use vegetables encompass 53 different botanical families and more than 60% of the species are pan African. About 75% of the primary use vegetables (275 species) are indigenous, 16% were introduced historically and now widely adapted. Only 8% were recently introduced and are regarded as exotic (*36*).

More than 150 species of indigenous vegetables from tropical Asia and Africa were collected, grown at AVRDC headquarters in Taiwan and evaluated for their nutritional values. Many species are high in one or more nutrients including vitamins A, C, E, folates, iron, calcium and antioxidants (*37–40*). Information on reported bioactive compounds, medicinal use, traditional uses and taboos of the African vegetables were reported as well (*40*).

For easier comparison of species varying in nutrient types and quantities, AVRDC designed a 'Nutritional Value' where:

Nutritional Value = sum of % RNI for selected nutrients

RNI is the recommended nutrient intake per person per day (*15*), and the nutritional values of vegetables in this report were calculated as the % coverage of the RNI values for eight important nutrients, including protein, vitamin A, C, E folate, iron, calcium and zinc. Table III presents the nutritional values of 80 African indigenous vegetable crops that include 59 species with the % RNI and the totals of eight major nutrients for each.

Many Indigenous vegetables are rich in vitamins A and C; consumption of a 100 g portion of either one of 20 IVs and 40 IVs out of the 80 IVs listed in Table III fulfill 100% or more of the RNI for vitamins A and C, respectively, for an adult. However, only 8 and 4 IVs out of the 80 IV crops could provide 15% of the RNI for iron and zinc, respectively, and most % RNIs were even lower ranging from only 5-10%.

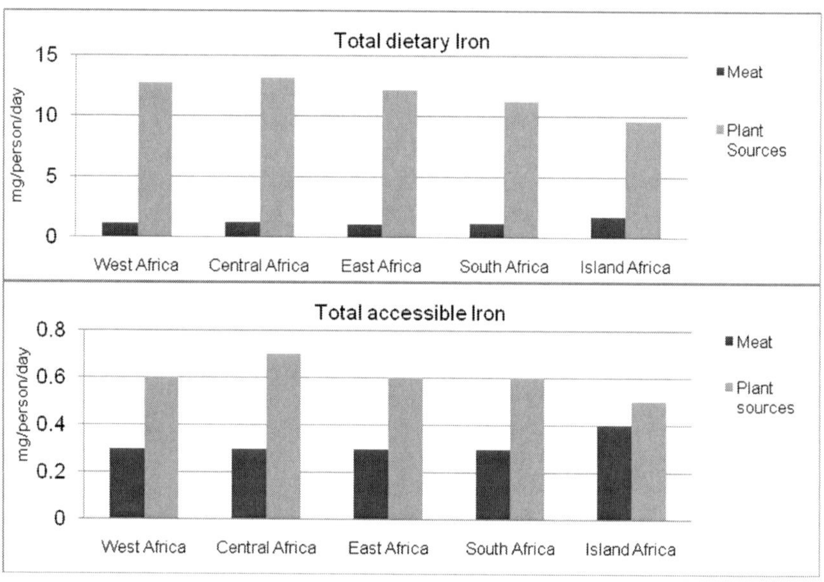

Figure 3. Total and accessible dietary iron from animal and plant sources. Data souce: (15). (see color insert)

The total and accessible dietary iron levels from animal and plant sources are compared in Figure 3. Plant foods are the main dietary iron source in tropical Africa even though iron bioavailability from plant sources is low (5-10% of the total iron content) (*41*). Though vegetables are not ideal dietary sources of iron, indigenous vegetables are still an important iron source in developing countries where diets are mainly plant-based. Certain vegetables (such as amaranth, jute mallow, young shoots of taro, moringa, and ceylon hydroleia in Table III) which have a relatively higher iron concentration could be consumed in order to have the best plant iron source when no better iron sources are available.

Indigenous vegetables, in fact most vegetables, are largely undcrused sources of nutrients relative to staple crops. To emphasize this point, we compared the nutritional values of indigenous vegetables versus several commonly consumed vegetables and a staple food (Figure 4). It is important to note that staple foods e.g. rice (*Oryza* spp.), cassava roots (*Manihot esculenta*) are a calorie source, and not considered a principal source of many nutrients when compared with vegetables.

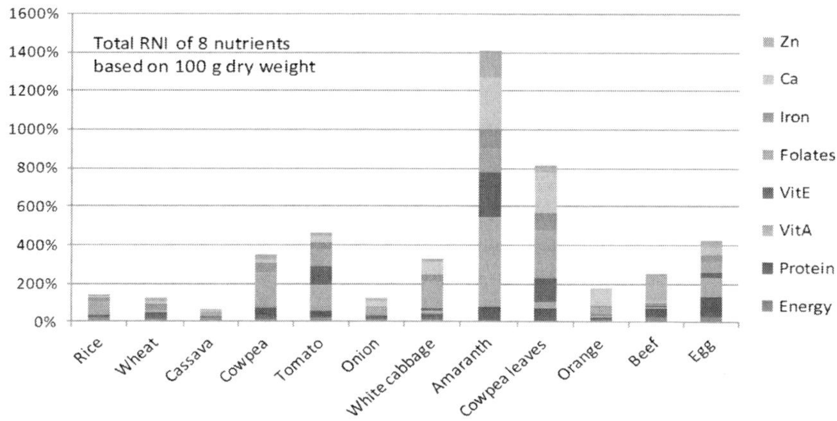

Figure 4. Comparison of total RNI of eight nutrients for staple, animal products, fruits and vegetables. Nutrient data source: AVRDC nutrient data (40) and USDA nutrient database (55). Note: RNI for vitamin C was not included. (see color insert)

In addition to nutritional values, we found that several African indigenous vegetables such as moringa (*Moringa oleifera*), amaranth (*Amaranthus* spp.), sweet potato leaves (*Ipomoea batatas*), spider plant (*Cleome gynandra*) and cassava leaves, have a high level of anti-inflammatory phytochemicals such as flavonoids and other antioxidants (*39*). Further more, most of the leafy vegetables contain 1-2 grams of oil in a 100 g fresh portion, which would also provide significant amounts of dietary omega-3 fatty acids (Yang, unpublished data). This because 20-50% of total fatty acids in green leafy vegetables are known to be in the form of α-linoleic acid (ALA, 18:3 n-3) (*42–44*), which has been strongly recommended in dietary guidelines for reduced risks of chronic inflammation, cardiovascular diseases, and cancers.

The following list represents a summary of some of the most nutritious indigenous vegetables:

- Amaranth (*Amaranthus cruentus, A. dubius, A. graecizans, A. hypochondriacu, A. lividus, A. blitum*): Amaranths are grown pan-tropically and in countries where the leaves are eaten, the crop is among the most nutrient-rich and most popular indigenous vegetables, even though the different species and types may have quite varied flavor and taste (*45*). Some amaranth species possess the C_4 photosynthetic system and are thus probably better adapted to higher temperatures and drought than C_3 species. Amaranth may become more important as farmers choose crops better adapted to conditions brought about by climate change.
- Jute includes three species (*Corchorus capsularis, C. trilocularis, and C. olitorius*). Each jute species is considered and consumed as a vegetable and very suited for planting in home gardens (*35*). While all three species are not distributed throughout the tropics, it is possible to find at least one species in almost every country of tropical Africa and Asia. Different parts of the plant can be eaten, including the stems and pods but the leaves are the most preferred and most nutritious items.
- Spider flower (*Cleome gynandra*): The plant is pan-tropic noteworthy for high folate content. Leaves, young shoots, and sometimes flowers are eaten. Sometimes the leaves are bitter, and are occasionally cooked together with amaranth, vegetable cowpea leaves, or nightshades (*35*). Like the amaranths, spider flower is a C_4 plant and adapted to tropical environments.
- Cassava (*Manihot esculenta*) is grown pan-tropically mainly for its root which is a starchy staple. However, cassava leaves are also extremely high in vitamin A and C but may contain toxic compounds which need to be removed during traditional preparation methods (*35*).
- Vegetable cowpea (*Vigna unguiculata*) is pan-tropic and eaten as pulse and a leafy vegetable. Cowpea leaves have high vitamin A and calcium contents.
- Drumstick tree (*Moringa oleifera*): The leaves, pods, branches and bark, of this pan-tropical plant are harvested and consumed. The leaves are rich in vitamins A, C, E, folates and iron, and the seeds are used for water purification. The leaf-powder is now used extensively to fight malnutrition in West Africa and can be added to special fortified food products for the extremely malnourished such as "Plumpy nut" paste (*35*) and with any source of energy consumed daily.
- Roselle (*Hibiscus sabdariffa*) is found growing throughout the tropics and sub-tropics. Its flowers, leaves and calices can all be consumed and notable for high levels of vitamin C which can be preserved in concentrated drink form which is highly suitable for and appreciated by young children.
- African eggplants (*Solanum aethiopicum, Solanum anguivi, Solanum macrocapon) and African nightshades (Solanum scabrum, Solanum villosum)* are also found Pan-tropically. The nutrient content of African eggplants are not as dense as many other indigenous vegetables, including African nightshades, yet they contain a great diversity of

nutrients. African eggplants and African nightshades are among the most popular and profitable vegetables in Africa (*45*, *46*).

The AVRDC Priority African Indigenous Vegetabless for Improving Nutrient Supplies

Adapted varieties of nutrient-rich vegetable crops enable farmers to supply high yields of good quality produce for markets. Consequently, variety choice is a critical factor affecting productivity and profitability of vegetable farmers. Hallmarks of a strong vegetable breeding and seed sector are the presence of a continuous stream of improved vegetable varieties linked to the production and marketing of healthy and high quality seed that is made available to farmers at affordable prices (*47*).

Increased global emphasis on the need to combat micronutrient malnutrition requires vegetable breeders to respond by developing nutrient-dense vegetables (*34*). This adds a major additional set of challenges to the breeding and postharvest research process and underlines its considerable complexity if successful products are to emerge. The AVRDC genebank (the world's largest public-domain collection of tropical vegetables) is a key resource to provide the variability in parental material needed for successful breeding programs. The AVRDC collection holds the largest collection of indigenous vegetables from Africa. All lines are freely available to vegetable breeders in both the public and private sectors (*46*).

In the past, IVs in Africa were grown mainly in small areas for home consumption or local sale using farmer landraces. Consumer demand for IVs has surged in Africa, leading to market requirements for IV varieties that are uniform for key quality traits. AVRDC researchers based in Sub-Saharan Africa have collaborated with public and private sector partners in the development of adapted IV varieties acceptable to farmers, markets, and consumers (*48*). The IV improvement process began with collection of diverse landraces from a wide array of IV crops from many African countries, and their inclusion as accessions in AVRDC and other germplasm collections; this ensures that landraces are maintained and available to future researchers. Because preferences for IV crops and traits vary among countries and regions, AVRDC staff coordinated national IV priority setting workshops where key players in the horticulture sector ranked IV crops in importance for breeding, and identified necessary traits for IV varieties. Subsequent IV breeding by AVRDC was carried out regionally and focused on selection within the genetic diversity present in landraces, emphasizing farmer participatory evaluation, purification and stabilization, multilocation testing, and national release. Released IV varieties bred by AVRDC are freely available to public and private seed sectors for large-scale seed production and commercialization. Examples of IV varieties developed by AVRDC include 'DB3' African eggplant, 'Tumaini' vegetable cowpea, 'Olevolosi' (=AVRDC code SS 49) African nightshade, and 'Arumeru' (=AVRDC code ST 3) Ethiopian mustard (*48*, *49*).

Table III. Nutritional Values (% RNI per 100 g fresh weight) of African Indigenous Vegetables

Scientific name	***Common name, Origin, Plant Part***	***% RNI (Recommended Nutrient Intake)***								***Total RNI***	***Yield t/ha***
		Pro	***A***	***C***	***E***	***F***	***Ca***	***Fe***	***Zn***		
Abelmoschus esculentus	Okra, Et., YP	4	12	78	4	8	8	5	0	118	10-20
Adansonia digitata	Baobab tree, YL	8	67	169	34	9	45	12	4	349	
Amaranthus cruentus	Purple amaranth, C.Am., YL	8	93	129	21	16	41	18	13	338	
Amaranthus dubius	Spleen amaranth, T.Am., YL	9	198	198	30	18	62	19	21	554	25-50
Amaranthus graecizans	Spreading pigweed, T.Af., YS	9	142	149	28	20	25	11	16	400	
Amaranthus hypochondriacus	Prince-of-Wales'-feather, N.Am., YS	10	175	198	21	17	37	13	25	496	
A. lividus; A. blitum	Livid amaranth, YS	9	164	202	36	10	41	12	38	511	25-50
Amaranthus mangostanus	Edible amaranth, E.A., YS	10	141	176	25	16	45	26	6	445	
Amaranthus tricolor	Joseph's Coat,Ind., YL	8	52	138	3	8	51	14	6	279	25-50
Amaranthus tricolor	Joseph's Coat, YS	11	182	236	11	47	46	10	5	548	
Asystasia gangetica	Chinese violet,Tropical, YS	9	135	111	18	7	26	11	0	317	25-50
Basella alba	Malabar spinach-green, E.A., V	4	87	109	11	18	7	6	0	244	20-80
Beta vulgaris var. cicla	Chard-light green, S.Eu., L	3	8	44	8	9	6	1	2	82	30-60
Beta vulgaris var. cicla	Chard-light green, YL	2	17	26	9	7	4	1	2	69	
Beta vulgaris var. cicla	Chard-red, S.Eu., YL	3	17	36	5	8	7	1	3	79	30-60
Beta vulgaris var. cicla	Chard-yellow, S.Eu., YL	2	13	24	3	6	3	1	2	54	30-60

Scientific name	Common name, Origin, Plant Part	% RNI (Recommended Nutrient Intake)								Total RNI	Yield t/ha
		Pro	A	C	E	F	Ca	Fe	Zn		
Beta vulgaris var. cicla	Swiss Chard, S.Eu., YL	4	28	44	3	17	10	7	2	115	30-60
Beta vulgaris var. crassa	Beet, S.Eu., R	3	0	11	1	2	1	4	3	25	30-60
Bidens pilosa	Hairy beggarsticks, N.Z., V	7	118	122	14		17	8		286	15
Brassica carinata	Ethiopian mustard, N.Af., YS	6	36	304	4		21	5		376	10-50
Brassica carinata	Ethiopian mustard, YL	7	15	371	14	22	23	9	7	466	35
Brassica juncea	Mustard, N.Eu., YL	4	30	160	15	10	20	4	2	245	10-50
Capsicum annuum	Chili pepper, C.Am., YL	9	62	207	3	13	27	12	6	339	1,5-18
Capsicum annuum	Chili pepper, YS	8	111	100	15	40	28	14	2	318	
Celosia argentea	Feather cockscomb, T.Af., L	4	60	4	3		9	8		87	47
Cichorium endivia	Endive, S.Eu., YL	4	61	47	15	2	11	10	3	152	20
Cleome gynandra	Spider-flower, T.Af., YL	10	128	291	9	57	30	13	6	543	30
Coccinia grandis	Ivy gourd, SE A., YS	7	53	109	10	25	4	10		217	10-13
Colocasia esculenta	Taro, M, L	1	5	11	0		11	2		30	5-6
Colocasia esculenta	Taro, S	2	9	7	0		2	2		23	
Corchorus capsularis	Jute, Sub & T., YS	12	284	161	23		28	13		522	3-10
Corchorus olitorius	Jute mallow, E.A., L	12	225	284	22	13	51	20		629	5-15
Corchorus trilocularis	Native jute, T.Af., YS	7	72	194	21	51	31	13	4	393	3-10

Continued on next page.

Table III. (Continued). Nutritional Values (% RNI per 100 g fresh weight) of African Indigenous Vegetables

Scientific name	*Common name, Origin, Plant Part*	*% RNI (Recommended Nutrient Intake)*								*Total RNI*	*Yield t/ha*
		Pro	*A*	*C*	*E*	*F*	*Ca*	*Fe*	*Zn*		
Cucurbita moschata	China squash, pumpkin, S.Am., YS	7	12	29	17	16	12	9		101	
Cucurbita moschata	China squash, pumpkin, FB	4	3	32	29	7	13	6	3	97	
Cucurbita moschata	China squash, pumpkin, YG	3	1	18	2	3	3	3	2	34	
Cucurbita moschata	China squash, pumpkin, MG	4	4	21	5	3	1	4	2	45	15
Glebionis coronarium	Edible chrysanthemum, garland chrysanthemum, V	7	61	71	8	6	14	9	3	179	
Hibiscus sabdariffa	Roselle, In., SE A., L	5	69	55	0		31	7		166	1-15
Ipomoea batatas	Sweet potato, S.A., L	5	75	43	9		8	6		145	7-30
Ipomoea batatas	Sweet potato	7	18	29	1		5	8		67	5-6
Lablab purpureus	Hyacinth bean, T.Af., B	6	4	28	2		4	8		52	
Lablab purpureus	Hyacinth bean, MF	4	23	69	0		4	4		104	
Lablab purpureus	Hyacinth bean, P	5	77	33	0	8	5	7	2	136	2.5-5
Lablab purpureus	Hyacinth bean, YP	10	0	31	0	9	15	7	8	81	7.5
Lablab purpureus	Hyacinth bean, MP	7	0	31	0	6	10	5	4	63	
Lagenaria siceraria	Bottle gourd, T.Af., F	1	0	12	0		1	2		16	25
Lycopersicon esculentum	Tomato, cherry, C.Am., F	2	5	67	8	1	1	3	1	89	8-27
Manihot esculenta	Cassava, S. Am., YL	17	414	822	127	14	57	19	13	1483	10

Scientific name	**Common name, Origin, Plant Part**	**% RNI (Recommended Nutrient Intake)**								**Total RNI**	**Yield t/ha**
		Pro	***A***	***C***	***E***	***F***	***Ca***	***Fe***	***Zn***		
Momordica charantia var. abbreviata	Kakorot, F	2	8	167	11		2	2		191	8-10
Momordica cochinchinensis	Spiny bitter gourd, SE A., IF	2	1	203	0		3	2		211	8-10
Momordica cochinchinensis	Spiny bitter gourd	2	2	431	32	2	5	4	2	480	
Momordica cochinchinensis	Spiny bitter gourd	7	94	279	18	35	19	6	5	463	
Moringa oleifera	Drumstick tree, In., YS	8	54	507	21	21	9	14	4	638	20
Nasturtium officinale	Water cress, Eu. & C. A., YS	4	71	101	7		21	4		208	50
Ocimum basilicum	Basil, T. Af., L	5	127	44	0		17	14		206	5-20
Ocimum basilicum	Basil, YS	6	60	98	15	1	48	19	5	253	
O. basilicum cv. purpurascens	Dark opal basil, T. Af., YS	6	99	94	13		42	10		263	5-20
Petroselinum crispum	Parsley, C./S. Eu., YL	8	83	389	32	21	16	14	5	568	10-20
Pisum sativum	Garden Pea shoot, E.A., YS	6	20	29	0		0	4		59	1.7
Portulaca oleracea	Purslane-red flower, S.Eu., V	3	70	11	3		7	6		100	12-17
Rumex acetosa	Garden sorrel, YL	3	0	40	7	4	7	1	2	63	
Sechium edule	Chayote, C.Am., F	1	0	11	0		2	2		16	20-30
Senna occidentalis	Coffee senna, L	10	522	290	36		19	13		890	
Senna occidentalis	Coffee senna, YL	11	95	747	14		3	7		878	

Continued on next page.

Table III. (Continued). Nutritional Values (% RNI per 100 g fresh weight) of African Indigenous Vegetables

Scientific name	*Common name, Origin, Plant Part*	*% RNI (Recommended Nutrient Intake)*								*Total RNI*	*Yield t/ha*
		Pro	*A*	*C*	*E*	*F*	*Ca*	*Fe*	*Zn*		
Sesbania grandiflora	Sesbania-red, SE A., F	3	1	127	4	7	2	5	3	152	55, leaf
Solanum aethiopicum	African scarlet eggplant, T. Af., MF	2	11	22	2	3	1	6		47	12-20
Solanum aethiopicum	African scarlet eggplant, YF	2	0	31	7	7	3	3	2	55	
Solanum anguivi	Anguivi, YF	3	3	40	9	16	7	6	3	86	
Solanum anguivi	Anguivi, MF	5	21	0	20	0	7	6	3	62	
Solanum macrocapon	African eggplant, T.Af., F	3	1	42	3	6	2	4	2	62	25
Solanum scabrum	African nightshade, In., YS	9	140	167	20	18	28	17	3	400	7-27
Solanum torvum	Water nightshade, C/S Am., YF	5	4	34	5		12	5		65	
Solanum villosum	African nightshades, T.Af., YS	8	71	175	18	15	25	18	6	336	20-25
Solanum zuccagnianum	Nakati, T.Af., L	10	140	97	57	2	34	14	6	360	40
Solanum zuccagnianum	Nakati, MF	3	43	25	28	17	7	2	3	128	
Solanum zuccagnianum	Nakati, YF	3	0	31	19	29	11	4	4	102	
Sphenoclea zeylanica	Chickenspike, wedgewort, T.Af., L	5	128	144	0		16	9		303	
Spinacia oleracea	Spinach, C/SW A., YL	4	45	24	9	8	11	10	2	113	10
Talinum triangulare	Potherb fame-flower, T.Af., V	4	78	6	8		6	5		107	10-60
Trichosanthes cucumerina	Snake gourd, E A., F	1	1	8	0		3	2		16	8-10

Scientific name	Common name, Origin, Plant Part	% RNI (Recommended Nutrient Intake)								Total RNI	Yield t/ha
		Pro	A	C	E	F	Ca	Fe	Zn		
Trichosanthes cucumerina	Snake gourd, YF	1	1	29	0		2	2		35	
Trichosanthes cucumerina	Snake gourd, YF	6	27	280	3		3	3		322	
Tropaeolum majus	Indian cress, S A., F	3	9	157	23		2	8		201	
Vigna unguiculata	Vegetable cowpea, T.Af., YL	10	220	182	63	40	77	10	3	605	0.4
Vigna unguiculata	Vegetable cowpea, YS	9	195	211	23	41	23	11	4	517	

Values used to calculate % coverage of RNI : Energy, 2300 kcal/day; Protein, 50 g/day; Vitamin A, 0,7 mg/day; Vitamin C, 45 mg/day; Vitamin E, 12 mg/day; Folate, 0,4 mg/day; Iron, 18 mg/day; Calcium, 700 mg/day; Zinc, 14 mg/day. Et., Ethiopia. C.Am., Central America. T.Am., Tropical America, T.Af. Tropical Africa. N. Am., North America. E.A., East Asia. In., India. S.Eu., Southern Europe. N.Z., New Zealand. N.Af., North Africa. N.Eu., Northern Europe. S.Eu., Southern Europe. SE A., Southeast Asia. M, Malaysia. Sub & T., Sub- and tropical. S.Am., South America. C.A., Central Asia. Eu., Europe. C./S. Eu., Central and South Europe. C/SW A., Central and South West Asia. Y, young. I, Immature, M, mature. B, Beans. FB, flower buds. G, gurds. L, leaves. P, pods. R, roots. S, shoots. S, stems. Data sources: RNI: (*15*); Nutrient values: Yang et al (*40*) and unpublished data AVRDC; origin and yield data: (*35*).

The breeding of AVIs in Africa is in its early stages and it is expected that future breeding will become more sophisticated with greater emphasis on selection for biotic and abiotic stress tolerances as well as yield and quality. Greater participation of the private sector in IV breeding will lead to more varieties, especially hybrids. As with other vegetables, there is concern that breeding (*50*) may result in IV varieties with diminished nutrient contents compared to landraces. Reduced contents of minerals and some vitamins may result from selection for higher yields (*50*). Unfortunately, vegetable breeding programs that directly select for higher nutrient content are not common (*51*), and many vegetables varieties bred for enhanced nutrient content have not been adopted (*52*). There are fortuitous cases when improved nutrient content is a consequence selection for an important quality trait such as deep red color in tomato and increased content of lycopene (an antioxidant) (*53*). High nutrient content can be bred into vegetable varieties as a long as traits such as high yield, quality, pest/disease resistance are not compromised. Selection for lower levels of a particular nutrient may occur if it is associated with negative quality traits such as strong smells or flavors; for example the association of high glucosinolate contents and bitterness in some crucifers. For many vegetable crops, breeders do not know how selection for yield or quality affects nutrient contents. Development of fast and inexpensive methods to measure nutrient levels, or molecular markers linked to genes conditioning high nutrient content would facilitate informed selection. Greater awareness of simple "rules of thumb" such as the association of high iron/provitamin A contents with deep green color in green leaf vegetables could help breeders visually select for some nutrients. Important to note that while the improvement of AIVs to enhance or maximize their nutrient composition is important, overarching breeding considerations may be focused on increasing consumption of the vegetable. As such, increased insect and disease resistance, slow flowering (e.g. with Spider flower) and longer-life that will ensure and increase the marketable and edible yields over time. These are among the key factors that will drive the market forward, increase profit at the farmer and community level and provide ample fresh product to the increasing market demand that is underway in Africa. A focus on proper processing to retain the nutrients and vitamins in the harvested AIV is also a key parameter to maximize the health benefits as well.

Research Gaps and Future Work

• Low food diversity in most sub-Saharan African countries has not improved in the past ten years. The situation is worsened by poverty and reduced crop diversity and the production of a few commodities over a larger number of nutritious and locally specific foods. Developing, empowering and revivifing local food systems to supply a rich diversity of local produce would help address the food, nutrition and poverty nexus (*22*, *34*).

• Much nutrition research attention has been focused on the health-promoting attributes of global vegetable crops, such as broccoli, tomato, onion, and cabbage. Yet most indigenous vegetables commonly consumed in tropical Asia and Africa have not been sufficiently studied for their positive and negative health factors. Some indigenous vegetables are integrated into local diets, but their health-promoting factors and potential hazards remain unknown. A better understanding of both the positive and negatives qualities of indigenous vegetables, particularly for underutilized species, would help to develop better evidence-based promotion and appropriate dietary strategies. Greater use and consumption of nutrient-rich and safe vegetables can be promoted, while precautions may need to be taken if certain indigenous vegetables contain critical amounts of anti-nutrient factors such as oxalates, alkaloids, and other substances deleterious to human health.

• Greater collaboration among agricultural, food, and health research institutions to understand the effect of regional and local food availability on public health, and to foresee and develop adaptive capacity that can help reduce climate change effects on food security at community and national levels. Research and development in the Africa agricultural sector should emphasize expanding crop diversity, food system productivity, and eco-friendly farming systems.

• Food, nutrition and health sectors should put more research efforts on food-based approaches that lead to diversification of crops, balanced diets, and ultimately, better health.

References

1. Mayer-Foulkes, D. A. *Global Econ. J.* **2011**, *11*, 1–26.
2. WHO. *Diet, Nutrition and the Prevention of Chronic Diseases*; WHO Technical Report Series 916. Report of a joint WHO/FAO expert consultation. World Health Organization: Geneva, Switzerland. 2003.
3. WHO. *Global status report on noncommunicable diseases 2010*; World Health Organization. http://www.who.int/nmh/publications/ncd_report_chapter1.pdf (accessed 07/01/11).
4. Ali, M.; Tsou, S. *Food Policy* **1997**, *22* (1), 17–38.
5. Keatinge, J. D. H.; Yang, R.-Y.; Hughes, Jd'A.; Holmer, R.; Easdown, W. J. The importance of ensuring both food and nutritional security in attainment of the Millennium Development Goals. *Food Secur.* **2011**, *3*, 391–401.
6. FAO. *State of food insecurity in the world, 2009*; Food and Agriculture Organization of the United Nations: Rome, 2010.
7. Bhutta, Z. A.; Ahmed, T.; Black, R. E.; Cousens, S.; Dewey, K.; Giugliani, E.; Haider, B. A.; Hirkwood, B.; Morris, S. S.; Sachdev, H. P.; Schekar, M. *Lancet* **2008**, *371*, 417–440.
8. WHO. *The World Health Report 2002. Chapter 4: Childhood and maternal undernutrition*; http://www.who.int/whr/2002/chapter4/en/index3.html (accessed 07/01/10).
9. Mclaren, D. S.; Frigg, M. *Sight and life manual on vitamin A deficiency disorders (VADD)*, 2nd ed.; Task Force SIGHT AND LIFE: Basel, 2001.
10. *Essentials of human nutrition*; Mann, J. I., Truswell, A. S., Eds.; Oxford: New York, 2000.

11. IOM. *Dietary reference intakes for vitamin A, vitamin K, arsenic, boron, chromium, copper, iodine, iron, manganese, molybdenum, nickel, silicon, vanadium and zinc*; National Academy Press: Washington, DC, 2002.
12. Yeum, K. J.; Russell, R. M. *Annu. Rev. Nutr.* **2002**, *22*, 483–504.
13. West, C. E.; Eilander, A.; van Lieshout, M. *J. Nutr.* **2002**, *132*, 2920S–2926S.
14. Howe, J. A.; Maziya-Dixon, B.; Tanumihardjo, S. A. *Br. J. Nutr.* **2009**, *102* (3), 342–349.
15. FAO; WHO. *Vitamin and Mineral Requirements in Human Nutrition*; World Health Organization: Geneva, 2004.
16. Hurrell, R.; Egli, I. *Am. J. Clin. Nutr.* **2010**, 1461–1467(2010).
17. Yang, R. Y.; Tsou, S. C. S. *J. Int. Cooperation* **2006**, *1* (1), 107–119.
18. Yang, R. Y.; Tsou, S. C. S.; Lee, T. C.; Chang, L. C.; Kuo, G.; Lai, P. Y. Moringa, a novel plant rich in antioxidants, bioavailable iron, and nutrients. In *Challenges in chemistry and biology of herbs*; ACS Symposium Series 935; American Chemical Society: Washington, DC, 2006; pp 224−239.
19. IZiNCG (International Zinc Nutrition Consultative Group). Assessment of the risk of zinc deficiency in populations and options for its control. Hotz, C., Brown, K.H., Eds.; *Food Nutr. Bull.* **2004**, *25*, S91−S202.
20. Caulfield, L. E.; Black, R. E. In *Comparative quantification of health risks: global and regional burden of disease*; Ezzati, M., Lopez, A. D., Rodgers, A., Murray, C. J. L., Eds.; WHO: Swaziland, 2004.
21. Hambidge, M. K.; Miller, L. V.; Westcott, J. E.; Sheng, X. Y.; Krebs, N. F. *Am. J. Clin. Nutri.* **2010**, *91.5*, 1478–1483.
22. UNSCN. *6th Report on the World Nutrition Situation*; United Nations Standing Committee on Nutrition: Geneva, 2010.
23. UNSCN. *Overweight and obesity – a new nutrition emergency?* SCN News 29. United Nations Standing Committee on Nutrition; Lavenham Press: U.K., 2005.
24. Popkin, B. M. *Asia Pac. J. Clin. Nutri.* **2001**, *10*, S13–S18.
25. Nugent, R. *Bringing Agriculture to the Table: How Agriculture and Food Can Play a Role in Preventing Chronic Disease*; Chicago Council: Chicago, 2011; 88 pp.
26. Johns, T.; Eyzaguirre, P. *Food Policy* **2007**, *32*, 1–24.
27. Ezzati, F.; Lopez, A. D.; Rodgers, A.; Hoorn, S. V.; Murray, C. J. L. *Lancet* **2002**, *360* (9343), 1347–1360.
28. Sanchez-Moreno, C.; Cano, M. P.; de Ancos, B.; Plaza, L.; Olmedilla, B.; Granado, F.; Martin, A. *J. Nutr.* **2004**, *134*, 3021–3025.
29. Lopez-Garcia, E.; Schulze, M. B.; Fung, T. T.; Meigs, J. B.; Rifai, N.; Manson, J. E.; Hu, F. B. *Am. J. Clin. Nutr.* **2004**, *80*, 1029–1035.
30. Nanri, A.; Moore, M. A.; Kono, S. *Asian Pac. J. Cancer Prev.* **2007**, *8*, 167–177.
31. Helmersson, J.; Arnlöv, J.; Larsson, A.; Basu, S. *Br. J. Nutr.* **2009**, *101*, 1775–1782.
32. Siemonsma, J. S.; Kasem, P. *PROSEA – Plant resources of South-East Asia 8 – Vegetables*, 2nd ed.; Prosea Foundation: Indonesia, 1996.
33. Engle, L. E.; Shanmugasundaram, S.; Hanson, P. M. *Acta Hortic.* **2003**, *623*, 263–274.

34. Keatinge, J. D. H.; Chadha, M. L.; Hughes, J. d'A.; Easdown, W. J.; Holmer, R.; Tenkouano, A.; Yang, R. Y.; Mavlyanova, R.; Neave, S.; Afari-Sefa, V.; Luther, G. C.; Ravishankar, M.; Ojiewo, C.; Belarmino, M.; Wang, J. F.; Lin, M. *Biol. Agric. Hortic.* **2012**, *28* (2), 71–85.
35. PROTA. In *Plant Resources of Tropical Africa 2: Vegetables*; Grubben, G. J. H., Denton, O. A., Messiaen, C. M., Schippers, R. R., Lemmens, R. H. M. J., Oyen, L. P. A., Eds.; PROTA Foundation: Wageningen, Netherlands, 2004.
36. PROTA. Comparative data on 275 vegetables. In *Vegetables of tropical Africa. Conclusions and recommendations based on PROTA 2: 'Vegetables'*; Bosch, C. H., Borus, D. J., Siemonsma, J. S., Eds.; PROTA Foundation: Wageningen, Netherlands, 2005.
37. Yang, R. Y.; Tsou, S. C. S.; Lee, T. C.; Wu, W. J.; Hanson, P. M.; Kuo, G.; Engle, L. M.; Lai, P. Y. *J. Sci. Food Agric.* **2006**, *86*, 2395–2403.
38. Yang, R. Y.; Wu, J.; Chen, D.; Lin, M.; Shiao, R.; Kuo, G. *Acta Hortic.* **2007**, *752*, 115–122.
39. Yang, R. Y.; Lin, S.; Kuo, G. *Asia Pac. J. Clin. Nutr.* **2008**, *17* (1), 275–279.
40. Yang, R. Y.; Keding G. B. Nutritional contributions of important African indigenous vegetables. In *African indigenous vegetables in urban agriculture*; Shackleton, C. M., Pasquini, M. W. Drescher, A. W., Eds.; Earthscan: London, 2009; pp 105−143.
41. FAOSTAT. *Food Security Statistics*; http://www.fao.org/economic/ess/food-security-statistics/en/ (accessed: 09/19/10).
42. Stuchlík, M.; Žák, S. *Biomed. Pap.* **2002**, *146* (2), 3–10.
43. Galli, C.; Marangoni, F. *Prostaglandins, Leukotrienes Essent. Fatty Acids* **2006**, *75*, 129–133.
44. van der Walt, A. M.; Ibrahim, I. M.; Bezuidenhout, C. C.; Loots, D. T. *Pub. Health Nutr.* **2009**, *12*, 525–530.
45. NRC (National Research Council). *Lost Crops of Africa. Volume II: Vegetables*; National Research Council; The National Academies Press: Washington, DC, 2006.
46. Ojiewo, C.; Tenkouano, A.; Hughes J. d'A.; Keatinge, J. D. H. Diversifying diets: Using African indigenous vegetables to improve nutrition and health. In *Using Agricultural Biodiversity to Improve Nutrition and Health*; Fanzo, J., Hunter. D., Eds.; Earthscan publications: London, 2013 (in press)..
47. Lenné, J. M.; Pink, D. A. C.; Spence, N. J.; Ward, A. F.; Njuki, J.; Ota, M. *Outlook Agric.* **2005**, *34*, 225–2322005.
48. Ojiewo, C.; Tenkouano, A.; Olouch, M.; Yang, R. Y. *Afr. J. Hortic. Sci.* **2010**, *3*, 1–23.
49. Afari-Sefa, V. A.; Tenkouano, A.; Ojiewo, C. O.; Keatinge, J. D. H.; Hughes, J. d'A. *Food Secur.* **2012**, *4* (1), 115–127.
50. Davis, D. R. *HortScience* **2009**, *44*, 15–19.
51. Simon, P. W.; Pollack, L. M.; Clevidence, B. A.; Holden, J. M.; Haytowitz, D. B. *Plant Breed. Rev.* **2009**, *31*, 325–392.
52. Munger, H. M. *HortScience* **1999**, *34*, 1158–1159.
53. Hanson, P. M.; Yang, R. Y.; Wu, J.; Chen, J. T.; Ledesma, D.; Tsou, S. C. S.; Lee, T. C. *J. Am. Soc. Hortic. Sci.* **2004**, *129*, 704–711.

54. WHO (World Health Organization). *WHO database on nutrition: NLiS (Nutrition Landscape Information System)*; http://apps.who.int/nutrition/landscape/report.aspx (accessed 09/16/10).
55. USDA. Nutrient Data Laboratory. *USDA Nutrient Database*; http://www.nal.usda.gov/fnic/foodcomp/search/ (accessed 09/21/10).

Chapter 16

Morphologic and Biochemical Diversity of Peasant's Baobab Tree (*Adansonia digitata* L.) Morphotypes in Senegal

Diaminatou Sanogo,[1] Babacar Diedhiou,[2] Abdoulaye Diop,[2] Haby Sanou,[3] Mactar Thiome,[4] Yaye Kène Gassama,[5] and Emmanuel Bassène[*,6]

[1]Institut Sénégalais de Recherches Agricoles, ISRA
[2]Department of Chemistry, Faculty of Sciences and Techniques, University Cheikh Anta DIOP Dakar (UCAD), Senegal
[3]Institut d'Economie Rurale, IER, Bamako, Mali
[4]Groupe de Laboratoires d'Analyses, Direction des Mines et de la Géologie, Dakar, Senegal
[5]Department of Biology Vegetal, Faculty of Sciences and Techniques, UCAD, Dakar, Senegal
[6]Department of Pharmacognosy and botany, Faculty of Medicine, Pharmacy and Odontology, UCAD, Dakar, Senegal
[*]E-mail: aynenut@hotmail.com.

Adansonia digitata L. is a multi-purpose species which plays an important role in the socio-economic and cultural life of Sub-Saharan populations. In Senegal, peasants distinguish various morphotypes of baobab according to their fruit traits. This study evaluates the morpho-biochemical significance of these local morphotypes using a scientific approach. Surveys were carried out in 2006, in thirteen villages of three various agro-ecological zone of Senegal, to identify criteria used in describing baobab morphotypes by peasants. Four morphotypes of baobabs trees were recognized by five ethnic groups according to the characteristics of the fruit like pulp taste, pulp quantity and the weight of the fruit. Eleven morphological markers (tree height and circumference, powder puff diameter, fruit circumference and length, pulp weight, fruit weight, the ratio of pulp weight and fruit weight, seed weight, fibers weight, thickness of endocarp) and two biochemical markers (total sugars and reducing sugars) were used to characterize these morphotypes through 96 sampled baobab trees and 1617 fruits in thirteen collections sites distributed over three different agroclimatic zones (Sudano-Guinean, Sudano-Sahelian and Sahelian). Only length of fruit, total sugars and reducing sugars contents allowed identifying four groups of *Andansonia digitata* which correlate with the traditional morphotypes. Characters used for the traditional classification are largely dependent on biochemical markers. Others characters such as mineral content (calcium, magnesium, iron) were also determined in pulp fruits. Thus, biochemical differentiation is found between the morphotypes of baobab in Senegal. This characterization of peasant's morphotypes according to morphobiochemical variables allowed for a precise typology of baobab varieties in Senegal.

Introduction

The baobab tree (*Adansonia digitata* L.) is a multi-purpose species which plays an important role in the socio-economic and cultural life of Sub Saharan populations. This species is largely used in Africa (*1*) and the classification of peasants' preferences put it among the most appreciable ligneous species in the Sahel (*2–4*). Research conducted on this species in the Sahel has provided information on its reproduction, agronomic potential (*5–9*), socio-economic importance in the Sahel (*5*, *10*, *11*) and physico-chemical characteristics (*12–18*).

Despite all of its socio-economic importance and scientific interest, its domestication, however, remains very limited because of sectorial approaches that provide only isolated information. We can name among them those in connection with local knowledge in Mali where the rural populations can distinguish the baobab morphotypes taking into consideration either the color of the bark which can be black, red or white, or the pulp taste and leaves or the scope of the tree (*6*). In Benin, a distinction is made according to the fruit-bearing potential of baobab and the form of the capsule (*19*). Furthermore, the variability of the species has been described utilizing physico-chemical characteristics in Senegal (*20*), and morphological and genetic characteristics of capsules in Benin (*21–23*).

But little information is available on the precise morpho-biochemical descriptions of baobab morphotypes as defined by peasants. Moreover, there is no information on their preferences, or reasons therefor. Thus, it is difficult to succeed in baobab domestication without effective participation of the populations in the overall management procedure and conservation of the species. The objective of this study is to conduct a detailed morpho-biochemical description of morphotypes of baobab known by the local populations in order to select those with preferable traits.

Materials and Methods

The study was conducted in three agro-climatic zones in Senegal (Table I) including 1) Eastern Senegal/Upper Casamance (ESUC) located in the Sudano-Guinean zone (rainfall 700-1000 mm, annual temperature 29.24°C, relative humidity 48.29%, ferruginous washed soil); 2) South Peanut Belt (SPB) located in the Sudano-Sahelian zone (rainfall 600-800 mm, annual temperature of 29.85°C, relative humidity 54.74%, washed tropical ferruginous soils with a sandy texture and clay washed and less graveled soils on a lateritic layer) and 3) Central-North Peanut Belt (CNPB) located in the Sahelian zone (rainfall 400-500 mm, annual temperature 27.24°C, relative humidity 64.96%, tropical ferruginous soils less washed named "Dior Soil", very sandy, often poor and Sub-arid brown soils named "Deck Soil", hydromorphic intergrades presenting a deficit of drainage). These areas of collection correspond to the natural baobab habitats in parks, forests and fallows. The predominant ethnic groups were Wolofs in Tambacounda, Pulaars in Kafrine and Sereres in Thies.

Thirteen natural baobab populations have been prospected, for every population a village was selected to determine the peasants' classification criteria of the baobab. The surveys were conducted through guided interviews with people of different ages and gender. These interviews were done with the predominant ethnic groups in the village, focusing on the state of the baobab population, the morphotypes of baobab they know, the ones they prefer, the organoleptic character of these morphotypes, the local management system and recommendations. Among thirteen villages, eight were selected for the identification and collection of the peasants' morphotypes. Villages were chosen on the basis of existence of a management system for the resource baobab related to collection of fruits and other plant parts.

The identification of trees of different morphotypes known by peasants was done in the presence of two to ten persons, generally one man, one woman and children. In each site, 30 to 36 adult trees were selected with a distance of at least 50 meters between the trees (a total of 96 trees). These selected trees were geo-referenced with a Garmin 2000 GPS, marked and identified by a number. At least 25 to 30 fruits per tree (a total of 1617 fruits) were collected and put in sealed bags, labeled with the name of the village, tree number, date of collection and organoleptic characteristics.

For each selected tree the following parameters were measured: the height by using a long pole, the circumference at 1.30 meters above ground and the diameter of the crown in two directions: East-West and North-South both using a ruban of 30 meters. For each fruit, the length, circumference, total weight; the thickness of the coat of the endocarp by using a caliper rule were measured. Then fruits were broken and pulp, seed and fibers were weighed. After maceration in water, seeds and fibers were separated, dried under the sun for 48 hours and then weighed. The quantity of the pulp was deducted. The variable ratio [weight of pulp/weight of fruit] was considered in this study taking into consideration one of the peasants' classification criteria which was abundance of the pulp (relative quantity of the pulp).

Table I. Geographical locations of sites of collection of the peasant morphotypes of baobab tree

Sites of collection	*Longitude/ Latitude*	*Soil Types*
ESUC[1], Tambacounda[2], Sudanian[3]		
Wally Babacar	14°02'03.0/ 13°19'19.7	Hydromorphic (argillaceous alluvia) and ferruginous tropical washed
Ndiaback	14°02'03.4/ 13°20'06.4	Hydromorphic and ferruginous tropical washed
Rabia	14°03'32.8/ 13°21'33.4	Hydromorphic (argillaceous alluvia) and ferruginous tropical washed
Touba Diagnène	14°04'06.0/ 13°20'07.5	Hydromorphic (argillaceous alluvia) and ferruginous tropical washed
SPB, Kafrine, Sudano-Sahelian		
Boye	15°36'986/ 14°26'362	Ferruginous tropical washed. fine gravels on lateritic armor
CNPB, Thiés, Sahelian		
Sindia Caphgoune	17°01'722/ 14°33'943	ferruginous tropical little washed (grounds dior) hydromorphic with a deficit of drainage
Soro Khassab	17°03'320/ 14°33'458	ferruginous tropical little washed (grounds dior) hydromorphic with a deficit of drainage
Ndiogoye	16°57'23.3/ 14°33'51.9	ferruginous tropical little washed (grounds dior) hydromorphic with a deficit of drainage

[1] NOTE: Agro-ecologic Zones: ESUC: Eastern Senegal- Upper Casamance; SPB: South Peanut Belt; CNPB: Center-North Peanut Belt. [2] Region, [3] Climatic Zone.

Twenty-four composite samples, composed of 5 to 10 fruits per sample, were collected on 43 trees. AOAC Methods (*24*) were used for the chemical analyses, namely: desiccation at 103°C to determine moisture, incineration at 550°C to determine ash, complexometry for calcium and magnesium, spectrophotometry for iron and phosphorus, the Luff-Schoorl method for sugars (reducing sugars, total sugars) and titrimetry for acidity.

The statistical analyses of the results were carried out using Statview and XL-Stat software. Variance analyses followed by average test comparisons by Newman-Keuls test to the threshold of probability $P < 5\%$ were used to characterize the morphotypes using physicochemical characteristics averages of pulp and to note variabilities observed in morphotypes.

Results

Peasant's Classification of Baobab Trees

The survey helped to determine the criteria used by the peasants to differentiate baobab trees in their natural habitats (Table II). The criterion taste or sweetness (very sweet, sweet, less sweet and bitter fruit) is used by all ethnic groups to classify the baobab. The abundance of pulp (abundant pulp and non abundant pulp) and the size of the fruit (big and small fruit) are two other criteria used mostly by 60% of the Pulaar groups in Tambacounda and Kaffrine and by 50% of the Serere groups in Thiès to classify the baobab. The two different ethnic groups in Tambacounda have established a link between the main criterion (taste) and the other criteria such as the shape and the weight of the fruit, the resistance of the endocarp to break, the stickiness and the color of the pulp.

Table II. Criteria mentioned by different ethnic groups to characterize the morphotypes of baobab tree existing in Senegal

Name of village	*Ethnic groups*	*Classification Criteria*
ESUC[1]	Tambacounda[2]	Sudanian[3]
Wally Babacar	Pulaars (Wolofs. Bambaras. Sarakolés)	Very sweet, sweet, less sweet, bitter or empty fruits [1]. Abundant pulp fruits. Less abundant pulp fruits [2]. Big fruits, small fruits [3].
Ndiaback		Very sweet, sweet, less sweet, bitter or empty fruits [1]
Rabia	Wolofs (Pulaars)	Very sweet, sweet, less sweet, bitter or empty fruits [1]
Touba Diagnène		Very sweet, sweet, less sweet, bitter or empty fruits [1]
Allagué		Very sweet, sweet, less sweet, bitter or empty fruits [1].abundant pulp. Less abundant pulp [2]. Non-sticky pulp [sweet taste] sticky pulp (less sweet taste) [4+1]. Long, easy to break fruit, with abundant pulp. Round, hard to break fruit, with less abundant pulp [5+6+2]. Color of pulp
Ndioum 1	Pulaars (Sarakolés)	Very sweet, sweet, less sweet, bitter or empty fruit [1]. Abundant pulp fruits. Sweeter and lighter. Less abundant pulp fruits less sweet and heavier [2+1+7]. Big fruits, small fruits [3]
SPB1	Kaffrine[2]	Sudano-Sahelian[3]
Ndony		Very sweet, sweet, less sweet, bitter or empty fruits [1]
Gniby	Pulaars (Sérères)	Very sweet, sweet, less sweet. bitter or empty fruits [1]

Continued on next page.

Table II. (Continued). Criteria mentioned by different ethnic groups to characterize the morphotypes of baobab tree existing in Senegal

Name of village	*Ethnic groups*	*Classification Criteria*
Boye		Very sweet, sweet, less sweet, bitter or empty fruits [1]. Abundant pulp fruits, less abundant pulp fruits [2]. Big fruits, small fruits [3]
SPB[1]	Kaffrine[2]	Sudano-Sahelian[3]
Ndony		Very sweet, sweet, less sweet, bitter or empty fruits [1]
Gniby	Pulaars (Sérères)	Very sweet, sweet, less sweet. bitter or empty fruits [1]
Boye		Very sweet, sweet, less sweet, bitter or empty fruits [1]. Abundant pulp fruits, less abundant pulp fruits [2]. Big fruits, small fruits [3]
CNPB	Thiès	Sahelian
Khoudi-adiène	Sérères	Very sweet, sweet, less sweet, bitter or empty fruits [1]. Big fruits, small fruits [3]. Abundant pulp fruits, less abundant pulp fruits [2]
Sindia Caphgoune		Very sweet, less sweet fruits, bitter or empty fruits, abundant pulp, less abundant pulp [2]. Big fruits, small fruits [3]
Soro Khassab	Sérères	Very sweet, sweet, less sweet, bitter or empty fruits [1]
Ndiogoye	Sérères	Very sweet, sweet, less sweet, bitter or empty fruits [1]

[1] Agro-ecologic Zones: ESUC: Eastern Senegal-Upper Casamance; SPB: South Peanut Belt; CNPB: Center-North Peanut Belt. [2] Region, [3] Climatic Zone. 1 = Taste of the pulp; 2 = Abundance of the pulp; 3 = Size of the fruit; 4 = Stickiness of the pulp; 5 = shape of the fruit; 6 = Resistance of the endocarp to break; 7 = weight; 8 = color of pulp.

The Pulaar group of this zone found that fruits with a sweet taste were those which were light, with high amounts of pulp, and fruits with little sweet taste were rather heavy with little pulp in it. The Wolof groups found that fruits with sticky pulp were less tasty than those with non-sticky pulp. They also observed that long fruits were easy to break, had much more pulp, whereas round fruits were hard to break and had less pulp. In short, four peasants' baobab morphotypes have been identified in the three agro-climatic zones: Morphotype 1: characterized by fruits with sweet pulp, these fruits are sometimes big, sometimes small, with little or much pulp. This morphotype 1 is preferred by peasants for local consumption and for sale. Morphotype 2: characterized by a very sweet pulp, its fruits are generally small, easy to break with much pulp. Peasants prefer morphotype 2 for local consumption. Morphotype 3: characterized by less sweet pulp, these fruits are generally big but contain less pulp and are difficult to break. Morphotype 3 is preferred by peasants for sale; Morphotype 4: characterized generally by very small fruits with bitter pulp or empty (sterile), morphotype 4 is generally not used by the populations. The peasants' preference for type I was mentioned in all agro-ecological zones for all ethnic groups.

Table III. Median values of morpho-biochemical characters of baobab tree peasant morphotypes

Morphotypes	*M 1*	*M 2*	*M 3*	*M 4*	*Probability*
Morphological characters: Tree					
Height (m)	19±0.4	17.5±0.6	18.8±0.6	18.2±0.6	0.2925 NS
Circumference	7.4±0.4	6.4±0.5	6.3±0.4	6.9±0.5	0.2292 NS
Powder puff diameter	19.1±0.7	16.5±0.5	17.1±1.0	18.3±0.8	0.155 NS
Morphological characters: Fruit					
Circumference (cm)	26.67 a	26.2 a	26.3 a	20.6 b	< 0.0001 S
Pulp weight (g)	48.6 a	47.7 a	51.4 a	23.0 b	< 0.0001 S
Fruit weight	239.4 b	220.6 b	265.4 a	134.5 c	< 0.0001 S
Seed weight	78.8 b	79.3 b	87.4 a	46.5 c	< 0.0001 S
Fibers weight	4.8b	4.8 b	5.6 a	2.8 c	< 0.0001 S
Pulp weight / fruit weight	0.21 a	0.2 a	0.2 b	0.19 c	0.0003 S
Thickness of endocarp (cm)	0.5 a	0.41 c	0.5 b	0.49 b	< 0.0001 S
Length	19.5 b	18.33 c	21.4 a	16.2 d	< 0.0001 S
Biochemical characters: Pulp					
Reducing sugars (%)	11.7 b	11.952 a	10.8 d	11.1 c	< 0.0001 S
Total sugars (%)	33.3 b	34.383 a	29.2 d	30.7 c	< 0.0001 S

NOTE: Averages followed by different letters are significantly different (P > 0.05 by Newman-Keuls method). NS = Non-significant. S = Significant.

Morphological and Organoleptic Characterization of Peasants' Morphotypes

The analysis shows that morphotypes of baobab described by peasants relate to the parameters measured on the tree (height, circumference and diameter of crown) (Table III). However, significant differences have been noted regards the parameters measured on the fruits. Morphotype 4 is different from the others morphotypes by a very low weight of fruits, contrary to morphotype 3 which is mainly different from others by high weight.

The PCA (Principal Component Analysis) highlighted two key areas that reflect 98.8% of the total variation (Figure 1). Axis 1 expresses 92.02% and is strongly correlated with the number of seeds, with the circumference and length of the fruit, the amount of pulp and weight of the fruits, seeds and fibers. Axis 2 with 6.82% is strongly correlated with the thickness of the endocarp, the relative amount of pulp (ratio pulp weight / fruit weight) and sugars (reducing sugars and total sugars).

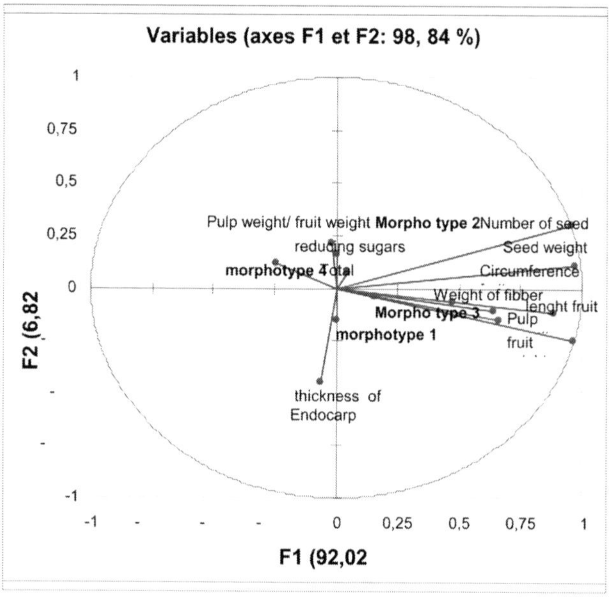

Figure 1. Morpho-biochemical fruits' parameters correlation circle.

Morphotype 2 is characterized by very sweet (reducing and total sugar quantity) and abundant pulp (high pulp weight or fruit weight) contrary to morphotype 1, which differs from others by a very thick endocarp.

The morphological and physical variables allow distinguishing four groups of baobab trees:

- group 1, characterized by long fruits, large circumference, with a very thick endocarp, a medium weight of fruits, fibers and seeds, high quantity and sweet pulp;
- group 2: characterized by medium to long fruits (18.33 cm), large circumference (26.20 cm), very thin endocarp (0.41 cm), medium weight of fruits, fibers and seeds, very sweet and high quantity pulp;

- group 3: characterized by very long fruits, large circumference, medium thickness endocarp, and a heavy weight of fruits, fibers and seeds, with medium quantity and less sweet pulp;
- group 4: characterized by very small fruits, with a medium thickness of endocarp, very low weight of fruits, fibers and seeds, less sweet and very low quantity of pulp.

Groups 1 and 2 as described morpho-biochemically correspond to peasant morphotypes 1 and 2, respectively. Groups 3 and 4 correspond to peasant morphotypes 3 and 4, respectively, but only on the morphological level. Indeed, the sugar contents (reducing and total) are higher in peasant morphotype 4 than in morphotype 3 contrary to peasants' classification.

Table IV. Medium values of physicochemical parameters of peasants' morphotypes pulp (M1-4)

Morphotypes	*M 1*	*M 2*	*M 3*	*M 4*
Reducing sugars(g/100g) (P≤0.0001)	11.7±0.5	11.952±0.181	10.8±1.0	11.1±0.3
Total sugars(g/100g) (P≤0.0001)	33.3±1.4	34.4±1.7	29.2±0.1	30.7±1.4
Magnesium (mg/100g) (P≤0.0001)	0.29±0.2	0.4±0.1	0.6±0.1	0.29±0.02
Acidity (mEq/100g) (P≤0.0001)	92.2±4.1	94.5±0.6	94.8±2.9	88.3±2.1
Calcium (mg/100g) (P=0.0069)	0.9±0.2	1.1±0.2	1.3±0.1	1.1±0.0
Ashes (g/100 g) (P=0.0014)	3.9±0.4	5.5±0.6	4.2±0.19	4.8±0.3
Moisture (g/100 g) (P=0.0748)	9.9±0.1	10.2±0.3	10.3±0.2	10.1±0.1

Physicochemical Characterization of the Morphotypes' Pulp

The evaluation of sugar contents (reducing and total sugar) in the pulp of the fruits reveals the existence of four distinct groups (table IV). The morphotype 2 presents the highest content of reducing sugars and total sugars of 11.952% and 34.383%, respectively. On the other hand, the lowest content of sugars is found in the pulp of the morphotype 3 with 10.774% for reducing sugars and 29.233% for total sugars. Morphotypes 1 and 4 present intermediate contents with 11.660% and 11.098%, respectively, for reducing sugars, 33.270% and 30.733%, respectively, for total sugars. In addition, three groups were distinguished according to acidity of the pulp, its contents in magnesium and ashes, and two groups according to calcium contents.

The highest acidity was noted for the morphotype 2 (94.500 mEq/100g) and morphotype 3 (94.833 mEq/100g), which are not significantly distinct. The magnesium content in morphotypes 1 and 4 was lowest at 0.290% and 0.287%, respectively, compared to morphotypes 2 and 3 (0.402% and 0.572%). The highest content of ashes was found in morphotype 2 (5.470%). The calcium contents of the pulp of morphotypes 2, 3 and 4 were relatively higher (1.107mg/100g; 1.252mg/100g and 1.095mg/100g, respectively) than those of morphotype 1 (0.908mg/100g). On the other hand, the moisture of pulp did not permit to discriminate between morphotypes. The low sensibility of our apparatus did not permit readings for iron and phosphorus on wavelengths of 508nm and 460nm, respectively.

Table V. Analysis of variance of morpho-biochemical parameters of baobab morphotypes' fruit

	Variance between morphotypes	*Variance within morphotypes*	*P>F*
Morphological parameters			
Fruit resistance to break (thickness of endocarp)	0.6	0.012	<0.0001
Abundance of pulp (pulp weight /fruit weight)	0.04	0.005	0.000
Fruit weight	794627.9	17110.7	<0.0001
Fibers weight	366.5	8.9	<0.0001
Seed weight	76439.4	2944.7	<0.0001
Number of seed	417344.4	14448.7	<0.0001
Fruit circumference	1903.6	78.9	<0.0001
Pulp weight	43411.5	827.5	<0.0001
Fruit length	1361.2	24.6	<0.0001
Biochemical parameters			
Reducing sugar	1.3	3.7	0.768
Total sugar	33	8.4	0.024

Table VI. Correlation between parameters of morphotypes' fruits

Morpho-biochemical parameters	*Morpho-biochemical parameters*									*Bio-chemical parameters*	
	Thickness of endocarp	*Pulp weight / Fruit weight*	*Fruit weight*	*Fruit length*	*Fibers weight*	*Seeds weight*	*Number of seed*	*Fruit circumfer-ence.*	*Pulp Weight*	*Reducing sugars*	*Total sugars*
Thickness of endocarp	1	-0.151	0.058	-0.008	0.007	-0.120	-0.188	0.086	-0.039	-0.130	-0.538
Pulp weight / Fruit weight	-0.151	1	-0.085	0.015	0.010	0.033	0.037	-0.014	0.189	-0.020	-0.322
Fruit weight	0.058	-0.085	1	0.679	0.644	0.910	0.848	0.475	0.875	-0.035	0.267
Fruit length	-0.008	0.015	0.679	1	0.579	0.607	0.589	0.241	0.590	-0.066	-0.275
Fibers weight	0.007	0.010	0.644	0.579	1	0.621	0.576	0.353	0.584	-0.049	-0.231
Seeds weight	-0.120	0.033	0.910	0.607	0.621	1	0.956	0.451	0.863	0.072	0.303
Number of seed	-0.188	0.037	0.848	0.589	0.576	0.956	1	0.433	0.802	0.109	0.399
Fruit circumference	0.086	-0.014	0.475	0.241	0.353	0.451	0.433	1	0.422	0.053	0.453
Pulp Weight	-0.039	0.189	0.875	0.590	0.584	0.863	0.802	0.422	1	0.059	0.320

Continued on next page.

Table VI. (Continued). Correlation between parameters of morphotypes' fruits

Morpho-biochemical parameters	*Morpho-biochemical parameters*									*Bio-chemical parameters*	
	Thickness of endocarp	*Pulp weight / Fruit weight*	*Fruit weight*	*Fruit length*	*Fibers weight*	*Seeds weight*	*Number of seed*	*Fruit circumfer-ence.*	*Pulp Weight*	*Reducing sugars*	*Total sugars*
Reducing sugars	-0.130	-0.020	-0.035	-0.066	-0.049	0.072	0.109	0.053	0.059	1	-0.335
Total sugars	-0.538	-0.322	0.267	-0.275	-0.231	0.303	0.399	0.453	0.320	-0.335	1

Variance of Morpho-Biochemical Parameters within and between Morphotypes' Fruits

The estimation of the variance of components between and within morphotypes shows that the variance between morphotypes is higher than the variance within morphotypes (Table V). Furthermore, it appears that the variance is higher for the fruits' quantitative morpho-biochemical parameters (fruit weight, circumference and length, fiber weight, seed weight, pulp weight and total sugars content) than the qualitative parameters (abundance of pulp and resistance of the endocarp to break). These results confirm the local population's classification which was mainly based on the pulp taste, which farmers determine by secondary quantitative (weight and or size of fruit), and qualitative characters (difficulty of fruit to break, abundant and sticky pulp).

Variance of Physicochemical Parameters of Fruit Pulps

The variance of biochemical characters between morphotypes, especially characters related to taste is higher than the variance within morphotypes (Table V). These results confirm the peasant's classification of the baobab tree which uses mainly the taste of the fruit's pulp to distinguish the morphotypes.

Correlation between Morphological and Physicochemical Parameters of Fruits

Significant correlations were found between some morpho-biochemical parameters of the fruit (Table VI). The fruit weight is positively correlated to the weight of fibers ($r = 0.64$), to the total weight of seeds ($r = 0.91$), to the number of seeds ($r = 0.84$), to the weight of the pulp ($r = 0.87$) and to the fruit length (0.67). The weight of the fibers correlated with fruit length ($r = 0.57$), with weight of seeds ($r = 0.62$), with number of seeds ($r = 0.57$) and with weight of the pulp ($r = 0.58$). Moreover, the thickness of the endocarp (comparable to resistance of the endocarp to break) is negatively correlated to total sugar content (-0.538), i.e., the lower the thickness of the endocarp, the higher the total sugar content in the fruit pulp. This is in line with the resistance of the endocarp to break being used by local populations as criterion to distinguish morphotypes.

Correlations between the Physicochemical Parameters of the Fruit Pulp

Table VII shows a significant correlation between reducing sugars and magnesium ($r = 0.52$), and moisture ($r = 0.66$). Total sugars are negatively correlated to calcium ($r = -0.61$) and to magnesium ($r = -0.63$). The matrix of correlation also shows that magnesium is positively correlated to calcium ($r = 0.59$) and to moisture ($r = 0.77$).

Table VII. Correlations between physicochemical parameters of the pulp of morphotypes' fruits

Variables	*Ca^{++}*	*Mg^{++}*	*Moisture*	*Ash*	*Acidity*	*Reducing sugars*	*Total sugars*
Calcium	1	0.59^s	0.26	- 0.3	- 0.06	0.12	$- 0.62^s$
Magnesium	0.59^s	1	0.77^s	- 0.41	0.03	0.53^s	$- 0.63^s$
Moisture	0.26	0.77^s	1	- 0.38	- 0.15	0.67^s	- 0.43
Ash	- 0.23	- 0.41	- 0.38	1	0.34	- 0.26	0.37
Acidity	- 0.06	0.03	- 0.149	0.34	1	- 0.003	- 0.15
Reducing sugars	0.12	0.53^s	0.67^s	- 0.26	- 0.003	1	- 0.34
Total sugars	$- 0.62^s$	$- 0.63^s$	- 0.43	0.37	- 0.15	- 0.34	1

S: significant at p > 0.05

Discussion

This constitutes the first morphological characterization of the peasants' morphotypes of baobab in Senegal. Previous studies did not take into account the knowledge of the local population. Several authors (Soloviev et al. (*20*), Gaye et al., (*25*). Sidibé et al. (*6*)) note that endogenous knowledge of the population on the traditional characterization of the baobab trees remains uninvestigated. This classification made by Senegalese peasants based on the characteristics of the fruit, in particular the quantity and taste of the pulp, the weight of the fruit, approaches that done in Benin, where the population distinguishes the morphotypes of baobab according to the shape of capsules (*19*). However, these authors classified varieties only according to the potential of fruits production and did not integrate the morphometric variables of the tree. Our investigation also differs from the one done in Mali where the population distinguishes several morphotypes of baobab according to the color of the bark, which can be black, red or white (*6*). Another study indicated that the criteria of distinction between the morphotypes vary according to regions and ethnic groups (*26*), e.g., in Southern Sudan and Southern Sahelian regions, the Malinkés and Kassonkes ethnic group distinguish according to the shape of the fruits, while in Northern Sudan Bambara distinguish according to the color of the trunk.

Studies relating to the morphological characterization of forest fruit trees in Africa showed that variability is especially noted on the fruits. This is the case for *Detarium microcarpum* (*27*), for *Vitelaria paradoxa* (*28*, *29*) and for *Adansonia digitata* (*20*, *21*, *25*). Our investigation identified four groups of baobab trees from the morphological and physical characters of the four morphotypes as described by local peasants. Only the variables measured were determinant in this classification. In our study, the most determinant variables of classification by morphological and physico-chemical parameters were the weight, the weight of fibers, the total weight of seeds, the relative quantity of pulp and the taste of the fruit. Results correspond to the peasants' classification which established a link between the taste of pulp and the weight of the fruit and/or the quantity of pulp.

In Benin, Assogbadjo (*22*) also found that the length of the capsule, the weight of the pulp, the total weight of the capsule, the weight of the almond, thickness of the capsule and the ratio [length/width] were determinant in the classification of the baobab tree.

In Mali (*30*), as well as in Benin (*19*), peasant typologies allowed identifying varieties of the baobab tree similar to our investigation in Senegal.

But none of these authors established a link between endogenous knowledge and the scientific knowledge to describe in detail the local varieties. Our study made it possible to partly validate the indigenous knowledge on the baobab tree. Indeed, there exists a correlation between taste (a biochemical parameter) and thickness of the endocarp of the fruit (a morphological parameter): the lower the thickness of the endocarp, the higher is the total sugar content in the fruits pulp. This aspect is used by the peasants to characterize the morphotype 2 (very sweet morphotype), recognized by its fruits which break easily as soon as they fall.

The morphological and physical characters of the fruit allow distinguishing four peasants' morphotypes. Sugar content (reducing and total) is the only biochemical character which differentiates peasants' morphotypes, contrary to acidity, which does not differentiate morphotypes 2 and 3, magnesium (morphotypes 1 and 4), calcium (morphotypes 2, 3 and 4), ashes (morphotypes 1 and 3 and morphotypes 3 and 4). It is interesting to note that morphotype 4, neglected by the peasants, has an undeniable nutritional value for its high content in calcium.

The variation of the physic-chemical parameters of the morphotypes' pulp may be explained by the environmental factors as well as by the plant-related factors. Indeed, periodic environmental variation (light, temperature, moisture, CO_2) can influence the distribution of various components within the plant, such as water and minerals absorption and migration of the saps (*31*). Potassium contents in the soil with a relatively low ratio of carbon/nitrogen may support the accumulation of sugars, thus partly explain the sweet taste of the pulp. It was shown that potassium activates the formation and the accumulation of sugar and starch (*32*). In addition, Heller et al. (*33*) find that mineral absorption from the ground is strongly influenced by the composition of the medium, but varies also according to species, age of the plant and the morphotype of cells concerned.

Conclusion

We identified in this study four morphotypes of baobab trees according to the peasants' selection criteria which are mainly based on the size of the fruit, its resistance to be broken, the abundance of the pulp, its taste, its color and its sticky character. These peasants' morphotypes are also different from the physico-chemical characteristics, such as sugar content (reducing and total). A study extended to other agro-ecological zones in Senegal would make it possible to complete the typing of the baobab tree by morphologic and biochemical aspects and select the most viable varieties.

Acknowledgments

This work was supported by grants from the African Forest Research Network (AFORNET). We thank this institution and its donors. We also thank the Senegalese Forest Service, the village peasants who allowed us to carry out this study and anonymous reviewers for their comments to improve the manuscript.

References

1. De Caluwé, E.; Halamova, K.; Van Damme, P. In *African Natural Plant Products: New discoveries and Challenges in Chemistry and Quality*; Juliani, H. R., Simon, J. E., Ho, C.-T., Eds.; ACS Symposium Series 1021; American Chemical Society: Washinton, DC, 2009; pp 51−84.
2. Cissé M. I. *Les parcs agroforestiers au Mali : état des connaissances et perspectives pour leur amélioration*; ICRAF/AFRENA: Bamako, 1995; 53 pp.
3. Ndour, B.; Gaye, A. In *Troisième atelier des centres et programmes nationaux de semences forestières africains*; Boye, A., Ndiaye, B., Bauwens, D., Eds.; PRASEF: Dakar, Sénégal, 1997; pp 45−52.
4. Ouedraogo, J. S. *Les parcs agroforestiers au Burkina Faso*; ICRAF: Bamako, 1995; 76 p.
5. Bonkoungou, E. G.; Ayuk, E.; Zoungrana, I. Les parcs agroforestiers des zones semi-arides d'Afrique de l'Ouest. *Actes du Symposium International Ouagadougou*; Burkina Faso, 1993; pp 25–27.
6. Sidibe, M.; Williams, J. T. *Baobab Adansonia digitata*; International Center for Underutilised Crops: Southampton, U.K., 2002; p 100, ISBN 0854327762.
7. Danthu, P. *Le Flamboyant* **2000**, *53*, 22–24.
8. Samba, N. A. S.; Gaye, A.; Fall, S. T.; Diouf, M.; Diallo, I. *Le Baobab Nouvelle plante maraîchère du Sahel. Fiche technique.* Unival / ISRA, 2003; Vol. 6, no. 3.
9. Diallo, I. *Etude de la biologie de la reproduction et de la variabilité génétique chez le jujubier (Zizyphus mauritiana Lam).* Ph.D. Dissertation, UCAD, Faculty of Sciences, Dakar, Senegal, 2002.
10. Nikiéma, S. In *Les parcs agroforestiers des zones semi-arides d'Afrique de l'Ouest*; Bonkoungou, E. T., Zoungrana, A. I., Eds.; ICRAF: Bamako, 1993; pp 36−50.
11. Codjia, J. T. C.; Assogbadjo, A. E.; Ekue, M. R. M. *Cahiers Agricultures* **2003**, *12*, 321–331.
12. Ighodalo, C. E. H.; Catherine, O. E.; Daniel, M. K. *Plant Foods Hum. Nutr.* **1991**, *41*, 151–154.
13. Scheuring, J. F. *Sight Life Newsl.* **1999**, *1*, 21–24.

14. Yazzie, D.; VanderJagt, D. J.; Pastuszyn, A.; Okolo, A.; Glew, R. H. H. *J. Food Compos. Anal.* **1994**, *7* (3), 89–193.
15. Sidibé, M.; Koné, M.; Scheuring, J. F. *Agroforestry Today* **1998**, *10*, 7–9.
16. Wickens, G. E. The Uses of the Baobab (*Adansonia digitata* L.) in Africa. In *Browse in Africa*; ILCA/FAO: Addis-Abeba, Ethiopia, 1982; Vol. 107, pp 151−154.
17. Gaiwé, R.; Nkulinkiye-Nfura, T.; Bassène, E.; Olschwang, D.; Ba, D.; Pousset, J. L. *Int. J. Crude Drug Res.* **1989**, *27* (2), 101–104.
18. Diop, A.; Boaretto, A.; Pasotore, P. *Soachim J.* **1996**, *1*, 30–37.
19. Codjia, J. T. C.; Fonton, B. K.; Assogbadjo, A. E.; Ekue, M. R. M. *Le baobab (Adansonia digitata), Une espèce à usage multiple au Bénin.*; CECODI, CBDD, VeCo, SNV, FSA, 2001; 44 p
20. Soloviev, P.; Niang , T. D.; Gaye, A.; Totte, A. *Fruits* **2004**, *59*, 109–119.
21. Assogbadjo, A. E.; Sinsin, B.; Van Damme, P. *Fruits* **2005**, *60* (5), 327–340.
22. Assogbadjo, A. E.; Kyndt, T.; Sinsin, B.; Gheysen, G.; Van Damme, P. *Ann. Bot.* **2006**, *97*, 819–830.
23. Assogbadjo, A. E.; Kyndt, T.; Chadare, F. J.; Sinsin, B.; Gheysen, G.; Eyog-Matig, O.; Van Damme, P. *Agroforestry Syst.* **2008**, *75*, 157–165.
24. A.O.A.C. *Official Methods of Analysis of the Association of Official Analytical Chemists*, 40th ed.; A.O.A.C.,1984
25. Gaye, A.; Samba, A. N. S.; Kaire, M.; Diallo, I.; Badiane, S.; Fall, T.; Ndiaye, A. *Domestication et valorisation de quelques fruitiers forestiers au Sénégal. Rapport technique final d'activités*; Ref. FNRAA: Project 009AP01SS310800, 2004; 43 pp.
26. Sanou, H.; Korbo, A.; Sanogo, D.; Jensen, J. S.; Raebild, A. *Plant Genet. Resour. Newsl.* **2008** (155), 1–10.
27. Kouyate, A. M.; Van Damme, P. *Fruits* **2002**, *57*, 231–238.
28. Maranz, S.; Wiesman, Z. *J. Biogeogr* **2003**, *30*, 1505–1516.
29. Sanou, H.; Picard, N.; Lovett, P. N.; Dembele, M.; Korbo, A.; Diarisso, D.; Bouvet, J. M. *Genetic Resour. Crop Evol.* **2005**, *00*, 1–17.
30. Sidibe, M.; Kone, M.; Scheuring, J. F.; Timbely, D.; Sidibe, M. M.; Hofman, P.; Frigg, M. *Agroforestry Today* **1996**, *8*, 13–15.
31. Moreau, F.; Roger, P. *Biologie et Multimédia - Université Pierre et Marie Curie - UFR des Sciences de la Vie*; http://www.snv.jussieu.fr/bmedia/Photosynthese-cours/06-facteurs.htm (10/03/2009).
32. Arunachalam, A.; Maithani, K.; Pandey, H. N.; Tripathi, R. S. *For. Ecol. Manage.* **1998**, *109*, 151–161.
33. Heller, R.; Esnault, R.; Lance, C. *Physiologie végétale. 1. Nutrition*, 6th édition de l'Abrégé; Dunod: Paris, 1998.

Chapter 17

The Role of African Indigenous Plants in Promoting Food Security and Health

Lorraine S. Cordeiro*

[1]Department of Nutrition, School of Public Health and Health Sciences, University of Massachusetts Amherst, 100 Holdsworth Way, Amherst, Massachusetts 01003-9282
***E-mail: lcordeiro@nutrition.umass.edu.**

Food insecurity and hunger afflict more than one billion people worldwide with negative implications for the health, productivity, and well-being of vulnerable members of our global population. African indigenous plants have the potential to play a central role in addressing food insecurity and associated health concerns in Sub-Saharan Africa. Understanding the historical and current uses of African indigenous plants during periods of food shortage, hunger, and disease is essential to developing effective programs and policies to promote the sustainable production and consumption of these local edible plants for population health. This work aimed to describe indigenous edible plants commonly consumed by rural populations during periods of food shortage in rural villages in Tanzania and to assess the associations between household food insecurity, health, and household consumption of indigenous edible plants. Indigenous edible plant consumption and household food security were independently associated with the number of self-reported morbidity symptoms among adolescents in this study ($p<0.05$). These findings suggest that indigenous plants may play a role in moderating the health status of individuals in both both food secure and food insecure households.

Introduction

Given rising food prices and the current global economy, the state of food insecurity in the world today is troubling (*1*, *2*). By definition, food insecurity "exists when all people, at all times, [*do not*] have physical, social and economic access to sufficient, safe and nutritious food that meets their dietary needs and food preferences for an active and healthy life" (*3*). The Food and Agriculture Organization (FAO) estimates that close to one billion people worldwide, most of whom live in low- and middle-income countries, lack sufficient calories to meet their dietary needs (*4*). It is likely that these food insecure individuals will experience poorer health and nutritional outcomes than their food secure peers (*5*, *6*).

Africa depends on a productive and healthy population for economic growth, making investments in agriculture, nutrition and health an optimal goal. Africa is the only region in the world where hunger is on the rise (*7*) and yet ironically it is the continent where future agricultural growth may need to be located to feed an ever growing global population. Food security remains one of the continent's most persistent challenges to economic growth (*7*). Demographic trends complicate matters; an estimated 41% of Africa's population is below 15 years of age (*8*) and a considerable proportion of these children are already undernourished or at risk (*9*). Identifying local solutions to food insecurity is the keystone to sustainable economic growth and the prevention of malnutrition, particularly in Sub Saharan Africa (*10*, *11*).

Figure 1

Food security is a localized concept, based on regional climatic conditions, coping strategies, and self-perceptions of targeted communities, among other indicators (*12*). While often portrayed as an immediate concern or crisis, food insecurity has a rich historical context that includes the use of indigenous plants to moderate the health and nutritional impacts of poor diet quantity and quality (*13*, *14*). African indigenous plants are important local and culturally-accepted sources of nutrition (*13*, *15*) and disease prevention (*11*, *16–18*), particularly during periods of food shortage (*18–20*). Many African edible plants have antioxidant and antimicrobial properties (*14*, *21*, *22*), thus corroborating in part some of the community-based perceptions of the health benefits of these plants (*10*). Cultivation and consumption of local edible plants offers a sustainable solution to food insecurity and associated diet-related diseases (*10*, *17*, *23*). The consumption of some indigenous plants can also pose a health risk and those antinutritive factors as well as the health promoting factors need to also be recognized. The objectives of this study were 1) to describe indigenous edible plants commonly consumed by rural populations during periods of food shortage in rural villages in Tanzania and 2) to assess the association between health and household consumption of indigenous edible plants.

Materials and Methods

Qualitative Data

Focus group discussions (FGD) were held in four villages in Kilosa District, Morogoro region, Tanzania in the dry season (see Figure 1). The villages of Rudewa, Gongoni and Madoto were situated along major transport routes, while Malangali and Tindiga were remote villages with a mixture of settled and nomadic populations. Each focus group included 6 to 12 male and female adults. Participants were actively involved in farming activities and discussion centered on coping strategies used during periods of food insecurity. FGDs lasted between 1.5-2 hours and were held in Swahili after obtaining informed consent from participants. Detailed notes from FGDs were translated into English and used to inform survey item design. The results of FGDs related to the use and consumption of African indigenous plants are presented in this chapter.

We also conducted in-depth interviews with four elderly men to elicit historical experiences with food insecurity, traditional food practices, and the use of indigenous plants to address health and nutritional deficiencies. While the number of participants in FGD were quite limited, the information gleaned was complimented by a quantitative household survey.

Quantitative Data

The research team administered surveys to 736 households across 28 rural villages in Kilosa District, Tanzania. Cluster sampling, using a standard population proportionate to size (PPS) approach, was employed to select villages from a master list of all villages in the district. Given the original study focus on adolescents, households selected for inclusion in this study had to have an adolescent in residence. The four villages in which FGDs were conducted were included in the survey sample.

Fourteen (14) interviewers were trained by a Harvard University/Muhimbili University research team on data collection procedures for household and individual level surveys. The research team administered surveys in Swahili to heads of households, main food preparers, and adolescents in their homes. The household head provided data on demographics, income and assets, food frequency, and food security. The main food preparer provided one 24 h dietary recall on the household's dietary intake.

Household Dietary Diversity Score (HDDS), is a proxy indicator of household food insecurity where households that consume fewer food groups in their diet are consider food insecure (*24*, *25*). Diet quality measures such as HDDS better reflect both the macro- and micronutrient composition of household diets. The methodology used to generate the HDDS is described in detail by Swindale and Bilinsky (*25*). For descriptive analyses, food insecure households (consuming ≤ 4 food groups) were coded '0', while food secure households (consuming 5+ food groups) were coded '1' (*26*). HDDS was entered as a continuous variable in multivariate analyses.

Classification of African Indigenous Plants

Since FGD participants and survey respondents provided vernacular names and descriptions of edible plant species, a literature review only was used to determine the scientific names and properties of these plants. Since the period of data collection coincided with the hot, dry season, when food was relatively scarce in rural Kilosa, the majority of plants described by participants were foraged from the wild. Plants that were determined to have origins in Africa were classified as indigenous, coded '1', and other food sources, such as commonly cultivated foods like maize and rice were coded '0'. Vernacular and tentative identification of scientific names of indigenous plants reported in the household diets, as well as nutritional and medicinal benefits reported in other studies, are presented in Table I.

Health Status

Morbidity data was asked of all adolescents, using a 30-day to a 12-month period of retrospection depending on the specific illness, disease, or symptom. Symptoms of morbidity included episodes of malaria, tuberculosis, acute respiratory infection, acute diarrhea, ear discharge, oral candidiasis, parotid swelling, and enlarged lymph nodes in two or more of the following sites: neck, groin, and axillae (*27*). Health status was reported as a continuous variable, number of morbidity symptoms, in multivariate analyses.

Statistical Analysis

Qualitative data are presented in absolute terms, through the use of quotations and by listing edible plants recorded from FGDs. For the purposes of this paper, commonalities in consumption patterns and uses of indigenous edible plants are presented in the results.

Quantitative data derived from household surveys was analyzed using SPSS version 20 (IBM), with $P < 0.05$ denoting statistical significance. Multivariate linear regression was used to examine the association between the number of morbidity symptoms reported by adolescents and their household's consumption of indigenous edible plants, after adjusting for household food security status, income, and number of household members.

The Tanzanian Commission for Science and Technology (COSTECH) granted the research permit for this study. Human subjects' research approval was obtained from the Tufts University Institutional Review Board. Data collection for this study was supported by UNICEF/Tanzania.

Results and Discussion

Of the 736 households included in this study, 40.9% (n=301) were regular consumers of African indigenous plants. Among household consumers of indigenous plants 66% were classified as food insecure and 33.7% were food secure ($p<0.05$), suggesting a heavy reliance on these plants among the poor. Approximately 20 indigenous plant foods were commonly consumed by the study sample during the dry season, a period of peak food insecurity and hunger (Table I). This number is similar to that reported by Lyimo et al. (*28*) who recorded 30 different species of indigenous vegetables in seven villages in Morogoro region.

Focus group discussions and in-depth interviews revealed a collective knowledge of indigenous plant use and shifts in the population's food experiences from colonial times to the current period. The FGD participants reported that several indigenous plants had both medicinal and nutritive properties. A review of the literature on nutritional and medicinal benefits of edible plants listed by FGD participants and recorded in household dietary data showed that many of these plants are excellent sources of micronutrients (Table I). All FGD participants felt that the time during which we were conducting this study, the dry season, was a period of peak food insecurity. Consumption of the *least preferred* indigenous plants was reported to be more frequent at times of severe or very severe food shortage across all villages studied. The FGD participants noted that indigenous plants were available year-round, and in one village, Malangali, *wild* indigenous plants were consumed even when households had sufficient food to eat. In contrast, consumption of *wild* indigenous plants was listed as a coping strategy to mitigate food shortage only during *moderate* and *very severe* periods of food insecurity in the villages of Rudewa Gongoni, Tindiga, and Madoto. In Madoto, women reported roasting capoc seeds, *Ceiba pentandra*, as a last resort during the periods of food shortage:

"We roast seeds from the Capoc fruit when there is no food and lots of hunger. We don't like to eat this but our children are hungry right now. We [women] try to save local leafy vegetables for the children, as these are available at all times because they grow in the wild. But when these plants dry up, then you know that our situation with food has become very, very bad."
~ *Young mother, Madoto village*

The FGD participants were also aware of the risks of consuming some local plants and noted that overconsumption could lead to adverse health outcomes.

"Sometimes our children swell up after eating roasted Capoc seeds; if they eat too much. And some of our people get poisoned from certain wild fruit, and yet they eat it when there is nothing else to eat. We do know how much to eat so that we do not get sick from too much of one food but when we are too hungry we have little choice."
~ *40 year old woman, Madoto village*

Rice and thick maize porridge (*ugali*) were considered to be preferred foods in Rudewa Gongoni. The FGD participants reported that these preferred foods were scarce at the time of the study, and most of the villagers incorporated their least preferred foods like *ndizi kicementi* or *ndizi moshi (Musa spp.)*, wild varieties of banana that are usually not harvested and typically left for the birds to eat. In some cases, villagers consumed unripe wild fruit and vegetables, including bananas, an indication of the severity of the food security experience in this village. In Malangali, FGD participants reported that they had already consumed most of the maize seed stock which had been designated for planting. Hence, they lacked both seed from last year's harvest as well as capital to purchase seeds for planting during the upcoming rainy season. Villagers in Malangali and Tindiga described active harvesting of wild vegetables such as *mchunga (Launaea cornuta)*, *mlenda (Corchonus tritocularis, C. tridens)*, and *mchicha pori* (*Amaranthus graecizans)* at the current time, as food stocks had ran very low.

In-depth interviews with Kilosa elders provided invaluable historical information. These elders indicated that the current population had markedly fewer traditional skills to address rising food insecurity, especially in regards to the knowledge and use of traditional food sources. Elders spoke about transformations in the behavior of local people who were autonomous in food production and foraging prior to being colonized by the British. Elders' observations over the past seven decades included the loss of land by the local communities, people waiting for their food rations under colonial rule, and their sense that community members had become used to having decisions made for them by a centralized administration. The elders believed that these long-lasting societal changes in food decision-making practices had an intergenerational impact. One elder in Rudewa Gongoni reported that heavy and extensive logging which commenced during colonial times drastically reduced forest coverage in Kilosa District. He spoke about historical use of forest products during periods of food shortage and hunger:

"We used to store more food in the old days. Nowadays, people do not seem to know how to store or protect their food from insects. We used to smoke some foods so that they stored well. If you look today, it seems like people are waiting by the side of the road for food. They don't know what to do when there is no food – who will give them food? The government? We used to know all the local foods, the wild foods, and our fathers would know which plants to eat. We have lost this knowledge and our forests. The crops today in the field are new to the world - they cannot take us through hunger during bad weather."
~ *88 year old man*

The loss of traditional knowledge of indigenous plants was also documented in FGDs in the four districts in Tanzania (*14*). According to elders in this study, traditional vegetables were locally available but the younger generation did not recognize these vegetables as food. They also found that younger villagers were unable to identify plants gathered from the wild (*14*). The lack of transfer of traditional knowledge across generations presents a particular challenge to identification and promotion of indigenous edible plants that could provide important nutrition and health benefits, particularly when food security is highly prevalent and for those that rely primarily on subsistence farming for their dietary intake. Without active knowledge of these indigenous plants and appropriate culinary preparation, younger generations are more likely to label them as unusual to the diet and find them less desirable (*14*).

In past 15 years, there has been an emergence of research examining the associations between household food insecurity, health status, and the consumption of indigenous vegetables (*11*, *13*, *18*, *23*, *33*). The extent to which consumption of indigenous vegetables can promote health, provide nutrition, and moderate food insecurity is of great interest to rural development initiatives in developing countries. The qualitative data presented above provides critical information on perceptions, utilization, descriptions, and knowledge of indigenous plants in Kilosa District. We also present the results of the associations between household food insecurity, health status, and the consumption of indigenous vegetables based on our analysis of household and adolescent survey data from Kilosa, Tanzania.

In a multivariate regression model, we examined the association between the number of morbidity systems reported by adolescents in this study - the dependent variable - and household consumption of indigenous edible plants, after adjusting for income, number of household members, and household food insecurity (Table II). Indigenous edible plant consumption and household food security were independently associated with the number of self-reported morbidity symptoms among adolescents in this study ($p<0.05$). Households that consumed indigenous edible plants had a -0.312 unit decrease in the number of self-reported morbidity symptoms among adolescents, after adjusting for income, number of household members, and household food insecurity status ($P<0.05$) (Table II).

Table I. Nutritional properties and purported medicinal uses of tentatively identified indigenous plants consumed during periods of food shortage as reported in FGD and derived from 24h dietary recalls of households in Kilosa District, Tanzania [1,2]

Common name	*Vernacular name*	*Tentative botanical name*	*Nutritional properties*	*Purported nutritional and medicinal uses*
1. Amaranth, Spinach	Mchicha kienyegi	*Amaranthus spinosus (28)*	Protein 4.6g Ca 43.2 mg/100g Fe 3.8 mg/100g Vitamin C 249 mg/100g (*28*)	Anemia, high blood pressure, diabetes, severe malnutrition, night blindness, headaches, and dizziness (*10*, *29*)
2. Amaranth, Spinach	Majani ya Ifene, mchicha pori	*Amaranthus graecizans (28)*	Protein 4.8g Ca 246 mg/100g Fe 3.0 mg/100g Vitamin C 46.3 mg/100g (*28*)	Anemia, high blood pressure, diabetes, night blindness, and headaches (*10*, *29*)
3. Banana (various)	Ndizi (types: nguruwe, moshi, kicementi)	*Musa spp. (30)*	High in potassium, rich source of micronutrients (*30*)	Stomachaches, high blood pressure, digestion, malaria[2]
4. African night-shade	Mnavu	*Solanum nigrum, S. scabrum, S. americum, S. villous (28)*	Protein 1g Ca 66.8 mg/100g Fe 2.5 mg/100g Vitamin C 235 mg/100g (*28*)	Diarrhea, anemia, high blood pressure, diabetes, problems of sight, peptic ulcers, stomachaches, and skin infections (*10*, *29*)
5. Bitter lettuce	Mchunga/Sunga	*Launaea cornuta (13)*	Fe 44.6 mg/100g Zn 0.26 mg/100g β-Carotene 2.69 mg/100g (*13*)	Malaria, measles, hookworms, diabetes, hemia, stomachaches, high blood pressure, temperature regulation, and infections (*29*)

Common name	***Vernacular name***	***Tentative botanical name***	***Nutritional properties***	***Purported nutritional and medicinal uses***
6. Wild soursop	Mtope tope, Mchekwa	*Annona senegalensis, A.chrysophylla (31)*	Mg 42.2 mg/100g Ca 28.9 mg/100g Fe 1.3 mg/100g Zn 0.64 mg/100g (*31*)	Diarrhea, stomach pain, and abscesses (*31*)
7. Capoc seed	Masufi, Msufi	*Ceiba pentandra (32)*	Crude protein: 324 g/kg Calcium 3.8 g/kg Phosphorous 9.3 g/kg Fatty acid: Palmitic acid (*32*)	Powdered fruit taken with water to treat intestinal parasites and stomaches (*32*)
8. Bulrush millet	Muhilile/mhilile	*Cleome hirta (Klotzsch) (13)*	Fe17.5 mg/100g Zn 0.32 mg/100g β-Carotene 0.95 mg/100g (*13*)	Painful menstruation, chest pain, and diarrhea[2]
9. Spider flower plant	Mgange	*Cleome glynandra (13)*	Fe 49.95 mg/100g Zn 0.41 mg/100g β-Carotene 2.10 mg/100g (*13*)	Headaches, fever, ear problems, common cold, blood pressure, diabetes, anemia, night blindness, toothaches and stomachaches (*10*, *29*)

Continued on next page.

Table I. (Continued). Nutritional properties and purported medicinal uses of tentatively identified indigenous plants consumed during periods of food shortage as reported in FGD and derived from 24h dietary recalls of households in Kilosa District, Tanzania [1,2]

Common name	*Vernacular name*	*Tentative botanical name*	*Nutritional properties*	*Purported nutritional and medicinal uses*
10. Wild mlenda/Jute mallow	Mlenda	*Corchonus tritocularis, C. tridens, C. fascicularis (13)*	Fe 4.20 mg/100g Zn 0.20 mg/100g β-Carotene 6.31 mg/100g Rich source of potassium, copper, manganese. (*13*)	Cough, eye diseases, and stomachache[2]
11. Marula	Majani ya mbwegele	*Sclerocarya birrea (31)*	Protein 3.31g/100g CHO 90.35g/100g (*31*)	None reported
12. Flameflower	Majani ya pwimbiji/Pwimbiji	*Talinum portulacifolium (13)*	Fe 36.45 mg/100g Zn 0.30 mg/100g β-Carotene 1.52 mg/100g (*13*)	Anemia[2]
13. Local mushroom	Uyoga	*Not identified*	n.a.	
14. Cassava leaves	Kisamvu, majani ya mihogo	*Manihot esculenta (13)*	Fe 3.65 mg/100g Zn 0.45 mg/100g β-Carotene 5.13 mg/100g (*13*)	Anemia, diarrhea (*29*)

Common name	***Vernacular name***	***Tentative botanical name***	***Nutritional properties***	***Purported nutritional and medicinal uses***
15. Pumpkin leaves	Majani ya maboga	*Cucurbita pepo (13)*	Fe 26.65 mg/100g Zn 0.20 mg/100g β-Carotene 5.34 mg/100g (*13*)	Eye diseases[2]
16. Sweet potato leaves	Matembele	*Ipomea batatas (13)*	Fe 8.35 mg/100g Zn 0.20 mg/100g β-Carotene 1.93 mg/100g (*13*)	Eye diseases, anemia, burns, asthma (*10*, *29*)
17. Cowpea leaves	Majani ya kunde	*Vigna unguiculata (13)*	Fe 17.90 mg/100g Zn 0.30 mg/100g β-Carotene 4.45 mg/100g (*13*)	Eye diseases, anemia, skin irritation (*10*, *29*)

[1] Indigenous plants were tentative identified using secondary sources. [2] Some medicinal properties were reported by focus group participants.

Table II. Associations between household food insecurity, household consumption of indigenous edible plants, and self-reported number of morbidity symptoms among adolescents in Kilosa, Tanania[1], [2]

	β value (95% CI)	*P* [2]
Indigenous Edible Plant Consumption	- 0.307 (-0.509 to -0.105)	0.003
Covariates		
Household size (no. of people)	-0.016 (-0.061 to 0.030)	0.502
Income (TSh in millions)	-0.320 (-2.083 to 1.444)	0.722
Household Dietary Diversity Score (HDDS)	- 0.07 (-0.147 to -0.0024)	0.044
Constant	1.437 (0.991 to 1.882)	0.000
R^2		0.032

[1] The β regression coefficient was calculated using general linear model (GLM) test. [2] Statistical significance was assessed as $p<0.05$.

These findings suggest that indigenous plants may play a role in moderating the health status of adolescents, and possibly other household members, across both food secure and food insecure households. Further research is needed to investigate the role of indigenous plant consumption on the health and nutritional status of food insecure individuals. The variety and composition of household and individual diets is likely to be a factor in the overall health and nutrition gains of rural populations in Africa. Furthermore, due to the loss of traditional knowledge of these important plants, special efforts need to be made to not only record current knowledge, past uses and preparation of indigenous plants for both medicinal and dietary purposes but to introduce this knowledge to the younger generations.

Conclusion

As African nations continue to expand cultivation efforts of high-yielding cash crops, protection and promotion of indigenous varieties of fruits and vegetables receive limited attention (*10*). Furthermore, in African urban centers the nutrition transition from traditional food sources towards Western-style diets has resulted in a rapid decline in the consumption of indigenous edible plants, which are often labeled 'low status' or 'least preferred' foods (*10*, *34*). Diminishing availability of indigenous plants due to cultivation of forested land and loss of knowledge of indigenous plant use in Africa suggests that more attention needs to be paid to the conservation and promotion of this important local food security safety net (*34*, *35*).

In communities that experienced long-term disadvantage in the distribution, access, and quality of food in Africa, the historical perspectives of food insecurity may shed light on previous and current food practices, behaviors, and health outcomes related to the use of African indigenous plants. Maxwell (*12*) argues that the various interpretations of food security "reflect the nature of the food problem as it is experienced by poor people themselves ... [and] understanding food security requires explicit recognition of the complexity and diversity, and that it necessarily privileges the subjective perceptions of the food insecure themselves." The experiences and perspectives of the local villagers in this study on food insecurity, as well as their knowledge, attitudes and practices related to indigenous edible plants, is a critical first step in understanding the potential role that these plants can play in promoting food security, nutrition and health in resource-constrained regions of Africa. Further research on the potential benefits of specific indigenous plants, their role in moderating food insecurity and associated health outcomes is essential for the development of programs and policies targeting rural development in Africa.

Acknowledgments

I would like to thank the Tanzania Commission for Science and Technology (COSTECH), the Tanzania Food and Nutrition Centre; UNICEF/Tanzania; Kilosa District Council, Ward Officers, Village Leaders and Health Officers; study participants; and the field research team. This study was funded by Tufts University and UNICEF/Tanzania.

References

1. FAO. The state of food insecurity in the world 2010. *Food and Agriculture Organization of the United Nations*; Rome, Italy, 2011.
2. Pinstrup-Andersen, P.; Herforth, A. Food security: achieving the potential. *Environment: Science and Policy for Sustainable Development* **2008**, *50*, 48–61.
3. FAO. The state of food insecurity in the world 2001. *Food and Agriculture Organization of the United Nations*, Rome, Italy, 2002.
4. FAO. The state of food insecurity in the world 2010. *Food and Agriculture Organization of the United Nations*, Rome, Italy, 2010.
5. Cordeiro, L. S.; Wilde, P. E.; Semu, H.; Levinson, F. J. Household food security is inversely associated with undernutrition among adolescents from Kilosa, Tanzania. *J. Nutr.* **2012**, *142*, 1741–7.
6. Hadley, C.; Lindstrom, D.; Tessema, F.; Belachew, T. Gender bias in the food insecurity experience of Ethiopian adolescents. *Soc. Sci. Med.* **2008**, *66*, 427–38.
7. FAO. The state of food insecurity in the world 2011. *Food and Agriculture Organization of the United Nations*, Rome, Italy, 2012.
8. Population Reference Bureau. World Population Data Sheet 2012. 2012; http://www.prb.org/Publications/Datasheets/2012/world-population-data-sheet.aspx.
9. World Health Organization. 2013. http://www.afro.who.int/en/clusters-a-programmes/frh/child-and-adolescent-health/programme-components/child-health.html.

10. Herforth, A. *Promotion of traditional African vegetables in Kenya and Tanzania: a case study of an intervention representing emerging imperatives in global nutrition.* Dissertation: Cornell University Library, 2010.
11. Cordeiro, L. S. Household dietary diversity, wild edible plants, and diarrhea among rural households in Tanzania. *J. Med. Act. Plants* **2012**, *1*, 4.
12. Maxwell, S. Food security: a post-modern perspective. *Food Policy* **1996**, *21*, 155–70.
13. Weinberger, K.; Msuya, J. Analítico: indigenous vegetables in Tanzania. *Technical Bulletin, AVRDC* **2004**.
14. Keller, G. B.; Mndiga, H.; Maass, B. L. Diversity and genetic erosion of traditional vegetables in Tanzania from the farmer's point of view. *Plant Genetic Resources: Characterization and Utilization* **2005**, *3*, 400–13.
15. Burlingame, B.; Charrondiere, R.; Mouille, B. Food composition is fundamental to the cross-cutting initiative on biodiversity for food and nutrition. *J. Food Compos. Anal.* **2009**, *22* (5), 361–365.
16. Irungu, B. N.; Rukunga, G. M.; Mungai, G. M.; Muthaura, C. N. In vitro antiplasmodial and cytotoxicity activities of 14 medicinal plants from Kenya. *S. Afr. J. Bot.* **2007**, *73* (2), 204–207.
17. Flyman, M.; Afolayan, A. The suitability of wild vegetables for alleviating human dietary deficiencies. *S. Afr. J. Bot.* **2006**, *72*, 492–7.
18. McGarry, D. K.; Shackleton, C. M. Children navigating rural poverty: rural children's use of wild resources to counteract food insecurity in the Eastern Cape, South Africa. *J. Child Poverty* **2009**, *15* (1), 19–37.
19. Maxwell, D.; Caldwell, R. The coping strategies index: a tool for rapid measurement of household food security and the impact of food aid programs in humanitarian emergencies. *Field Methods Manual*, 2nd ed.; USAID/CARE/WFP/Tufts U./TANGO, 2008.
20. Paumgarten, F. The role of non-timber forest products as safety-nets: a review of evidence with a focus on South Africa. *GeoJournal* **2005**, *64* (3), 189–197.
21. Johns, T.; Chapman, L. Phytochemicals ingested in traditional diets and medicines as modulators of energy metabolism. *Recent Adv. Phytochem.* **1995**, *29*, 161–188.
22. *The biodiversity of traditional leafy vegetables*; Chweya, J. A., Eyzaguirre, P. B., Eds.; International Plant Genetic Resource Center: Rome, Italy, 1999.
23. Kaschula, S. Wild foods and household food security responses to AIDS: evidence from South Africa. *Popul Environ.* **2008**, *29* (3), 162–185.
24. Hoddinott, J.; Yohannes, Y. Dietary diversity as a household food security indicator*Food and Nutrition Technical Assistance Project*; AED: Washington, D.C., 2002.
25. Swindale, A.; Bilinsky, P. Household Dietary Diversity Score (HDDS) for measurement of household food access: indicator guide. *Food and Nutrition Technical Assistance Project*; AED: Washington, D.C., 2005.
26. Steyn, N.; Nel, J.; Nantel, G.; Kennedy, G.; Labadarios, D. Food variety and edietary diversity scores in children: are they good indicators of dietary adequacy? *Public Health Nutr.* **2006**, *9*, 644–50.

27. World Health Organization. *Interim WHO Clinical staging of HIV/AIDS and HIV/AIDS case definitions for Surveillance: Africa Region*; 2005.
28. Lyimo, M.; Temu, R.; Mugula, J. Identification and nutrient composition of indigenous vegetables of Tanzania. *Plant Food Hum. Nutr.* **2003**, *58* (1), 85–92.
29. Keller G. *African nightshade, eggplant, spiderflower et al.–production and consumption of traditional vegetables in Tanzania from the farmers point of view.* Masterarbeit im wissenschaftlichen Studiengang Agrarwissenschaften an der Georg-August Universität Göttingen, Fakultät für Agrawissenschaft; 2004.
30. Davey, M. W.; Stals, E.; Ngoh-Newilah, G.; Tomekpe, K.; Lusty, C.; Markham, R.; Swennen, R.; Keulemans, J. Sampling strategies and variability in fruit pulp micronutrient contents of West and Central African bananas and plantains (Musa species). *J. Agric. Food Chem.* **2007**, *55*, 2633–44.
31. Eromosele, I. C.; Eromosele, C. O.; Kuzhkuzha, D. M. Evaluation of mineral elements and ascorbic acid contents in fruits of some wild plants. *Plant Food Hum. Nutr.* **1991**, *41* (2), 151–154.
32. Narahari, D.; Asha Rajini, R. Chemical composition and nutritive value of kapok seed meal for broiler chickens. *Br. Poult. Sci.* **2003**, *44* (3), 505–509.
33. Lockett, C. T.; Calvert, C. C.; Grivetti, L. E. Energy and micronutrient composition of dietary and medicinal wild plants consumed during drought: study of rural Fulani, northeastern Nigeria. *Int. J. Food Sci. Nutr.* **2000**, *51*, 195–208.
34. Jaenicke, H.; Höschle-Zeledon, I. Strategic framework for underutilized plant species: with special reference to Asia and the Pacific, and to Sub-Saharan Africa. *Bioversity International* **2006**.
35. Shackleton, C; Shackleton, S The importance of non-timber forest products in rural livelihood security and as safety nets: a review of evidence from South Africa. *S. Afr. J. Sci.* **2004**, *100*, 658–664.

Chapter 18

An Assessment of the Essential Oil and Aromatic Plant Industry with a Focus on Africa

Ramu Govindasamy,[*,1] Surendran Arumugam,[1] and James E. Simon[2]

[1]Department of Agricultural, Food and Resource Economics,
Rutgers, the State University of New Jersey,
New Brunswick, New Jersey 08901
[2]New Use Agriculture and Natural Plant Products Program (NUANPP)
Department of Plant Biology and Plant Pathology,
Rutgers, the State University of New Jersey,
New Brunswick, New Jersey 08901
[*]E-mail: govindasamy@AESOP.Rutgers.edu.

Africa is a producer and consumer of essential oils. Yet, its emergence as a major global source of most essential oils has been limited by many constraints largely related to the weakness of the value chain and supply. Those essential oils that are exclusively sourced from Africa are usually linked to a limited number of aromatic plants indigenous to the continent and which still are only found there. As the use of natural products such as essential oils provide high value niche crops and plant-based products that can provide income generating opportunities for African communities, a review of the major African essential oils in context to the global industry could provide insight into opportunities. According to the United Nations Comtrade statistics, the size of essential oil fragrance and flavor global market was estimated at US$ 24 billion in 2011, growing at an annual rate of 10 %. The major consumers in the multi-billion dollar global essential oils market are United States (40%), Western Europe (30%) and Japan (7%), with trade in essential oils and related products increasing at about 10% per year. The United States is the largest importer (US$ 2,721 million) and consumer of essential oils, with consumption equaling about 40% of the total production. Essential oils are used in the food industry for flavoring or flavor enhancing and in the aromatic industry directly in flavor and fragrance applications and formulated into a wide range of personal care and industrial products. In this paper, the overall global market for essential oils and aromatic plants is examined with a focus on the major internationally traded ones from Africa, as is their impact on the trade and development including discussions relating to economic returns and distributions. The sources of information used for this examination were secondary data such as various organizations and international country trade statistics.

1. Introduction

The international essential oil industry is highly complex and tightly knit circle of processors, traders, dealers and producers. Globally, over 100 countries produce essential oils to varying degrees and many of these countries have been producing essential oils for decades if not centuries. Although commonly known as fragrance materials for the aromatic industry, essential oils are also used for flavoring or as flavor enhancing agents. Primary markets for essential oils include the flavor and fragrance industries such as soft drink companies, food and beverage companies, perfume and fragrance, cosmetic, personal care and pharmaceutical companies. These commercial markets require reliable supplies of consistent, high quality, and price competitive products. Since the use of essential oils in both flavor and fragrance industries is interrelated, few separate production or market statistics are available to characterize each industry relative to essential oil sourcing.

As these products are niche high value products yet not major agricultural commodities, few countries keep track of import/export of individual essential oils or even the aggregated group of products. Many essential oils are produced in one country, sold to another where the oils are further rectified, blended, and aggregated and then sold to another larger essential oil buyer who then may in turn re-sell the same product to another or export it from their country into yet another country so that even when statistics are available, it is different to track the actual origin of a particular oil. Often the country of export is listed as the source. Export products are sold directly to buyers in Europe, Asia and America and through agents in Europe and North America. Exact export statistics are also not directly available for the international market; as the products are used in various industries such as cosmetic industry, food industry as well as pharmaceutical industry. Furthermore, many oils don't have separate trade codes to track them, which make the analysis of the industry very difficult from a trade perspective.

Overall, essential oils number over a 1,000, from which approximately 300 are of commercial importance (*1*). Plants accumulate these aromatic compounds such as terpenes and phenylpropanoids in specialized tissues which are distinct and characteristic of the plant families and as such the essential oils can be found in different parts such as leaves, flowers, fruits, roots, rhizomes, and wood. Different plant parts may also express different aromas. The essential oils are generally distilled using fresh, partially dry, or dry tissues using hydrodistillation, steam distillation though additional technologies including solvent extraction, super critical fluid extraction are also employed. Different technologies result in different qualities and characteristics and also have different cost implications resulting from the specific systems employed. While so many essential oils are in commerce, there are about ten major aromatic plants that account for 80% of the world market for essential oils. These include orange, corn mint (Japanese mint, Mentha arvenis), eucalyptus cineole-type, citronella, peppermint, lemon, eucalyptus citronellal-type, clove leaf, cedar wood (US) and Litsea cubeba (Table I). Many essential oils are still collected in commercial quantities from wild sources. The citrus and woods oils are generally by-products (*2*).

Table I. The most utilized essential oils and major producers

Product	*Botanical Origin*	*Country*
Orange	*Citrus* x *sinensis*	Australia, Brazil, Dominican Republic, Israel, Italy and U.S.A.
Corn mint	*Mentha arvensis*	Brazil, China, India, Japan, North Korea, Paraguay, Taiwan and Thailand
Eucalyptus (cineole-type)	*Eucalyptus* spp.	Australia, Austria, Brazil, China, India, Paraguay, Portugal, South Africa and Spain
Citronella (lemongrass)	*Cymbopogon* spp.	China, India and Vietnam
Peppermint	*Mentha* x *piperita*	Australia, China, Italy, Japan and U.S.A.
Lemon	*Citrus* x *limon*	Argentina, Australia, Brazil, Greece, Spain, Italy, U.S.A. and Peru
Clove leaf	*Syzygium aromaticum*	Brazil, Indonesia, Madagascar, Sri Lanka and Tanzania
Cedar wood	*Juniperus* spp., *Thuja occidentalis*	U.S.A. and China
Litsea cubeba	*Litsea cubeba*	China
Sassafras	*Cinnamomum camphora, Ocotea pretiosa*	Brazil and U.S.A.
Lime	*Citrus aurantifolia*	Brazil, China, Cuba, Ghana, Haiti, Ivory Coast, Jamaica, Mexico and Peru
Spearmint	*Mentha spicata, M.* x *gentilis* (form. *M. cardiaca*)	Argentina, Australia, Brazil, Bulgaria, China, Egypt, France, Hungary, Japan, Korea, Morocco, New Zealand, Paraguay, Romania, Russia, Taiwan, UK, U.S.A. and Yugoslavia

SOURCE: Directorate Marketing (*3*); botanical listings added.

The United Nations Comtrade statistics department denoted that the total size of the global market was approximately $24 billion in 2011. According to the Directorate Marketing, Department of Agriculture, Forestry & Fisheries, Republic of South Africa (*3*), the major consumers in the multi-billion dollar global essential oils market are United States (40%), Western Europe (30%) and Japan (7%), with trade in essential oils and related products increasing at about 10% per year.

According to MacTavish and Harris (*4*), China, India, Indonesia and Brazil are the top traditional essential oils producers in the world. The key factors that kept these countries as the highest producers include a large population and available field workers, low labor costs, sizable internal consumption, investment in scientific and technical training, strong economic position, and a well-developed export sector. France, United States and Australia are the top producers in improved varieties of essential oils (*4*). France holds the top position in lavandin production, whereas, United States holds the top position in the production of peppermint and spearmint. These countries have favorable factors that contribute to their production capacity in improved varieties such as selection and introduction of improved varieties, establishment of intensive and mechanized systems of production, harvesting and distillation, a strong historical basis of production with oil quality thus being defined largely based upon these producing countries, and sound research infrastructures.

In 2011, the size of global exports relating to essential oils averaged US$24 billion a year and is growing at an annual rate of 10 % (*5*). The United States is the largest importer (US$ 2,721 million) and consumer of essential oils (*5*), with consumption equaling about 40% of the total production. European Union member countries import essential oils amounting to US$2,614 million dollars (2011), of which France is the leading importer, reported at US$2,892 million (32%). Among all countries, China, India, Indonesia, and Brazil are the top essential oil producers. Africa's share is less than 1% of the world market (*6*). Strong growth at the beginning of the 21st century in the pharmaceutical, cosmetics and nutraceutical industries have led to continued growth in demand for essential oils and plant extracts.

According to Harris (*7*), a small number of essential oil suppliers are now labeling their products as having Generally Recognized as Safe (GRAS) status and implying not too subtly that this means that they are of therapeutic quality and also safe for internal use. GRAS however does not have anything to do with therapeutic quality and only refers to the safety of the product. Other essential oils are targeted for the certified organic markets, recognizing that different countries require different standards and certifying organizations. Alternatively, others focus on newer innovative certifications seeking to brand their products as being sourced ethically, sustainably, using environmentally friendly systems of collection. The primary markets for essential oils are the flavor and fragrance industries that include beverage, food flavor, cosmetic, and perfume and fragrance companies. These commercial markets require reliable supplies of consistent high quality, price competitive products. Aromatherapy is an emerging economic industry, which is manufacturing a wide range of newer products based upon natural essential oils, and other aromatic materials for the purpose of changing a person's mind, mood, cognitive function or health.

According to In Consult Ltd (*8*), the market for aroma therapeutic/essential oil products in France, Germany, and the United Kingdom was valued at US$ 532 million. The aromatherapy market in the U.S. has grown from US$ 316 million in 1996 to US$ 454 million in 2001. The compound annual growth rate (CAGR) for aromatherapy sales in the U.S. for 1996-2001 is 7.5%. Natural personal care (NPC) markets in the U.S. grew to US$ 4.1 billion in consumer sales in 2002 (*9*).

2. Methodology

In this paper, we use the available secondary sources of data from various organizations such as the Food and Agriculture Organization of the United Nations (FAO), the World Trade Organization (WTO), the International Trade Center (ITC), and the United States Department of Agriculture (USDA) and other regional associations and trade organizations. Estimates focused on size and trade of global essential oil and aromatic industry with a focus on African essential oils. A major objective of this review is to characterize the overall market size of essential oils and to predict the growth opportunities and their impact on trade and development, including economic returns and distributions.

3. Results

3.1. Global Market

Egypt, France, Italy, Morocco, Madagascar and the US are prominent countries in the use of improved varieties of organic essential oils in flavor, fragrances and cosmetic industry (*10*). With over 300 essential oils, only about 20 dominant the international market (*1*).

According to 2011 UN Comtrade statistics, the United States is the single largest trader of essential oils with its exports and imports reported at US$ 2,056 and US$ 2,721 million, respectively. The same statistics showed that the United States, France and the United Kingdom were the most involved in both exports and imports compared to other nations, which primarily did one or the other function. The United States and France have primary functions in importing essential oils and processing the materials to re-exporting to other countries. North American and Western European countries use most of the global essential oil resources.

The average imports from France were higher than exports between years 2001 and 2011, suggesting that both countries also use significant quantities of essential oils. These figures are not surprising given than so many essential oils come from tropical and subtropical countries and thus cant be produced in Europe (or the US). Both countries do purchase essential oils from other countries, repackage those oils and export a further processed product. Japan is one of the largest consumers of essential oils in the world, with importation continuously increasing since 2001, with insignificant exportation. Germany is also one of the largest consumers of the essential oils among the European community. German essential oil imports have remained higher when compared to its exports.

Based upon the United Nations commodity statistics (*5*), Ireland was the largest exporter of essential oils with US$ 6,944 million. Clearly, this export reflects a major food processing industry that has developed given that no essential oils of any quantity are produced there. The United States was the largest importer with US$ 2,721 million. The same sources reported global exports and imports to be US$ 24 billion and US$ 21 billion, respectively.

The overall global trade in essential oils, perfume and flavor materials between 2001 and 2011 shows that the average variation between imports and exports of essential oils, perfume and flavor materials were US$ 1 billion (Figure 1). The overall global export value of essential oils, perfume and flavor materials was US$ 9 billion in 2001, growing to US$ 24 billion in 2011, an increase of 167% in 11 years; whereas, the global import value of essential oils, perfume and flavor materials were reported at US$ 8 billion in 2001, reaching US$ 21 billion in 2011, an increase of 162% in the same period.

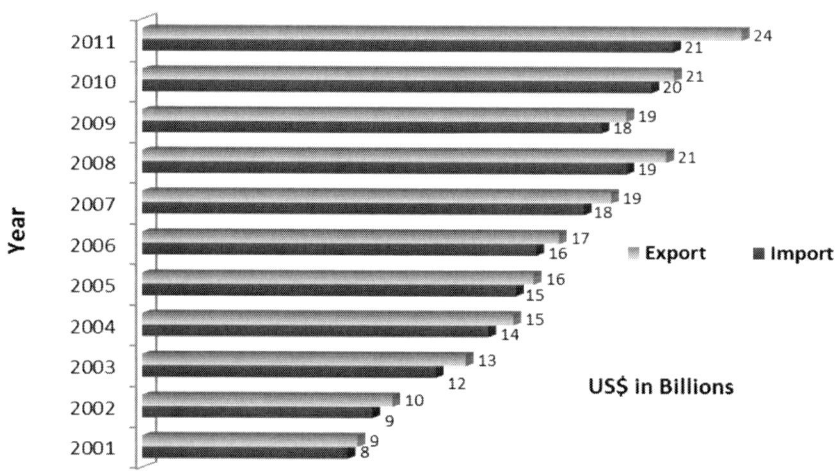

Figure 1. Global exports and imports of essential oil, perfume and flavor Materials from 2001 to 2011. SOURCE: UN Comtrade (5).

The top ten countries exporting essential oil, perfume and flavor materials from 2001 to 2011 shows that only a few actually produce essential oils (Table II). During the year 2001, Ireland exported US$ 2,213 million in essential oils, perfume and flavor materials, and it has increased to US$ 7262 million in 2011, an increase of 214%. According to the International Trade Center, Ireland is now the world's top exporter in essential oils sector. France exported US$ 1,007 million in essential oils in 2001, increasing to US$ 2,298 million in 2011, showing it to be the second largest exporter of essential oils sector. The United States is currently the third largest exporter of essential oils in the world. The United States in 2001 exported essential oil, perfume and flavor materials totaling about US$ 990 million and this grew 108% to US$ 2,056 million in 2011. According to UN Comtrade statistics, Germany is the fourth largest country in exporting US$ 2,032 million of essential oil, perfume and flavor materials with 207% growth between 2001 and 2011. Switzerland exported US$ 809 million of essential oils, perfume and flavor materials during 2001, increasing to US$ 2,025 million in 2011 with 150% growth. Singapore also reported over a US$ 171 million of essential oil, perfume and flavor material exports in 2011. India (1,056%), Singapore (821%) and Spain (226%) showed the highest growth (relative to % point) in terms of export of essential oils, perfume and flavor materials (Table II). Overall, European countries exported more quantities of essential oils, perfume and flavor materials compared to all the other continents.

The top ten essential oil, perfume and flavor materials importing countries in the global market during the 2001 to 2011 include the US, France and the United Kingdom (Table III). The United States is the largest essential oil, perfume and flavor materials importer among all the countries globally with US$ 2,721 million in 2011, an increase of 554% over a decade. France is the second largest importer of essential oils, perfume and flavor materials among the top ten countries with US$ 2,892 million in 2011, an increase of 186% over 2001. The United Kingdom is the third largest importers in essential oils, perfume and flavor materials, importing over US$ 1,177 million in 2011, with 85% growth over 2001. During the same time period, Germany imported US$ 972 million of essential oils, perfume and flavor materials followed by Mexico with US$ 922 million, Spain US$ 845 million and Italy with US$ 814 million. In comparison, Japan, Canada and the Netherlands reported importation of more than a half billion US dollars of essential oils, perfume and flavor materials. In terms of percentage growth, the United States was the highest, with 554% of essential oils, perfume and flavor materials imports, followed by Russian federation (422%), Mexico (287%), France (186%), Spain (129%), Germany (114%), Italy (112%), Netherlands (102%) and United Kingdom (85%). Japan had the lowest import growth rate, though still significant at 73%, when one compares that to other foods and manufactured products, among top ten importers of essential oils, perfume and flavor materials.

Many counties including the United States, France, Germany, United Kingdom, Spain and Netherlands participate both as exporters and importers of essential oils (Tables II and III). Ireland did not report any significant imports of essential oils in the top ten lists. The United States is the single largest net importer of essential oils in the world.

The U.S.A. and Russian federation reported the highest growth in the essential oils sector with 554% and 422% increases, followed by Mexico with 287% and France with 186% (Figure 2). In addition, Spain, Germany, Italy, Netherlands, UK and Japan reported an import growth rate of essential oil, perfume and flavor materials that is less than the world average of 162%.

Table II. Top 10 Global exporting countries of essential oil, perfume and flavor materials from 2001 to 2011 (figures in Million $US)

Year	*Ireland*	*France*	*U.S.A.*	*Germany*	*Switzerland*	*Singap-ore*	*UK*	*Netherla-nds*	*India*	*Spain*	*World*
2001	2213	1007	990	661	809	171	584	338	61	172	8566
2002	2400	1164	1087	728	991	220	621	407	78	202	9944
2003	4015	1417	1253	941	1086	332	719	484	123	267	13094
2004	5223	1556	1334	1123	1172	366	741	530	124	299	15109
2005	5781	1467	1269	1228	1254	408	744	576	225	312	15382
2006	5922	1604	1357	1350	1421	479	823	579	282	348	16649
2007	6534	1864	1539	1559	1592	557	891	602	321	433	18618
2008	7293	1994	1681	1716	1744	664	867	687	449	430	20636
2009	6604	1795	1686	1728	1473	716	787	612	385	340	18919
2010	6489	2034	1906	1849	1839	1071	986	568	457	505	21193
2011	6944 7262	2298	2056	2032	2025	1576	1090	992	705	561	23910
Growth %	214	128	108	207	150	821	87	194	1056	226	179

SOURCE: UN Comtrade (*5*).

Table III. Top 10 Global importing countries of essential oils, perfume and flavor materials from 2001 to 2011 (figures in million US$)

Year	*U.S.A.*	*France*	*UK*	*Germany*	*Mexico*	*Spain*	*Italy*	*Russian*	*Japan*	*Netherlands*	*World*
2001	416	1011	637	454	238	369	383	111	351	256	8079
2002	467	1157	735	488	300	433	443	137	356	244	9048
2003	1407	1355	820	590	396	596	618	167	382	320	11626
2004	2122	1489	992	653	434	616	703	197	435	327	13535
2005	2548	1562	1033	765	524	633	760	235	444	304	14746
2006	2608	1764	1053	767	632	691	704	268	382	312	15602
2007	2581	2073	1145	920	685	743	831	355	430	367	17513
2008	2830	2335	1108	990	761	823	871	431	457	436	19386
2009	2439	2183	1057	892	716	572	773	451	477	369	17791
2010	2595	2512	1111	1052	812	775	759	529	494	449	20340
2011	2721	2892	1177	972	922	845	814	579	607	517	21162
Growth (%)	554	186	85	114	287	129	112	422	73	102	162

SOURCE: UN Comtrade (*5*).

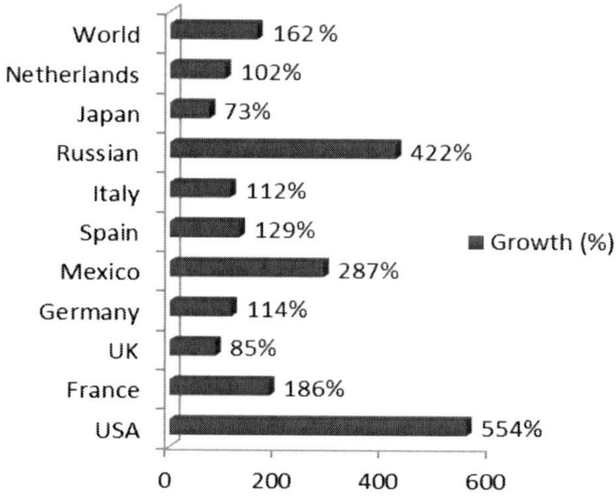

Figure 2. The fastest growing global markets in the essential oils sector based on imports from 2001 – 2011. SOURCE: UN Comtrade (5).

3.2. European Union Essential Oils Trade

During 2008, essential oil imports by European Union countries totaled about €810 million, that is a €187 million increase when compared to 2003 (Table IV). During this time period (2003-2008), trade figures showed a significant difference between intra-trade and extra-trade which was higher than the former. The United States continued as the leading supplier of essential oils to the European Union, valued at about 20% of total imports.

Table IV. European Union essential oil trade (€Millions)) from 2003-2008

Trade	*2003*[1]	*2004*	*2005*	*2006*	*2007*	*2008*
Total Trade	623	635	659	725	749	810
Intra-Trade	254	249	243	283	284	307
Extra-Trade	138	150	166	173	164	173
Developing Countries	232	237	249	269	301	330

[1] NOTE: Value (€Millions)   Source: Eurostat (*11*).

However, within the EU community, France was the leading importer as well as supplier providing from 7% up to 26% (Table V) of the EU nations' essential oil needs. China, Brazil and United Kingdom were the largest suppliers of essential oils after United States and France, accounting for about 30%. France, the United Kingdom and Germany were the leading importers, accounting for 64% within the European Union. In 2011, about 127,534 tones of essential oils valued at US$ 2,614 million were imported into the European Union.

Table V. European Union essential oil imports from leading suppliers in 2011

Importing EU countries	*2011 (Total Value) US$ In Million*	*2011 (Total Quantity Imported) Volume in Tonnes*	*Leading Suppliers*
France	2892	121103	United States (10%), Morocco (8%), Italy (8%), India (7%), China (7%)
United Kingdom	1177	66135	United States (28%), Argentina (18%), France (10%), China (7%), Brazil (5%)
Germany	972	68246	France (21%), United States (10%), The Netherland. (9%), China (9%), India (6%)
Netherland	516	78157	Brazil (22%), United States (20%), France (7%), India (5%), Spain (4%)
Spain	845	50627	France (16%), China (14%), Indonesia (10%), United States (7%), Germany (7%)
Italy	814	34420	France (26%), United Kingdom (25%), The Netherlands (14%), Germany (7%)
Total EU	2614	127534	United States (20%), France (10%), China (6%), Brazil (5%), United Kingdom (5%),

Source: UN- Comtrade (*5*).

According to the UN Comtrade database, the total export of essential oils, perfume materials and cosmetics totaled US$1,949 million in 2001 and reaching US$ 6,467 million in 2011, an increase of 232% (Figure 3). The total import of essential oil, perfume materials and cosmetics were reported to be US$ 1,015 million in 2001, growing to US$ 2,614 million in 2011, resulting in 158% growth rate.

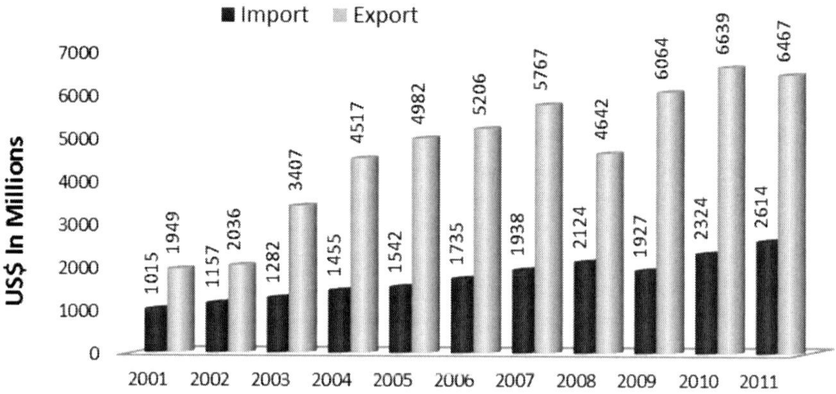

Figure 3. European Union (EU27) essential oil, perfume materials and cosmetics trade from 2001-2011. SOURCE: UN Comtrade (5).

The United States led as the highest trader of essential oil, perfume materials and cosmetics with the European Union compared to other leading trading countries. The European Union exported US$ 1,889 million of essential oil, perfume materials and cosmetics to the United States during 2011 and at the same period they imported US$ 436 million of essential oil, perfume materials and cosmetics from United States. In this region, Switzerland was the second largest trading partner to the European Union in 2011, exporting US$ 268 million and importing US$ 1,342 million. The European Union exported US$ 320 million of essential oil, perfume materials and cosmetics to Turkey in 2011. The European Union exported about US$ 200 million of essential oil, perfume materials, and cosmetics to Japan and Mexico.

All these figures, trends and economic assessments address the significant value that processing contributes to the value-addition and economic impact of this industry. As most of the essential oils exported from the EU for example are based upon 'raw materials' originating in the developing world including Africa (Figure 4). To the extent that the essential oil industry can grow in Africa, we suggest that growth can occur in strengthening the value chain and in developing as a processor of raw essential oils and carrier oils. Africa has a long way to go to be viewed by the industry as a reliable supplier of competitively priced essential oils (as a global supplier of raw/crude semi-processed (e.g. distilled) essential oils. This is in particular reference to those essential oils that are or can be procured in other regions. In addition, Africa, with notable exceptions, in general has lacked the additional secondary processing, fractionation and manufacturing base for essential oils that could contribute significantly to an array of blended and prepared essential oil products for the private sector to use in final product manufacturing.

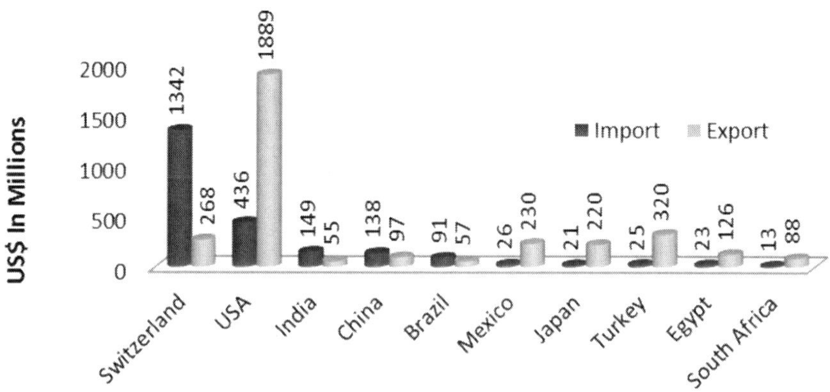

Figure 4. European Union (EU27) essential oil, perfume materials and cosmetics trade by country in 2011. SOURCE: UN-Comtrade (5).

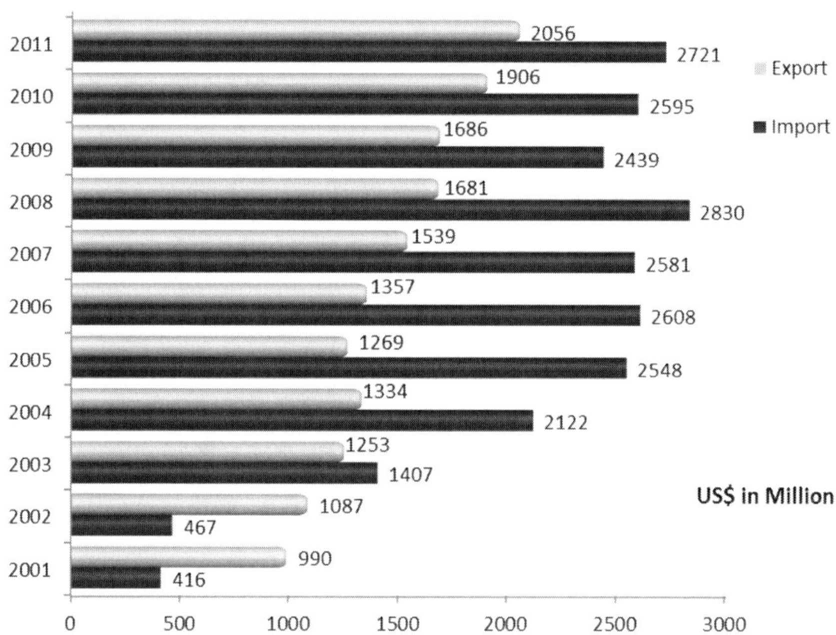

Figure 5. United States global essential oil trade from 2001 to 2011. SOURCE: UN-Comtrade (5).

3.3. United States Essential Oils Trade

The market for essential oils has been growing rapidly for the past several years in the United States. Due to wide variety of plants, forms and uses, definitions of goods and products vary based on the market. The United States imported a total of US$ 416 million worth of essential oils in 2001, which increased to US$ 2,721 million in 2011 (UN Comtrade, 2011; see Figure 5). This represents a six-fold increase in U.S. essential oil imports in ten years. The period of highest growth occurred between 2005, 2006, 2008 and 2011. This also occurred then during the time of a major economic collapse in the U.S.A. indicating that even when most all other industry sectors faced declining or stagnant growth, the essential oil sector continued to grow.

3.4. Developing Countries Essential Oils Trade

Growth in essential oils also occurred in the developing world and it is here where additional lessons or illustrative examples of potential opportunities in Africa can be noted. For example, the essential oils industry in India is significant. This country has had a long and vibrant history in its use of aromatic and medicinal plants. India exported US$ 61 million worth of essential oil, perfume and flavor materials in 2001, increasing to US$ 705 million in 2011 with a growth of 1,055% over the past ten years. The Indian import of essential oil, and perfume and flavor materials was US$ 33 million in 2001 and now is about six times higher a decade later (Figure 6). The export of Indian essential oil always exceeded imports in this same time period. When considering the high consumption and utilization of essential oils within India itself, the size of the industry becomes significantly higher.

Figure 6. Indian global essential oil and perfume and flavor materials trade from 2001-2011. Source: UN-Comtrade (5).

Brazil can serve as a second example of an emerging developing country and its essential oil trade. Brazil exported US$ 77 million worth of essential oil, perfume and flavor materials in 2001, which increased to US$ 320 million in 2011. Its' imports totaled US$ 81 million in 2001, rising to US$ 183 million in 2011, an increase of 126% (Figure 7). India and Brazil had a positive trade balance in the reference period.

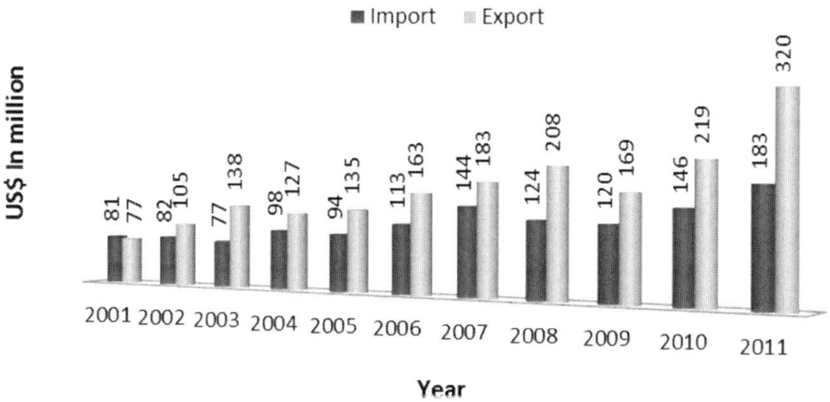

Figure 7. Brazilian global essential oils and perfume and flavor material trade from 2001-2011. Source: UN-Comtrade (5).

As a final example of emerging markets for essential oils, Malaysian imports of essential oil, perfume and flavor materials have always been much higher than their exports of the same commodities (Figure 8). During 2001, Malaysian essential oil exports were valued at US$ 5 million compared to US$ 75 million in imports in the same period. Malaysia exported US$ 17 million worth of essential oil, perfume and flavor materials in 2011, whereas, it imported US$ 114 million during the same period.

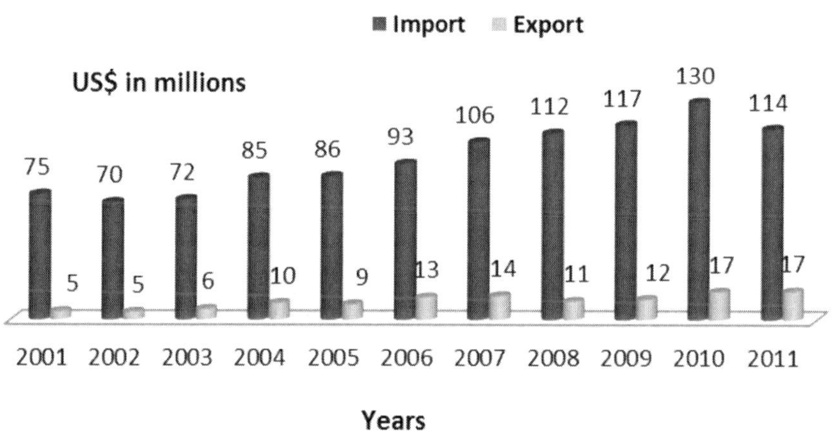

Figure 8. Malaysian global essential oil and perfume and flavor materials trade from 2001-2011. SOURCE: UN-Comtrade (5).

3.5. African Countries and the Essential Oil Trade

The continent of Africa is famous for a wide range of indigenous aromatic plants- from the North African to Southern Africa and the island of Madagascar. The use of these aromatic plants are stepped in history, culture and tradition. Over the centuries and continued today, essential oils from Africa have been used in foods, flavorings, religious and spiritual ceremonies and traditions and for health and medicinal applications. Across the many African countries, are a few key African players in the international market. Other African countries produce small quantities, use such products locally and regionally, or market them to larger buyers in Africa where then they may finally appear statistically as exports to Europe and the U.S.A..

South Africa

South African export of essential oils, perfume and flavor materials accounted for US$ 19 million in 2001, US$ 63 million worth of imports reported during the same period (Figure 9). By 2011, the South Africa import of essential oil perfume and flavor materials accounted for US$ 175 million, with exports totaling US$ 62 million. South African imports were consistently higher than the exports of essential oil, perfume and flavor materials, which was similar to what was reported for Malaysia. Though South Africa has abundant natural resources, exports were very limited, because of poor technology and trade knowledge, deficiency of capital investment and insufficient organizational support (*3*) and limited number of larger-scale buyers and processors.

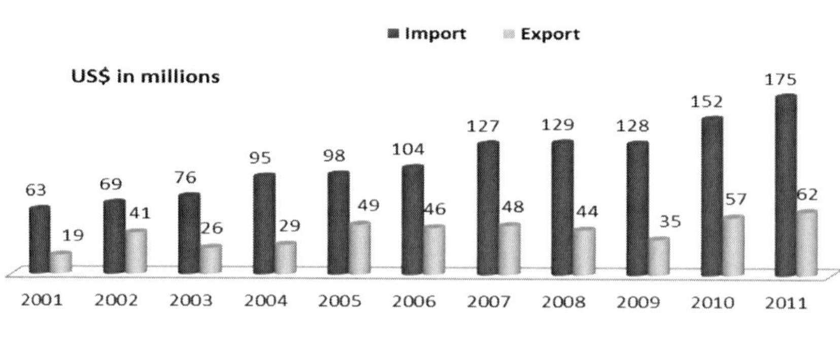

Figure 9. South African global essential oil and perfume and flavor materials trade from 2001-2011. SOURCE: UN-Comtrade (5).

While South Africa is considered the economic leader in the essential oils in sub-Sahara Africa, these products are produced and/or traded in a wide number of additional southern African countries, albeit generally in lower volumes and trade. Export value of essential oils, perfume and flavor materials from southern African countries into the international marketplace from over a decade, 2001-2011 period surprisingly shows Swaziland as a leader for a number of year, exporting US$ 169 million worth of essential oil, perfume and flavor materials in 2001, increasing to US$ 319 million in 2007 with a growth of 88% over seven years period (Table VI). In the case of South Africa, the exports value of essential oil, perfume and flavor materials was 18.9 million in 2001, increasing to 62 million in 2011. Madagascar, internationally recognized and respected for a wide range of essential oils (cloves, cinnamon, geranium, ylang ylang and more) exported essential oils, perfume and flavor materials valued at $9.6 million in 2001, increasing to $41.6 million in 2011. Moreover Seychelles, Congo, Malawi, Mozambique, etc., reported exporting very low levels of exports of essential oils, perfume and flavor materials during this same time frame.

As this overview focuses on the major essential oils that enter into international trade, its noteworthy that many indigenous and introduced aromatic plants are traded all over sub-Sahara Africa in smaller quantities and often are not specified in commerce making its impact difficult and in understanding the supply and availability of such oils over a long time period. As reported by the South African Essential Oil Producer Association, just in South Africa (in 2010) and coming in from different provinces, essential oils of limited quantity came from Artemisia (assuming Artemisia afra and not A. annua), buchu, chamomile, citronella, Eriocephalus, eucalyptus, geranium, lavender, lavendin, lemonbalm (Melissa), lemon grass, lemon tea tree, lippia, marjoram, rosemary, spearmint, tagetes, Tea tree, thyme and vetiver (*3*). As some of these arise from development projects, it is unclear how many of these will be produced over longer time periods and meet buyer expectations.

Importation of essential oils, perfume and flavor materials imports by Southern African countries from other regions during this same time period (2001 and 2011) indicated significant commerce (Table VII). The major importing markets for essential oils, perfume and flavor materials within Southern African countries were South Africa followed Zimbabwe, Swaziland, Tanzania, Malawi, and Namibia. It is in the partial replacement of importation of essential oils for locally produced ones that could contribute to the economic development of a developing country. That would be contingent upon several factors including but not limited to the ability to successfully introduce and grow the specific aromatic plant in the new country. Yet, the local use of imported oils for internal markets may also offer greater flexibility in quality of essential oils manufactured and sold locally and regionally. The requirement to meet international norms need not always be an issue when used in local soaps, cosmetics, shampoos and other products.

Table VI. Southern African Countries Exports of Global Essential Oils, Perfume and Flavor Materials Trade from 2001-2011 (US$ in thousands)

Period	*2001*	*2002*	*2003*	*2004*	*2005*	*2006*	*2007*	*2008*	*2009*	*2010*	*2011*
Swaziland	169538	502487	573583	551802	445106	356180	319162	NA	NA	NA	NA
South Africa	18915	40888	26215	29389	49100	46285	48439	44384	35259	56773	62302
Madagascar	9669	8372	6600	6861	7226	10749	13770	14378	17333	27036	41675
Zambia	250	93	5	23	73	64	315	2183	8434	15271	15757
Namibia	67	150	350	1981	1932	119	36	472	810	2985	359
Zimbabwe	4	1356	NA	478	768	313	373	44	215	498	310
Mauritius	41	19	39	53	40	38	38	73	23	77	220
Botswana	1	551	20	66	49	62	12	7	29	9	121
Tanzania	5	10	11	69	19	191	181	385	430	2564	1469
Lesotho	36	7	4	1	NA	NA	NA	7	NA	NA	NA
Malawi	NA	6	58	99	143	156	97	33	51	79	79
Mozambique	NA	NA	76	8	NA	NA	69	NA	34	179	71
Congo	NA	NA	NA	NA	NA	NA	4	6	20	19	NA
Seychelles	NA	NA	2	7	1	NA	4	3	NA	NA	NA

Note: NA- Data not available Source: UN-Comtrade (*5*).

Table VII. Southern African Countries imports of Global Essential Oils, Perfume and Flavor Materials Trade from 2001-2011 (US$ in thousands)

Period	*2001*	*2002*	*2003*	*2004*	*2005*	*2006*	*2007*	*2008*	*2009*	*2010*	*2011*
South Africa	62781	69083	75708	95262	97558	104167	127303	128714	127887	152433	174620
Zimbabwe	16896	13965	4476	7162	2395	1991	821	1870	5859	23593	NA
Swaziland	14861	76043	109180	284610	140394	29839	18092	NA	NA	NA	NA
Tanzania	7642	13368	14835	13346	21534	26327	30345	32747	35014	39465	52783
Malawi	4486	5067	5231	6500	7665	8693	8643	13176	12155	22912	20667
Namibia	4340	8093	6684	10118	4509	3547	6185	6241	5004	11818	14807
Mozambique	3421	4062	7638	10344	11493	13343	11120	18936	19302	14688	11311
Mauritius	3072	3088	4321	4362	5089	4320	5102	5633	4799	5573	5709
Zambia	1966	1387	4607	7648	7849	6179	10235	11129	9077	14301	15759
Madagascar	1682	1143	2740	4012	3641	5651	7360	8566	6284	8170	7880
Seychelles	935	1169	1093	852	1142	1742	1203	1349	NA	NA	NA
Botswana	375	3711	9957	9753	6294	3113	5015	4265	4938	5997	8858
Lesotho	45	439	542	863	NA	NA	NA	1147	2567	NA	NA
Congo	NA	NA	NA	NA	NA	NA	4633	9551	5156	5717	NA

Note: NA- Data not available   Source: UN-Comtrade (*5*).

Table VIII. Western African Countries Exports of Global Essential Oils, Perfume and Flavor Materials Trade from 2001-2011 (US$ in thousands)

Period	*2001*	*2002*	*2003*	*2004*	*2005*	*2006*	*2007*	*2008*	*2009*	*2010*	*2011*
Senegal	1429	1726	1798	2749	4185	4344	5535	5411	6435	6237	5739
Togo	232	298	1186	381	171	NA	NA	19	30	NA	76
Ghana	16	NA	265	NA	16	16	3	17	90	184	29
Niger	NA	NA	NA	NA	3	NA	NA	2	7	2	1
Nigeria	1	NA	65	NA	NA	NA	167	345	NA	10	265
Mali	3	63	43	85	18	187	277	448	707	NA	NA
Benin	NA	NA	10	NA	NA	1	NA	NA	NA	NA	NA
Guinea	3	1	NA	NA	1	40	4	NA	NA	NA	NA

Note: NA- Data not available   Source: UN-Comtrade (*5*).

Table IX. Western African Countries imports of Global Essential Oils, Perfume and Flavor Materials Trade from 2001-2011 (US$ in thousands)

Period	*2001*	*2002*	*2003*	*2004*	*2005*	*2006*	*2007*	*2008*	*2009*	*2010*	*2011*
Nigeria	57112	58896	58350	NA	NA	151763	254394	104158	158575	249154	258374
Ghana	13620	25170	24190	NA	28528	32563	35686	27999	22857	21765	41270
Senegal	5382	6627	7687	8306	8293	8519	11115	10870	10602	8800	13905
Mali	2958	1982	2135	2133	2366	3406	3771	4662	NA	5874	NA
Togo	1794	2434	3369	3207	1126	NA	410	844	1439	1503	8264
Burkina Faso	747	202	331	730	21	211	583	689	738	928	5352
Benin	567	515	381	214	59	171	137	197	503	1127	NA
Niger	182	62	177	98	150	73	203	100	70	206	231
Mauritania	123	114	106	45	165	347	31	79	114	281	229
Cape Verde	122	520	912	1062	1252	945	915	1008	1417	1200	1410

Note: NA- Data not available   Source: UN-Comtrade (*5*).

The imports value of essential oils, perfume and flavor materials in South Africa was US$ 62.8 million in 2001 and increased to about US$ 174.6 million in 2011, while in the case of Zimbabwe imports value was about US$ 16.9 million in 2001 and increasing to US$ 23.6 million in 2010. In the case of Swaziland, the import was around US$ 14.9 million worth of essential oils, perfume and flavor materials in 2001, increasing to US$ 18 million in 2007. The exports of Swaziland essential oils always exceeded imports in the reference period. In 2011, the imports market for essential oils, perfume and flavor materials declined in Malawi, Mozambique and Madagascar as compared to previous years.

From Western Africa

Western African countries also export and import values of essential oil, perfume and flavor materials. From 2001 through 2011, the major player in Western African exports was Senegal which exported US$ 1.4 million worth of essential oils, perfume and flavor materials in 2001, increasing to US$ 5.7 million in 2011 with a growth of 301 % over ten-year period (Table VIII). In the case of Togo, the exports were valued about US$ 232,000 in 2001, decreasing to US$ 76,000 in 2011. Moreover Guinea, Benin, Mali, Nigeria, etc., have very low levels of exports of essential oils, perfume and flavor materials during the period under study.

Relative to imports, Nigeria accounted for about US$ 57 million worth of essential oils, perfume and flavor materials in 2001, increasing to US$ 258.4 million in 2011 (Table IX). Ghana imported essential oils, perfume and flavor materials valued at US$ 13.6 million in 2001, increasing to US$ 41.3 million in 2011 followed by Senegal, Mali, Togo, and other West African countries. Several West African countries such as Burkina Faso, Mauritania and Cape Verde were net importers only. Local production of essential oils is not captured by these data sources.

From Eastern Africa

Eastern Africa Countries also exhibited growth in essential oils, perfume and flavor materials between 2001 and 2011 yet with far greater variability by specific countries in the region (Tables X and XI). The exports value of essential oils, perfume and flavor materials from Kenya for example was about US$ 3.9 million in 2001, decreasing to US$ 775,000 in 2010. In contrast during this same time frame, exports from Uganda and Tanzania was about US$17,000 and US$ 5,000, respectively in 2001, increasing to US$ 702,000 and US$ 1.5 million in 2011. Tanzania, in comparison, showed significantly stronger growth with exports of essential oils, perfume and flavor materials compared to other countries in this region.

Table X. Eastern Africa Countries Exports of Global Essential Oils, Perfume and Flavor Materials Trade from 2001-2011 (US$ in thousands)

Period	*2001*	*2002*	*2003*	*2004*	*2005*	*2006*	*2007*	*2008*	*2009*	*2010*	*2011*
Kenya	3911	66	667	393	837	611	772	1080	859	775	NA
Uganda	17	12	37	NA	7	90	195	10	380	398	702
Tanzania	5	10	11	69	19	191	181	335	430	2564	1469
Ethiopia	3	NA	9	NA	NA	NA	NA	101	103	14	234
Rwanda	3	15	6	NA	NA	NA	NA	34	1	23	49
Burundi	NA	NA	NA	NA	NA	12	NA	NA	NA	NA	NA
Sudan	NA	29	NA	2	NA	42	NA	NA	NA	NA	NA

Note: NA- Data not available   Source: UN-Comtrade (*5*).

Table XI. Eastern Africa Countries Imports of Global Essential Oil and Perfume and Flavor Materials Trade from 2001-2011 (US$ in thousands)

Period	*2001*	*2002*	*2003*	*2004*	*2005*	*2006*	*2007*	*2008*	*2009*	*2010*	*2011*
Kenya	13465	26624	27851	39924	36015	43771	49797	49836	52638	58071	NA
Uganda	8177	9416	9900	17539	19164	26138	24014	30816	25445	25236	32494
Tanzania	7642	13368	14835	13346	21534	26327	30345	32747	35014	39465	52783
Ethiopia	5354	6358	7782	9249	12682	14865	16192	23382	27748	33820	30225
Sudan	1822	2708	4497	5168	6391	15022	15220	12517	11179	NA	NA
Rwanda	585	937	492	392	478	439	638	1325	4034	NA	8291
Burundi	69	42	158	178	554	1782	2858	2345	6987	3372	NA

Note: NA- Data not available Source: UN-Comtrade (*5*).

Table XII. Northern Africa countries exports of Global Essential Oils, Perfume and Flavor Materials Trade from 2001-2011 (US$ in thousands)

Period	*2001*	*2002*	*2003*	*2004*	*2005*	*2006*	*2007*	*2008*	*2009*	*2010*	*2011*
Algeria	14791	17774	18310	20775	14945	20171	26786	35283	24097	29914	32584
Egypt	9370	8867	8338	10251	6879	9670	11493	49038	46381	65000	78290
Morocco	4430	4169	5236	6150	8405	10285	13181	14248	10095	12414	13167
Tunisia	29	35	2	3	6	9	166	1008	189	107	191

Note: NA- Data not available   Source: UN-Comtrade (*5*).

Table XIII. Northern Africa countries imports of Global Essential Oils, Perfume and Flavor Materials Trade from 2001-2011 (US$ in thousands)

Period	*2001*	*2002*	*2003*	*2004*	*2005*	*2006*	*2007*	*2008*	*2009*	*2010*	*2011*
Algeria	16460	21430	31103	40091	38374	38373	45079	54006	51780	60925	78964
Morocco	15827	17116	23929	25596	28729	46876	65335	77586	74954	74954	76854
Tunisia	12251	13222	14332	13335	14777	16241	17589	20684	22309	21094	24312
Egypt	10786	14326	18134	12727	16108	12969	14625	84891	71269	119336	138089
Libya	NA	NA	465	342	NA	NA	788	2375	3993	6086	NA

Note: NA- Data not available; nes= not specified. Source: UN-Comtrade (*5*).

Kenya is a leading importer of essential oils, perfume and flavor materials among the Eastern Africa at nearly US$ 13.5 million in 2001, increasing to about US$ 58 million in 2010. While in Uganda and Tanzania, the imports value was around US$ 8.2 million and US$ 7.6 million in 2001, increasing to US$ 32.5 million and US$ 52.9 million in 2011 followed by Ethiopia, Sudan, Rwanda and Burundi.

From Northern Africa

Northern African countries have been long recognized for the production and/or wild harvesting of a wide range of aromatic plants. Selected statistics reflecting the export and import values of essential oils, perfume and flavor materials during the period between 2001 and 2011 are illustrated in Tables XII and XIII. A major export market for essential oil, perfume and flavor materials was Algeria accounting about US$ 14.8 million in 2001, increasing to US$ 32.5 million in 2011. Egypt, a major producer of culinary herbs, and aromatic plants exported US$ 9.4 million worth of essential oil, perfume and flavor materials in 2001, increasing to US$ 78.3 million in 2011. Tunisia was reported as having very low levels of export markets for essential oil, perfume and flavor materials accounting US$ 29,000, increasing about US$ 191,000 during the period under scrutiny. The tables were not normalized for population differences.

North African countries also serve as major import markets for essential oil, perfume and flavor materials from various regions. Importations in the region steadily increasing from 2001 to 2012. Algeria and Morocco, for example, imported a significant value of essential oil, perfume and flavor materials in 2001 valued at US$ 16.5 million and US$ 15.8 million, increasing to US$ 78.9 million and US$ 76.8 million in 2011, respectively.

3.6. Types of Essential Oils Exported from African Countries

Table XIV provides the list of essential oils commodities exported by African countries during the period 2005 to 2011. The table further indicates that Madagascar's primary exported essential oil commodities are Essential oils, clove, Ginger, Niaouli, and non specified (nes). The reported data indicates that the export value of Essential oils, clove, Ginger, Niaouli, nes, was about US$ 13.33 million in 2007, increasing to about US$ 41.33 million in 2011.This indicates that there has been significant growth in the value of Essential oils, clove, Ginger, Niaouli, nes, under the period of study. In the case of Egypt, the major exported commodities of essential oil are essential oils of citrus fruits, nes,. The exported value was about US$ 6.2 million in 2008, increasing to of US$ 12.8 million in 2011. The important essential oil commodities produced by South Africa such as Essential oils of lemon. The reported data indicates that the export value of Essential oils of lemon was about US$ 4.6 million in 2005, increasing to about US$ 8.5 million in 2011.

Table XIV. Top five export commodities of essential oils from African countries during the period of 2005-2011

Country or Area	*Year*	*Commodity*	*Trade (USD)*	*Weight (kg)*
Madagascar	2011	Essential oils, clove, Ginger , Niaouli, nes	41339321	1863914
Egypt	2011	Essential oils of citrus fruits, nes	12807952	101031
South Africa	2011	Essential oils of lemon	8542800	374265
Tunisia	2011	Essential oils of orange	6428309	2403
South Africa	2011	Essential oils, terpenic by-products etc., nes	6045054	694315
Madagascar	2010	Essential oils,clove, Ginger , Niaouli, nes	26934748	2280553
Egypt	2010	Essential oils of citrus fruits, nes	14031471	225505
Morocco	2010	Essential oils, terpenic by-products etc., nes	9959207	379546
South Africa	2010	Essential oils of lemon	9216995	465208
Tunisia	2010	Essential oils of orange	5428873	3191
Madagascar	2009	Essential oils,clove, Ginger , Niaouli, nes	17185784	1841454
Morocco	2009	Essential oils, terpenic by-products etc., nes	8396503	276345
Egypt	2009	Essential oils of citrus fruits, nes	7935991	325321
South Africa	2009	Essential oils of lemon	4907453	274084
Tunisia	2009	Essential oils of orange	3801898	1340
Morocco	2008	Essential oils, terpenic by-products etc., nes	17234842	399014
Morocco	2008	Essential oils, nes	15388336	366771

Country or Area	*Year*	*Commodity*	*Trade (USD)*	*Weight (kg)*
South Africa	2008	Essential oils of lemon	9140830	441521
Tunisia	2008	Essential oils of orange	6201300	45139
Egypt	2008	Essential oils of citrus fruits, nes	6155989	2324674
Madagascar	2007	Essential oils,clove, Ginger , Niaouli, nes	13333226	1852434
Morocco	2007	Essential oils, terpenic by-products etc., nes	12993545	364494
Tunisia	2007	Essential oils of orange	6084951	22340
South Africa	2007	Essential oils of lemon	4564475	297643
South Africa	2007	Essential oils of citrus fruits, nes	2635177	86631
Morocco	2006	Essential oils,clove, Ginger , Niaouli, nes	10674024	355586
Morocco	2006	Essential oils, terpenic by-products etc., nes	8145178	360589
Tunisia	2006	Essential oils of orange	3519618	8042
South Africa	2006	Essential oils of lemon	3431770	286743
South Africa	2006	Essential oils of citrus fruits, nes	2233982	88142
Swaziland	2005	Essential oils, terpenic by-products etc., nes	9453574	241811
Morocco	2005	Essential oils, nes	7372314	267639
South Africa	2005	Essential oils of citrus fruits, nes	4803692	193981
South Africa	2005	Essential oils of lemon	4635164	787852
Tunisia	2005	Essential oils of orange	3142262	3456

Source:UN data (*12*); nes=nonspecified.

3.7. Essential Oils Value Chain Analysis

Buyer requirements play a major role in determining the commodity to be produced, and the end-user feedbacks act as a major factor in establishing their needs. Value chain analysis along the channels from the producer to the end-user in conjunction with realistic production programs is vital for establishing a profitable enterprise in the essential oils industry. To maintain consistent demand, producers must be able to supply consistent quality and quantity in an efficient and reliable manner. The value chain for essential oils can be disaggregated under production, processing and marketing (Figure 10).

Along the production value chain, common issues relate to inadequate knowledge and information about correct variety selection and lack of extension support for production of biomass. The quality of oil depends on the genetic material used, local soil, climate, and system of production, timeliness of harvest, method of postharvest handling and distillation as well as the cleanliness of biomass for distillation. Contaminated biomass often leads to poor quality oils largely due to off odors from weeds and extraneous materials. Along the processing value chain, common issues relate to poor distillation processes stemming from inadequate infrastructure as well as poor techniques, leading to inefficient production of essential oils, lower recoverable yields and variable final quality due to small-sized distillations commonly used across Africa.

The smaller practical distillation units provide practical solutions for locally collected materials in remote locations and reflect the traditional smaller land holdings in sub-Sahara Africa yet causes complication in ensuring larger bulked essential oils reflect a consistent quality. Establishment of multiple distillation units lead to differences in the quality of essential oils. This will increase the cost of production of essential oils and increase variability for many oils. Location of distilleries in relation to production areas also plays a major in terms of cost of transportation. In developing countries, extension support relating to establishment of distilleries and up to date distillation techniques are also inadequate, leading to poor quality and ineffective production of essential oils. Lack of expertise and quality control laboratories lead to increased risk of producing contaminated essential oils. For most small producers, marketing uncertainty is the major issue as is the difficulty in procuring low interest reasonable loans for inputs and purchasing of needed supplies. Most end-users and institutional buyers demand consistent quality and quantity. Due to the nature of small production volumes, small and mid-sized farmers find it difficult to maintain quality and supply consistently. Moreover, small farmers often sell to intermediaries who aggregate the products and supply to institutional buyers. Under such a scenario, if for some reason, when the intermediaries such as traders and brokers disappear, markets also vanish. Producers tend to have limited knowledge about market opportunities and product prices for essential oils and have limited access to those same markets. For most producers, these niche market opportunities are sporadic and there is often high volatility in output prices due to volume fluctuations.

Although it is difficult to quantify the cost of transaction from the farm-gate to the ultimate consumer, past studies indicate that about 10 to 35% of farm-gate costs can be attributed to harvesting, with an average of 22%. About 11 to 37% of costs can be attributed to pests, diseases and weed control, with an average of 22%. Plant nutrient costs and water, as needed, could be estimated at about 10% of farm-gate costs. Cost of extraction is estimated to be about 60 to 75% of processor costs (*3*). This presents a challenge because in many areas of Africa investment in agriculture, agroforestry or NonTimber Forest Species by the private sector (at the producer/collector or processor/buyer levels) are minimal resulting often in technological limitations in distillation and processing and in on-site purchasing of collected and/or cultivated materials. At the global level, as a commodity, trading of essential oils takes place on a larger-scale and is highly competitive. Many developed countries import essential oils from neighboring developing countries and re-export them after adding value through further processing. These re-exports, not only reduces the supply irregularities but also any local imbalances caused by climate, diseases and pests, and sporadic changes in supply and demand.

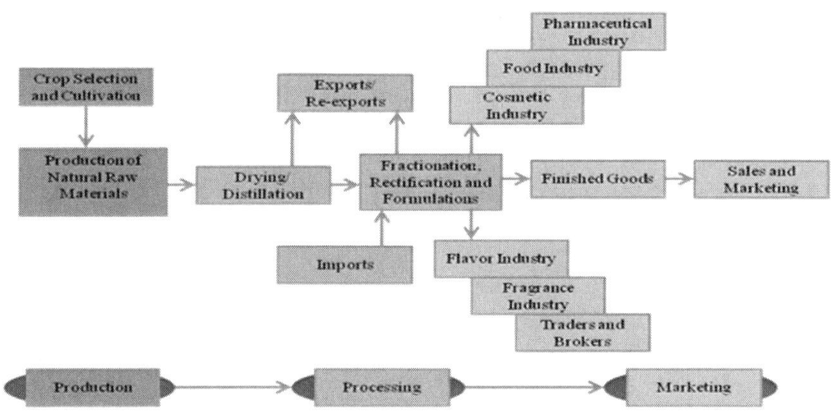

Figure 10. Value chain mapping for the essential oil industry.

4. Conclusion

Essential oils and perfume materials are the key important components of the flavors and fragrances industries. This review estimated the overall size of essential oils market based on available secondary data from various countries and international organizations. The discrepancies in reported data were due mainly to non-reporting or missing information and the data to separate out essential oils from perfume and fragrance materials was not available. Additionally, variations in product category definitions resulted in inclusions and exclusions of products from various categories, making country trade data difficult to compare. Specialty high value resins, niche local/regional aromatic oils and emerging new essential oils were not included in this discussion. The purpose of this analysis was to give a broader picture of the international global essential oil industry despite incomplete and/or unavailable data.

According to US Comtrade (*5*) statistics, China, India, Indonesia and Brazil are the top traditional essential oils producers worldwide because of large population and available field workers, low labor costs, sizable internal consumption, investment in scientific and technical training, strong economic position, and a well-developed export sector. According to the 2011 figures, Ireland became the world's top exporter in the essential oil sector, with US$ 6,944 million followed by France with US$ 2,298 million, the United States with US$ 2,056 million, Germany with US$ 2,032 million and Switzerland with US$ 2,025 million. Each of these leading countries have favorable factors that contribute to their production capacity in improved varieties, such as selection of improved varieties, establishment of intensive methods, simplification of production systems and sound research infrastructures. Countries such as the United States, France, Germany, United Kingdom, Northern Ireland, Spain and the Netherlands participated in both the export and import of essential oils. In terms of value, the United States is the single largest net importer of essential oils in the world, whereas, Ireland is the largest net exporter of essential oil in the world. The European Union reports the highest trade value in essential oils with the United States.

In case of developing economies, Brazil, Malaysia and South Africa were highly dependent on the imports of essential oils during 2001-2011. Essential oils represent a value-added product in the agricultural sector of developing economies. They provide additional income in developing economies, particularly, to the rural people where the natural resources are abundant (*3*). For the strengthening of the essential oils industry, a coalition of stakeholders such as producers, private sector, NGOs, and government could come together to support the growth of the industry and establish cooperation and transparency. However, in reality that often does not drive a market. Using a market-first science driven approach led by the private sector working in partnership with producing/collecting communities could foster economic growth. Regardless of the model(s) employed, coordinated efforts are needed among the actors in the value chain to optimize production techniques involving variety selection, agronomic practices and input application resulting in consistent quantity, control quality of oil along the value chain and more importantly establish market linkages to improve efficiency. African countries would also be served well by a continual investment in scientific and technical trainings coupled to quality assurance programs and a priority toward further value-addition and processing for local and regional sales and products.

References

1. Van de Braak, S. A. A. J.; Leijten, G. C. J. J. *Essential Oils and Oleoresins: A Survey in the Netherlands and Other Major Markets in the European Union*; CBI, Centre for the Promotion of Imports from Developing Countries: Rotterdam, The Netherlands, 1999; p 116.
2. Simon, J. E. Essential oils and culinary herbs. *Advances in New Crops*; Timber Press, Portland, OR, 1990; pp 472–483.
3. Directorate Marketing. *A Profile of the South African Essential Oils Market Value Chain*; Department of Agriculture, Forestry & Fisheries: Republic of South Africa, 2011; www.daff.gov.za/docs/AMCP/EssOilsMVCP2010-2011.pdf .
4. MacTavish; Harris. *An Economic Study of Essential Oil Production in the UK: A Case Study Comparing Non-UK Lavender/ Lavandin Production and Peppermint/spearmint Production with UK Production Techniques and Costs*; ADAS Consulting Ltd, Jan. 2002; http://www.uicnmed.org/nabp/web/documents/Essential%20oils%20study%20-%20UK.pdf.
5. UN Comtrade. *United Nations Commodity Trade Statistics Database*; N.p., n.d. 2012; http://comtrade.un.org/.
6. AfricaBiz. *Introduction To Essentials Oils Industry And Trade*; N.p., 15 Apr. 2003; http://businessafrica.net/africabiz/arcvol1/is47front.php.
7. Harris, B. Editorial. *Int. J. Aromather.* **2006**, *16* (2), 55.
8. In Consult Ltd. *Pre-Feasibility Study For Floriculture, Farming, Production, Processing and Marketing*; Government of Pakistan, May 2006; http://www.pc.gov.pk/usefull%20links/Studies%20for%20P&DD/Agriculture%20and%20Agro%20Based/Establishment%20of%20Floriculture%20Farm.pdf>.
9. Nutrition Business Journal. *U.S. Nutrition Industry Sales Grow 7.3% in 2002 to $58 Billion*; New Hope 360. N.p., 24 Aug. 2003; Web. 29 Dec. 2012, http://m.newhope360.com/us-nutrition-industry-sales-grow-73-2002-58-billion.
10. ITC. *Marketing Manual and Web Directory for Organic Spices, Culinary Herbs and Essential Oils*; International Trade Centre, 2006; http://www.intracen.org/uploadedFiles/intracenorg/Content/Exporters/Sectoral_Information/Agricultural_Products/Organic_Products/Marketing_Manual_for_Organic_Spices_Culinary_Herbs_Essential_Oils.pdf.
11. Eurostat. European Commission, 28 Dec. 2012; http://epp.eurostat.ec.europa.eu/portal/page/portal/statistics/search_database.
12. UN data. *A world of information*; data.un.org/.

Editors' Biographies

H. Rodolfo Juliani

Rodolfo Juliani obtained his Ph.D. in plant biology from the National University of Cordoba in Argentina (internationally recognized institution founded in 1613). Dr. Juliani currently serves as Assistant Research Professor at Rutgers University. He is working on the biology and chemical diversity of plants and quality (effectiveness and safety) of natural plant-based products for human (food, nutrition, health, cosmetic and industrial uses) and environmental well-being. Dr. Juliani has published over 80 research papers, books, book chapters and book reviews. He has contributed to develop the natural plant product industry for income generation, by expanding research infrastructure and human capacity of non-profit, research and higher education institutions in Africa and other parts of the world.

James E. Simon

Professor James Simon received his Ph.D. in environmental stress physiology and plant science from University of Massachusetts, Amherst, MA. Dr. Simon was a Professor at Purdue University for 17 years and in 2000 relocated his program to Rutgers, The State University of New Jersey, where he is the Director of the New Use Agriculture and Natural Plant Products Program and a Professor in the Dept. Plant Biology & Plant Pathology. Dr. Simon's research focuses on plant genetics, genetic diversity, plant breeding, natural products discovery and characterization, botanical authentication, standardization and quality control. Dr. Simon has published over 200 scientific papers, books and book chapters and has been the recipient of many academic and industry awards, most recently being the 2012 award recipient for Scientific Excellence by a researcher in a USAID Collaborative Support Research Program given by the Board for International Food and Agricultural Development (BIFAD), USAID.

Professor Chi-Tang Ho

Professor Chi-Tang Ho received his Ph.D. in chemistry from Washington University in St. Louis. Currently he is a Distinguished Professor in the Department of Food Science, Rutgers University. Dr. Ho has published over 590 refereed journal articles and 215 book chapters, and edited over 30 professional monographs. He has won numerous awards including the ACS Award for the Advancement of Application of Agricultural and Food Chemistry and IFT Stephen S. Chang Award in Flavor and Lipid Science, and is an elected Fellow of American Chemical Society and Institute of Food Technologist. His current research interests focus on functional food chemistry.

Indexes

Author Index

Aboagye, F., 79
Achel, G., 79
Adjei, S., 79
Amekuse, L., 33
Appiah, A., 79
Arumugam, S., 289
Asante-Dartey, J., 33, 63
Asomaning, W., 79
Bai, N., 115
Bassène, E., 255
Clement, J., 79
Cordeiro, L., 273
Diedhiou, B., 255
Diop, A., 255
Duker-Eshun, G., 79
Edoh, D., 79
Ekanem, A., 93
Fischer, S., 231
Gassama, Y., 255
Giordano, L., 33
Govindasamy, R., 289
Gqaleni, N., 185
Gustafson, K., 63
Gyampo, O., 79
Hanson, P., 231
Hastilestari, B., 3
He, K., 115
Ho, C., xi, 115
Juliani, H., xi, 33, 127, 167, 209
Katerere, D., 51
Keatinge, J., 231
Koffa, S., 33
Koroch, A., 33, 127
Kotze, M., 103
Lai, C., 115
Marnewick, J., 103
Mwine, J., 3
Nahm, H., 167
Neethling, H., 103
Nyarko, A., 79
Obiekezie, A., 93
Ojiewo, C., 137
Okine, L., 79
Oppong, I., 79
Pan, M., 115
Papenbrock, J., 3
Prinsloo, G., 19, 185
Rautenbach, F., 103
Roller, M., 115
Sanogo, D., 255
Sanou, H., 255
Shao, X., 115
Simon, J., xi, 33, 63, 127, 167, 209, 289
Street, R., 19, 185
Thiome, M., 255
Van Damme, P., 3
Venter, I., 103
Villani, T., 209
Wu, Q., 63, 209
Yang, R., 137, 231

Subject Index

A

African indigenous vegetables, 231
  AVRDC priority, improving nutrient supplies, 243
  children under 5 in Sub-Saharan Africa, nutritional status, 236*t*
  classification, 276
  dietary pattern, 237*f*
  dietary pattern and diversity, 237
  Morogoro region, 274*f*
  nutritional values, 239, 244*t*
  nutritious indigenous vegetables, 241
    Amaranth, 242
    Drumstick tree, 242
    Spider flower, 242
  overweight and non-communicable diseases, 235
  research gaps and future work, 250
  role in promoting food security and health, 273
    health status, 276
    household food insecurity, 284*t*
    indigenous plants, traditional knowledge, 279
    interviews with Kilosa elders, 278
    multivariate regression model, 279
    nutritional properties and purported medicinal uses, 280*t*
    preferred foods, 278
    qualitative data, 275
    quantitative data, 275
    results and discussion, 277
    statistical analysis, 277
  share of vegetable availability, 238*f*
  total and accessible dietary iron, 240*f*
  total RNI of eight nutrients, comparison, 241*f*
  undernutrition, 232
    iron, 233
    micronutrient deficiencies, regional trends, 234
    vitamin A, 233
    zinc, 233
  women in Sub-Saharan Africa, nutritional status, 235*t*
African nightshade and eggplant
  African Solanum vegetables, distribution and utilization, 151*t*, 153*t*
  eggplants, types, 139*f*
    Solanum aethiopicum L, 139
    Solanum anguivi Lam, 140
    Solanum macrocarpon L, 140
  genebank material, morphological and molecular characterization, 162
  glycoalkaloids, 160
    biosynthetic pathway, 161
    types, 162
  grown in fields at AVRDC, 150*f*
  nightshades, types
    S. americanum, 142
    S. florulentum, 142
    S. grossidentatum, 142
    S. hirsutum, 143
    S. tarderemotum, 143
    S. villosum, 144
  nutritional values, 148, 149*f*
  positive and potential negative health outcomes, 162
  selected African *Solanum* crops, fruits and leaves
    flavonoid content, 158*t*
    glycoalkaloid content, 159*t*
    nutrient content, 154*t*, 156*t*
  taxonomy, 138
  variety development and quality seed availability, 162
African potato *(Hypoxis hemerocallidea)*
  antinociceptive and anti-inflammatory activity, 55
  Benign Prostatic Hyperplasia (BPH), 55
  biological activity, 54
  botanical information, 52
  ethnobotanical use, 55
  phytochemistry, 53
  research, future directions, 57
  toxicology and adverse effects, 56
  traditional and contemporary use, 52
African traditional medicines, quality control, 185
  contamination and adulteration, 193
  fingerprinting and marker compounds, 196
    bioactive extract, 197
    use, 197
  good agricultural practice (GAP)
    impact and application, 186
    social impact, 188
  good collection practice (GCP), impact and application, 189
    medicinal plant, long-term survivability, 190
    medicinal plants, sustainable use, 190

good manufacturing practice (GMP), impact and application, 191
drying methods, 192
medicinal plant extracts, preparation, 195
medicinal plants, scientific validation, 195
product-tracking system, 200
propagation, 187
and safety control, good practice, 194

C

Crop management
African eggplants
seed germination and nursery establishment, 144
transplanting and crop management, 145
African nightshades
seed germination and nursery establishment, 146
transplanting and crop management, 146
Croton membranaceus
artefact in formation of amides, 83*f*
chemistry, 81
1,8-cineole or α-terpineol to cis-terpine, hydration, 83*f*
compounds isolated from roots, structures, 82*f*
ethnomedical uses, 81
manool, chemical structure, 85*f*
pharmacological properties
AChE inhibitory activity, 86
anticancer activity, 85
antimicrobial activity, 87
antioxidant activity, 86
5α-reductase and testosterone-induced growth, 84
smooth muscle relaxant and antihypertensive activity, 85
plant parts, 80*f*
as remedy for BPH, 87
acetylcholinestrase (AChE) inhibition, 89
alpha adrenergic receptor antagonism, 88
antioxidant action, 89
cholesterol absorption, inhibition, 90
julocretine, 88*f*
phytosterols and cholesterol, chemical structure, 90*f*
5α-reductase inhibition, 88
scopoteletin and synthetic alpha-blocker terazosin, 89*f*
other pharmacological properties, 87
terpenes and amides isolated from roots, 83*f*
toxicity, 87
trace elements, 84

E

Essential oil and aromatic plant industry, 289
African countries and essential oil trade, 304
Brazilian global essential oils and perfume and flavor material trade, 303*f*
commodities of essential oils from African countries, 316*t*
developing countries essential oils trade, 302
developing economies, 320
from Eastern Africa, 310
Eastern Africa countries exports, 311*t*
Eastern Africa countries imports, 312*t*
essential oil industry, value chain mapping, 319*f*
essential oils exports, 292
essential oils producers, 292
essential oils value chain analysis, 318
European Union (EU27) essential oil, perfume materials and cosmetics trade, 2001-11, 300*f*
European Union (EU27) essential oil, perfume materials and cosmetics trade, 2011, 301*f*
European Union essential oil imports, 299*t*
European Union essential oils trade, 298
from 2003-2008, 298*t*
exported from African countries, 315
exporters, 295
fastest growing global markets, 298*f*
global exporting countries, 296*t*
global exports and imports, 294*f*
global importing countries, 297*t*
global market, 293
importation, 305
Indian global essential oil and perfume and flavor materials trade, 302*f*
Malaysian global essential oil and perfume and flavor materials trade, 303*f*
methodology, 293

most utilized essential oils and major producers, 291*t*
from Northern Africa, 315
Northern Africa countries exports, 313*t*
South African global essential oil and perfume and flavor materials trade, 304*f*
Southern African countries exports, 306*t*
Southern African countries imports, 307*t*
traditional essential oils producers, 320
United States essential oils trade, 301
United States global essential oil trade, 2001 to 2011, 301*f*
from Western Africa, 310
Western African countries exports, 308*t*
Western African countries imports, 309*t*
Euphorbia tirucalli L. (Euphorbiaceae), 3
chemical composition, 8
classification, 5
disuses, 14
ecology and distribution, 8
pests and disease, 9
plant description, 7
tree sap latex, chemical substances, 9*t*
twig, 6*f*
uses, 10*t*
annual energy yields after conversion to electricity, 12*t*
conservation and agroforestry, 13
ornamental, 11
pesticides, 13
source of energy, 11
source of rubber, 13
traditional medicines, 9
vernacular names, 4*t*

F

Fonio *(Digitaria exilis)* from senegal, nutritional value
elemental composition, 131*t*
future prospects, 131
inflorescence, 128
materials and methods, 129
morphology, 128
proximate analysis, 130*t*

H

Harvesting and post-harvest handling
African eggplant, 146
African nightshades, 147
Hibiscus sabdariffa, 209
medicinal uses and biological studies, 217
antibacterial/antifungal/antiparasitic, 221
anti-cholesterol/anti-obesity, 219
antihypertensive, 218
anti-inflammatory/diuretic, 221
antioxidant activity, 220
antitumor, 220
methods of analysis
chromatographic methods, 216
nonchromatographic assays, 217
phytochemicals, 213*t*
anthocyanidin species, structures, 211*f*
fatty acids, oils, and aromatic volatiles, 212
nutrition, 216
organic acids, 212
polyphenols, 210
uses in food, 222

M

Marula [*Sclerocarya birrea* (A.Rich) Hochst], 19
commercial use, 28
income generator, resource poor communities, 27
leaves and immature fruits, 21*f*
medicinal uses, validation
amoebic dysentery, 23
antibacterial activity, 22
anti-inflammatory effect, 23
anti-malarial activity, 24
anti-mycobacterial activity, 24
hyperglaecemia, 22
hypertension, 24
muscle relaxation, 24
polydipsia, 23
secondary metabolites and nutrient composition, 25
fatty acid, 26
flavonoids, 26
vitamin C, 26
toxicity, 25
tree, 19

P

Peasant's baobab tree (*Adansonia digitata* L.), 255
baobab populations, 257
classification, 259

criteria mentioned, different ethnic groups, 259*t*
different morphotypes, trees identification, 257
geographical locations, 258*t*
morpho-biochemical characters, median values, 261*t*
morpho-biochemical description, materials and methods, 257
morpho-biochemical fruits' parameters correlation circle, 262*f*
morpho-biochemical parameters analysis of variance, 264*t*
variance, 267
morphological and organoleptic characterization, 261
morphological and physicochemical parameters, correlation, 267
morphotypes' pulp, physicochemical characterization, 263
parameters of morphotypes' fruits, correlation, 265*t*
physicochemical parameters, medium values, 263*t*
physicochemical parameters of fruit pulp correlations, 267
variance, 267
physicochemical parameters of pulp of morphotypes' fruits, correlations, 268*t*
Pycnanthus angolensis
ABTS free radical scavenging assay, 73*t*
biological activity
cetyl myristoleate, 72
selected other compounds, 74
tocotrienol derivatives, 72
ecology and ethnobotany, 63
phytochemistry
cetyl myristoleate and tocotrienol derivatives, 66*f*
fatty acids, 65
kombo butter, 67
myristoleic acid, enrichment, 68*f*
other compounds, 71*t*
seeds, fatty acid composition, 65*t*
seed with aril, 64*f*
tocotrienol derivatives, 68
compounds, interconversion, 70
proposed metabolic scheme, 69*f*

R

Rooibos
fermented Rooibos tea preparations, HPLC analysis
chemicals, 105
chromatographic conditions, 107
dietary intake analyses, 107
fermented/traditional, contents, 106*t*
genetic analyses, 107
HPLC chromatogram, 106*f*
sample analyses, 107
statistical analyses, 108
materials and methods
study design, 105
subjects, 104
physiological factors, 112
results
daily nutrients intake by female participants, 110*t*
effect of traditionally fermented Rooibos, 109*t*

S

Secoiridoid glucosides from Fraxinus Excelsior, 115
materials and methods
calibration curves, 117
cell culture and chemicals, 118
cell viability, determination, 119
extraction and isolation, 118
general experimental procedures, 117
HPLC analysis conditions, 117
nitrite assay, 119
nuzhenide and GI 3 compounds, HPLC calibration curve data, 119*t*
plant material, 117
results and discussion, 119
compounds 1-3 and 6-10 (CD_3OD), ^{13}C NMR data, 120*t*
compounds 4 and 5 (CD_3OD), ^{13}C NMR data, 121*t*
compounds structure, 122*f*
growth of various human cancer cells, effect of 1-11, 124*t*
LPS-induced nitrite production, effect of 1-11, 124*t*
Shea Butter *(Vitellaria paradoxa)*
benefits, 170
botany, 167
chemical composition
and fractionation, 173
triglyceride fraction, 171
unsaponifiable fraction, 171
unsaponifiables, amounts, 172*t*
currently available standards, 178
extraction, 168

locally processed West African unrefined, quality analysis, 176, 178*t*
fatty acid composition, 177
moisture content, 177
unsaponifiables, 177
production with consistent quality and extend shelf-life
antioxidants addition to butter, 180
locally processed shea butter, quality improvement, 179
quality, factors causing variability
fatty acid composition, regional variability, 175*t*
post-harvest processing and quality characteristics, 175
regional variability and chemical composition, 174
unrefined butter, quality characteristics and grading system, 179*t*
uses, 169

T

Traditional medicines
pharmacovigilance, 198
safety assessment, 198
Treatment and control of disease in fish
African medicinal plants, 93
aquaculture development in Nigeria, 93
Artemisia Annua L., ethanol extracts, 98
parasite mortality, 98*f*, 99*f*
showing host (fish) tolerance, 101*t*
standard threshold concentrations, 101*t*
toxicity test of concentrations, 100*t*
gills of goldfish, extract of Mucuna pruriens, 97*f*
goldfish, extracts of *Mucuna pruriens* and *Carica papaya*, 95
mortality of goldfish, 95*t*
skin and gill monogenean parasites, 96
skin of goldfish, extract of Carica papaya, 96*f*

U

Utilization
African eggplant, 147
African nightshades, 147

W

West African black pepper, 33
chemical composition of essential oils, comparison, 38*t*
functional properties
antifungal and antimicrobial, 42
antioxidant activity, 42
insecticidal, 40
other properties, 43
material and methods, 35
plant products, use, 41
proposed quality control standards, 38*t*
sensory, foreign matter and chemical analysis, 37*t*